Spectral Theory and Applications

CONTEMPORARY MATHEMATICS

720

Centre de Recherches Mathématiques Proceedings

Spectral Theory and Applications

CRM Summer School
Spectral Theory and Applications
July 4–14, 2016
Université Laval, Québec, Canada

Alexandre Girouard
Editor

AMERICAN MATHEMATICAL SOCIETY

2010 *Mathematics Subject Classification.* Primary 58Jxx, 35Pxx, 65Nxx.

Library of Congress Cataloging-in-Publication Data

Names: Girouard, Alexandre, 1976- editor.
Title: Spectral theory and applications / Alexandre Girouard, editor.
Description: Providence, Rhode Island : American Mathematical Society ; Montreal, Quebec, Canada : Centre de Recherches Mathématiques, [2018] | Series: Contemporary mathematics. Centre de recherches mathématiques proceedings ; volume 720 | "CRM Summer School, Spectral Theory and Applications, July 4-14, 2016, Universite Laval, Quebec City, Canada." | Includes bibliographical references.
Identifiers: LCCN 2018032989 | ISBN 9781470435561 (alk. paper)
Subjects: LCSH: Spectral theory (Mathematics) | AMS: Partial differential equations – Spectral theory and eigenvalue problems – Spectral theory and eigenvalue problems. msc | Numerical analysis – Partial differential equations, boundary value problems – Partial differential equations, boundary value problems. msc
Classification: LCC QC20.7.S64 S63945 2018 | DDC 515/.7222–dc23
LC record available at https://lccn.loc.gov/2018032989

DOI: https://doi.org/10.1090/conm/720

Visit the AMS home page at https://www.ams.org/

10 9 8 7 6 5 4 3 2 1 23 22 21 20 19 18

Contents

Preface

The *2016 CRM Summer School in Québec City* took place at Université Laval from July 4–14. The topic of the meeting was *Spectral Theory and Applications*. The event was sponsored by the Centre de Recherches Mathématiques (CRM), Institut des Sciences Mathématiques (ISM), the National Science Fondation (NSF), Groupe Interdisciplinaire de Recherche en Éléments Finis (GIREF), and Université Laval.

The summer school brought together students and internationally renowned experts from several subfields of spectral theory. The program consisted of six minicourses introducing fundamentals of spectral theory, applications to physics and partial differential equations, as well as spectral geometry and numerical methods. The minicourses were complemented by exercise sessions and computer labs. There where also some shorter presentations touching upon various related research topics, including classical integral operators, network analysis, and random matrices. The school featured about 50 participants from 12 different countries. These lectures should be useful to graduate as well as advanced undergraduate students.

The lecture notes by Thomas Ransford (Université Laval) cover the fundamentals of spectral theory in Hilbert spaces, up to the spectral theorem for compact self-adjoint operators. Applications to Sturm–Liouville differential equations are also presented. The presentation is streamlined and will appeal to any student who wants to quickly learn the most useful and concrete aspects of spectral theory on Hilbert spaces.

The lecture notes by Richard Laugesen (University of Illinois at Urbana-Champaign) focus on spectral theory of partial differential operators. The emphasis is on basic examples, such as the Laplace operator on bounded Euclidean domains. It starts from separation of variables for the Dirichlet and Neumann Laplacians and builds up to a proof of the spectral theorem for these operators, using weak solutions and quadratic forms on Hilbert spaces. The variational characterization of eigenvalues is then presented, with several applications, such as the monotonicity of Dirichlet eigenvalues under inclusion. Applications to the stability of reaction-diffusion equations are also discussed.

The lectures by Richard Froese (University of British Columbia) provide a mathematical background on classical and quantum mechanics. A novel aspect of the lectures is the continual blend and comparison of ideas from classical and quantum mechanics. An introduction to Lagrangian submanifolds and Legrendre transforms is included, and the Hamilton–Jacobi equation is used to construct solutions to the Schrödinger equation. Hidden variables are discussed as a means to illustrate the essential strangeness of the quantum description of nature. This is an area that is neither easy to access nor widely known.

The lectures by Felix Kwok (Hong Kong Baptist University) introduce numerical methods for approximating the eigenvalues and eigenfunctions of partial differential operators. Finite difference methods are first presented, followed by finite element methods. Applications to vibrating plates and their nodal patterns are presented.

The lectures by Yaiza Canzani (University of North Carolina at Chapel Hill) give an overview of spectral geometry on Riemannian manifolds and start with an informal discussion on the ubiquity and importance of the Laplacian. This serves as a road map to the lectures, which cover isospectrality, spectral asymptotics, and heat equation methods. A review of basic Riemannian geometry is included.

The paper by Ram Band (Technion) and Sven Gnutzmann (University of Nottingham) teaches basic quantum graph theory through a well-chosen set of exercises. It covers various vertex conditions, the secular function, and scattering, as well as a trace formula related to periodic orbits.

The paper by Dmitry Khavinson (University of South Florida) presents several open problems in the study of spectral and geometric properties of classical integral operators, such as the Cauchy operator, and single and double layer potentials.

The proceedings of the *2016 CRM Summer School in Québec City* on *Spectral Theory and its Applications* cover a large variety of topics and methods, combining geometric, analytic, and numerical ideas. We hope that this volume will serve as a reference for young mathematicians who are eager to learn the basics of this fascinating area of mathematics.

The editor would like to express his gratitude to all the contributing authors, as well as to all the speakers at the summer school.

Contemporary Mathematics
Volume **720**, 2018
https://doi.org/10.1090/conm/720/14520

Fundamentals of spectral theory

Thomas Ransford

ABSTRACT. The mini-course developed the elements of spectral theory needed for the other lectures at the conference. No prior knowledge of functional analysis was assumed. The notes below are a copy of what was presented during the lectures, together with some exercises.

Introduction

Two fundamental theorems from spectral theory of matrices:

- Every $n \times n$ matrix has eigenvalues (at least 1 and at most n).
- Every normal $n \times n$ matrix has an orthonormal basis of eigenvectors.

Our aims:

- Establish analogues of these theorems in infinite dimensions.
- Apply these new results to the theory of differential equations.

Plan of the mini-course:

(1) Normed spaces and operators
(2) Invertible operators
(3) The spectrum
(4) Hilbert spaces
(5) Operators on Hilbert spaces
(6) Compact operators
(7) Spectral theorem
(8) Sturm–Liouville equation

1. Normed spaces and operators

1.1. Normed spaces.

DEFINITION. A *normed space* is a vector space X equipped with a *norm*, namely a function $\|\cdot\| : X \to \mathbb{R}$ satisfying:

- $\|x\| \geq 0$ for all $x \in X$, with equality iff $x = 0$;
- $\|\lambda x\| = |\lambda|\|x\|$ for all $\lambda \in \mathbb{C}$ and all $x \in X$,
- $\|x + y\| \leq \|x\| + \|y\|$ for all $x, y \in X$.

2010 *Mathematics Subject Classification.* Primary 47A10; Secondary 34B05, 47B05, 47B15.

Note that X is a metric space, with metric $d(x,y) := \|x-y\|$.

X is *complete* if every Cauchy sequence in X converges to a limit in X. A complete normed space is called a *Banach space.*

1.2. Examples of normed spaces.

EXAMPLE 1. $X = \mathbb{C}^n$ with the Euclidean norm:

$$\|(\xi_1,\dots,\xi_n)\|_2 := \Big(\sum_{k=1}^n |\xi_k|^2\Big)^{1/2}.$$

EXAMPLE 2. $X = C[a,b]$ with the sup norm:

$$\|f\|_\infty := \sup_{t\in[a,b]} |f(t)|.$$

EXAMPLE 3. $X = \{\text{polynomials}\}$ with the coefficient norm

$$\Big\|\sum_{k=0}^n a_k t^k\Big\| := \sum_{k=0}^n |a_k|.$$

Examples 1 and 2 are Banach spaces, Example 3 is not.

1.3. Absolute convergence implies convergence.

THEOREM. *Let (x_k) be a sequence in a Banach space X. If $\sum_k \|x_k\| < \infty$, then $\sum_k x_k$ converges in X.*

PROOF. Set $s_n := \sum_{k=1}^n x_k$. For $n > m$, we have

$$\|s_n - s_m\| = \Big\|\sum_{k=m+1}^n x_k\Big\| \le \sum_{k=m+1}^n \|x_k\|.$$

As $\sum_k \|x_k\| < \infty$, the last term is small if m,n are large enough. Thus (s_n) is a Cauchy sequence, so it converges in X. □

1.4. Operators.

THEOREM. *Let X, Y be normed spaces and let $T : X \to Y$ be a linear map. The following are equivalent:*

(1) *T is continuous on X;*
(2) *T is bounded on the unit ball of X;*
(3) *there is a constant C such that $\|Tx\| \le C\|x\|$ for all $x \in X$.*

DEFINITION. A map T satisfying (1)–(3) above is called a *(bounded) operator.* The smallest constant C such that (3) holds is called the *operator norm* of T, denoted $\|T\|$.

1.5. Examples of operators.

EXAMPLE 1. Let $A = (a_{jk})$ be an $n \times n$ matrix, and define $T : \mathbb{C}^n \to \mathbb{C}^n$ by $Tx := Ax$. Then, writing $x = (\xi_1,\dots,\xi_n)$,

$$\|Tx\|_2^2 = \sum_{j=1}^n \Big|\sum_{k=1}^n a_{jk}\xi_k\Big|^2 \le \sum_{k=1}^n\sum_{j=1}^n |a_{jk}|^2 \|x\|_2^2.$$

So T is a bounded operator with $\|T\| \le (\sum_{j,k} |a_{jk}|^2)^{1/2}$.

EXAMPLE 2. Define $T : C[0,1] \to C[0,1]$ by $Tf(t) := tf(t)$.

$$\|Tf\|_\infty = \sup_{t\in[0,1]} |tf(t)| \leq \sup_{t\in[0,1]} |f(t)| = \|f\|_\infty.$$

So T is bounded with $\|T\| \leq 1$. In fact $\|T\| = 1$ (take $f \equiv 1$).

EXAMPLE 3. Define $T : C[0,1] \to C[0,1]$ by $Tf(t) := \int_0^t f$.

$$\|Tf\|_\infty = \sup_{t\in[0,1]} \left|\int_0^t f\right| \leq \int_0^1 |f| \leq \|f\|_\infty.$$

So T is bounded with $\|T\| \leq 1$. In fact $\|T\| = 1$ (take $f \equiv 1$).

EXAMPLE 4. Let $X = \{\text{polynomials}\}$ with the coefficient norm, and define $T : X \to X$ by $Tp := p'$. If $p_n(t) = t^n$, then

$$\|p_n\| = 1 \quad \text{and} \quad \|Tp_n\| = n.$$

Conclusion: this T is unbounded.

1.6. The space of bounded operators. We write $B(X,Y)$ for the set of bounded operators $T : X \to Y$.

THEOREM. *$B(X,Y)$ is a normed space with respect to the operator norm. Further, if Y is complete, then so is $B(X,Y)$.*

Important special cases:

- $Y = \mathbb{C}$. We write $X^* := B(X,\mathbb{C})$, the *dual space* of X. Elements of the dual are called (continuous) *linear functionals.*
- $Y = X$. We write $B(X) := B(X,X)$, which is now an algebra. If $S,T \in B(X)$, then also $ST \in B(X)$ and

$$\|ST\| \leq \|S\|\|T\|.$$

Exercises

1A Justify the completeness or incompleteness each of the spaces X in §1.2.
1B Let X be a Banach space and let (x_k) be vectors in X such that $\sum_{k\geq 1} \|x_k\| < \infty$. Prove that

$$\left\|\sum_{k\geq 1} x_k\right\| \leq \sum_{k\geq 1} \|x_k\|.$$

1C Prove the theorem in §1.4.
1D Prove the theorem in §1.6.
1E Prove that, if $S,T \in B(X)$, then $\|ST\| \leq \|S\|\|T\|$.

2. Invertible operators

2.1. Invertible operators. Let X be a normed space. As usual, $B(X)$ denotes the space of bounded operators on X.

We say that $T \in B(X)$ is *invertible* if there exists $S \in B(X)$ such that $ST = TS = I$. This S is unique, and is denoted by T^{-1}.

The set of all invertible operators on X is a group, denoted $GL(X)$.

2.2. The fundamental lemma.

LEMMA. *Assume that X is a Banach space. If $T \in B(X)$ and $\|T\| < 1$, then $(I - T)$ is invertible, and $(I - T)^{-1} = \sum_{k\geq 0} T^k$.*

PROOF. Since $\sum_{k\geq 0} \|T^k\| \leq \sum_{k\geq 0} \|T\|^k < \infty$, the series $\sum_{k\geq 0} T^k$ is convergent in the Banach space $B(X)$. Further

$$(I-T)\sum_{k\geq 0} T^k = \lim_{n\to\infty} (I-T)\sum_{k=0}^{n} T^k = \lim_{n\to\infty} (I - T^{n+1}) = I.$$

Likewise $\sum_{k\geq 0} T^k(I-T) = I$. □

2.3. The group of invertible operators is open. Recall: $GL(X)$ denotes the group of invertible operators on X.

THEOREM. *Assume that X is a Banach space. Then $GL(X)$ is open in $B(X)$.*

PROOF. By the fundamental lemma, I lies in the interior of $GL(X)$. Given $S \in GL(X)$, the map $T \mapsto ST$ is a homeomorphism of $B(X)$ onto itself taking $GL(X)$ onto itself and mapping I to S. Hence S lies in the interior of $GL(X)$ as well. Thus $GL(X)$ is open in $B(X)$. □

2.4. Continuity of inversion.

THEOREM. *Assume that X is a Banach space. Then the map $T \mapsto T^{-1}$ is continuous on $GL(X)$.*

PROOF. By the fundamental lemma, if $S_n \to I$, then $S_n^{-1} \to I$. Indeed, writing $S_n = I - T_n$, we have $\|T_n\| \to 0$, so for n large

$$\|(I-T_n)^{-1} - I\| = \Big\|\sum_{k\geq 1} T_n^k\Big\| \leq \sum_{k\geq 1} \|T_n\|^k \to 0.$$

Hence, for a general $S \in GL(X)$, we have

$$\begin{aligned} S_n \to S &\Rightarrow S_n S^{-1} \to I \\ &\Rightarrow (S_n S^{-1})^{-1} \to I \\ &\Rightarrow S S_n^{-1} \to I \\ &\Rightarrow S_n^{-1} \to S^{-1}. \end{aligned}$$

□

2.5. A generalization of the fundamental lemma.

LEMMA. *Assume that X is a Banach space. If $T \in B(X)$ and $\|T^k\| < 1$ for some $k \geq 1$, then $(I - T)$ is invertible.*

PROOF. By the fundamental lemma, $(I - T^k)$ is invertible. Also,

$$I - T^k = (I-T)\Big(\sum_{j=0}^{k-1} T^j\Big) = \Big(\sum_{j=0}^{k-1} T^j\Big)(I-T),$$

whence it follows that $(I - T)$ is invertible. □

2.6. Application to an initial-value problem.

THEOREM. *Let $b > 0$, let $n \geq 1$, let $a_0, \dots, a_{n-1} \in C[0,b]$, let $g \in C[0,b]$ and let $x_0, \dots, x_{n-1} \in \mathbb{C}$. Then there exists a unique solution f to*

$$(*)\begin{cases} f^{(n)}(t) + a_{n-1}(t)f^{(n-1)}(t) + \cdots + a_0(t)f(t) = g(t), \\ f^{(j)}(0) = x_j \quad (0 \leq j < n). \end{cases}$$

Idea for the proof. Convert into a system of first-order ODEs. For example, if $n = 2$, then $(*)$ is equivalent to

$$(**)\begin{cases} F'(t) + A(t)F(t) = G(t), \\ F(0) = X_0, \end{cases}$$

where

$$A := \begin{pmatrix} 0 & -1 \\ a_0 & a_1 \end{pmatrix}, \quad F = \begin{pmatrix} f_0 \\ f_1 \end{pmatrix}, \quad G = \begin{pmatrix} 0 \\ g \end{pmatrix}, \quad X_0 = \begin{pmatrix} x_0 \\ x_1 \end{pmatrix}.$$

So it is enough to prove

THEOREM. *Let $A : [0,b] \to B(\mathbb{C}^n)$ and $G : [0,b] \to \mathbb{C}^n$ be continuous maps, and let $X_0 \in \mathbb{C}^n$. Then there exists a unique solution F to*

$$(**)\begin{cases} F'(t) + A(t)F(t) = G(t), \\ F(0) = X_0. \end{cases}$$

PROOF. F solves $(**)$ iff $(I+V)F = H$, where

$$(VF)(t) := \int_0^t A(s)F(s)\,ds \quad \text{and} \quad H(t) := \int_0^t G(s)\,ds + X_0.$$

Here $V : Y \to Y$, where $Y := C([0,b], \mathbb{C}^n)$, a Banach space.
So it is enough to show that $V \in B(Y)$ and $(I + V)$ is invertible.

(1) Proof that V is a bounded operator on Y: Let $F \in C([0,b], \mathbb{C}^n)$. For each $t \in [0,b]$,

$$\begin{aligned} \|VF(t)\|_{\mathbb{C}^n} &= \left\| \int_0^t A(s)F(s)\,ds \right\|_{\mathbb{C}^n} \\ &\leq \int_0^t \|A(s)F(s)\|_{\mathbb{C}^n}\,ds \\ &\leq \int_0^t \|A(s)\|_{B(\mathbb{C}^n)} \|F(s)\|_{\mathbb{C}^n}\,ds \\ &\leq M \int_0^t \|F(s)\|_{\mathbb{C}^n}\,ds, \end{aligned}$$

where $M := \sup_{s \in [0,b]} \|A(s)\|_{B(\mathbb{C}^n)} < \infty$. Hence

$$\|VF\|_Y = \sup_{t\in[0,b]} \|VF(t)\|_{\mathbb{C}^n} \leq Mb \sup_{s\in[0,b]} \|F(s)\| = Mb\|F\|_Y.$$

Conclusion: V is a bounded operator on Y and $\|V\| \leq Mb$.

(2) Proof that $(I+V)$ is invertible: Repeat the previous computation with F replaced by $V^{k-1}F$ to get

$$\|V^k F(t)\|_{\mathbb{C}^n} \leq M \int_0^t \|V^{k-1}F(s)\|_{\mathbb{C}^n}\,ds.$$

Hence, by induction on k,

$$\|V^k F(t)\|_{\mathbb{C}^n} \leq \frac{M^k t^k}{k!}\|F\|_Y.$$

Hence

$$\|V^k F\|_Y \leq \frac{M^k b^k}{k!}\|F\|_Y.$$

Hence

$$\|V^k\| \leq \frac{M^k b^k}{k!}.$$

If k is large enough, then $M^k b^k/k! < 1$, so $\|V^k\| < 1$. By the (generalized) fundamental lemma, $(I+V)$ is invertible. Done! □

Exercises

2A Let (T_n) be a sequence of invertible operators in $B(X)$ converging to a non-invertible operator T. Show that $\|T_n^{-1}\| \to \infty$. [*Hint*: $\|I - T_n^{-1}T\| \leq \|T_n^{-1}\|\|T_n - T\|$.]

2B Let $n \geq 2$. Determine A, F, G, X_0 so that the systems $(*)$ and $(**)$ in §2.6 are equivalent.

2C Show that the solution F to $(**)$ in §2.6 satisfies

$$\|F\|_\infty \leq B\Big(\|X_0\| + \int_0^b \|G(s)\|\,ds\Big),$$

where B is a constant depending on A, but not on X_0 or G.

2D Combine the two preceding exercises to show that the solution f to $(*)$ in §2.6 satisfies

$$\max_{0\leq j\leq n-1} \|f^{(j)}\|_\infty \leq C\Big(\max_{0\leq j\leq n-1} |x_j| + \int_0^b |g(s)|\,ds\Big),$$

where C is a constant depending on $a_0, \ldots, a_{n-1}$, but not on $x_0, \ldots, x_{n-1}$ or g.

3. The spectrum

Throughout this section, X is a Banach space and $T \in B(X)$.

3.1. Eigenvalues and eigenvectors. $\lambda \in \mathbb{C}$ is an *eigenvalue* of T if there exists a non-zero $x \in X$ such that $Tx = \lambda x$. The vector x is a called an *eigenvector*.

Warning: If $\dim X = \infty$ then T may have no eigenvalues!

EXAMPLE. $X := C[0,1]$ and $Tf(t) := tf(t)$ (see Exercise 3A).

Two ways round this problem:

- impose extra conditions of T (shall explore this later);
- 'broaden' the definition of eigenvalue (subject for this section).

3.2. Spectrum.

DEFINITION. The *spectrum* of T is defined by

$$\sigma(T) := \{\lambda \in \mathbb{C} : T - \lambda I \text{ not invertible}\}.$$

If $\dim X < \infty$, then $\sigma(T) = \{\text{eigenvalues of } T\}$.

If $\dim X = \infty$, then $\{\text{eigenvalues of } T\} \subset \sigma(T)$, but inclusion may be strict.

EXAMPLE. Let $X := C[0,1]$ and $Tf(t) := tf(t)$.

- If $(T-\lambda I)$ is invertible, then there exists $f \in C[0,1]$ such that $(t-\lambda)f(t) = 1$ for all $t \in [0,1]$, so $\lambda \notin [0,1]$.
- Conversely, if $\lambda \notin [0,1]$, then $t \mapsto 1/(t-\lambda) \in C[0,1]$, and multiplication by this function is an inverse to $(T - \lambda I)$.
- Conclusion: $\sigma(T) = [0,1]$.

3.3. Spectral radius formula.

THEOREM. *The spectrum $\sigma(T)$ is a non-empty compact set, and*

$$\max\{|\lambda| : \lambda \in \sigma(T)\} = \lim_{n\to\infty} \|T^n\|^{1/n} = \inf_{n\geq 1} \|T^n\|^{1/n}.$$

PROOF. In three steps:

(1) $\rho(T) := \lim_{n\to\infty} \|T^n\|^{1/n}$ exists and equals $\inf_{n\geq 1} \|T^n\|^{1/n}$;
(2) $\sigma(T)$ is a closed subset of the disk $|z| \leq \rho(T)$;
(3) there exists $\lambda \in \sigma(T)$ with $|\lambda| = \rho(T)$.

Note: $\rho(T)$ is called the *spectral radius* of T.

Step 1: $\lim_{n\to\infty} \|T^n\|^{1/n} = \inf_{n\geq 1} \|T^n\|^{1/n}$.

Set $\eta := \inf_{n\geq 1} \|T^n\|^{1/n}$. It is enough to show that

$$\limsup_{n\to\infty} \|T^n\|^{1/n} \leq \eta. \qquad (*)$$

Let $\epsilon > 0$ and choose $m \geq 1$ such that $\|T^m\|^{1/m} < \eta + \epsilon$.

Given $n \geq 1$, write $n = qm + r$, where $0 \leq r < m$. Then

$$\|T^n\| \leq \|T^m\|^q \|T^r\| \leq (\eta+\epsilon)^{mq} \|T^r\| = (\eta+\epsilon)^{n-r} \|T^r\| \leq (\eta+\epsilon)^n M,$$

where M is a constant independent of n. Hence

$$\limsup_{n\to\infty} \|T^n\|^{1/n} \leq \eta + \epsilon.$$

Letting $\epsilon \to 0$, we get $(*)$.

Step 2: $\sigma(T)$ is a closed subset of the disk $|z| \leq \rho(T)$.

Define $F : \mathbb{C} \to B(X)$ by $F(\lambda) := T - \lambda I$. Clearly F is continuous. Also, recalling that $GL(X)$ denotes the invertible operators on X,

$$F^{-1}(GL(X)) = \mathbb{C} \setminus \sigma(T).$$

As $GL(X)$ is open in $B(X)$, we have $\mathbb{C} \setminus \sigma(T)$ is open in $\mathbb{C}$, whence $\sigma(T)$ is closed in $\mathbb{C}$.

Now suppose that $|\lambda| > \rho(T) = \inf_{n\geq 1} \|T^n\|^{1/n}$. Then there exists $n \geq 1$ such that $\|(T/\lambda)^n\| < 1$. By the generalized form of the fundamental lemma, it follows that $(I - T/\lambda)$ is invertible, and therefore $\lambda \notin \sigma(T)$. Conclusion:

$$\lambda \in \sigma(T) \implies |\lambda| \leq \rho(T).$$

Step 3: There exists $\lambda \in \sigma(T)$ with $|\lambda| = \rho(T)$.

Two cases to consider:

Case 1: $\rho(T) = 0$.

We need to show that T is not invertible. Suppose, on the contrary, that T is invertible. Then, for all $n \geq 1$, we have

$$1 = \|I\| = \|T^n (T^{-1})^n\| \leq \|T^n\| \|T^{-1}\|^n.$$

Then $\|T^n\|^{1/n} \geq 1/\|T^{-1}\| > 0$ for all n, contradicting $\rho(T) = 0$.

Case 2: $\rho(T) > 0$.

Without loss of generality $\rho(T) = 1$. Then, by Step 2, $\sigma(T) \subset \overline{\mathbb{D}}$. We need to show that $\sigma(T) \cap \partial\mathbb{D} \neq \emptyset$. Suppose, on the contrary, that $\sigma(T) \subset \mathbb{D}$. Then $(I - zT)$ is invertible for all $z \in \overline{\mathbb{D}}$.

As $z \mapsto (I - zT)^{-1} : \overline{\mathbb{D}} \to B(X)$ is continuous on a compact set, it is uniformly continuous. So, given $\epsilon > 0$, there is $\delta > 0$ such that

$$z, w \in \overline{\mathbb{D}},\ |z - w| \le \delta \quad \Rightarrow \quad \|(I - zT)^{-1} - (I - wT)^{-1}\| \le \epsilon.$$

Using the identity $(I - S^2)^{-1} = ((I - S)^{-1} + (I + S)^{-1})/2$ and induction on n, it follows that, for all $n \ge 0$,

$$|z - w| \le \delta \quad \Rightarrow \quad \|(I - (zT)^{2^n})^{-1} - (I - (wT)^{2^n})^{-1}\| \le \epsilon.$$

In particular, taking $z = 1$ and $w = (1 - \delta)$, we get

$$\|(I - T^{2^n})^{-1} - (I - ((1 - \delta)T)^{2^n})^{-1}\| \le \epsilon.$$

To summarize, $\forall \epsilon > 0\ \exists \delta > 0\ \forall n \ge 0$

$$\|(I - T^{2^n})^{-1} - (I - ((1 - \delta)T)^{2^n})^{-1}\| \le \epsilon. \qquad (*)$$

Now $\rho((1 - \delta)T) = (1 - \delta) < 1$, so $((1 - \delta)T)^{2^n} \to 0$ and hence

$$(I - ((1 - \delta)T)^{2^n})^{-1} \to I.$$

Together with $(*)$, this shows that, for all sufficiently large n,

$$\|(I - T^{2^n})^{-1} - I\| \le 2\epsilon.$$

Thus $(I - T^{2^n})^{-1} \to I$, whence $(I - T^{2^n}) \to I$ and $T^{2^n} \to 0$.
This contradicts the supposition that $\rho(T) = 1$. Done! □

3.4. Application: Fundamental theorem of algebra.

THEOREM. *Every polynomial $p(z) = z^n + a_1 z^{n-1} + \cdots + a_n$ has a root.*

PROOF. We have $p(z) = \det(zI - A)$, where

$$A := \begin{pmatrix} 0 & 1 & 0 & \dots & 0 \\ 0 & 0 & 1 & \dots & 0 \\ \vdots & \vdots & \vdots & \ddots & \vdots \\ 0 & 0 & 0 & \dots & 1 \\ -a_n & -a_{n-1} & -a_{n-2} & \dots & -a_1 \end{pmatrix}.$$

Take $\lambda \in \sigma(A)$. Then $p(\lambda) = \det(\lambda I - A) = 0$. □

3.5. Some further developments.

- λ is an *approximate eigenvalue* of T if there exists a sequence (x_n) of unit vectors in X such that $\|(T - \lambda I)x_n\| \to 0$. One can show every $\lambda \in \partial\sigma(T)$ is an approximate eigenvalue. Hence every $T \in B(X)$ has approximate eigenvalues.
- *Spectral mapping theorem*: If p is a polynomial then $\sigma(p(T)) = p(\sigma(T))$.
- *Holomorphic functional calculus.* One can define an operator $f(T)$ for every function f holomorphic on a neighborhood of $\sigma(T)$. Idea: use Cauchy integral formula

$$f(T) = \frac{1}{2\pi i} \int_\Gamma f(z)(zI - T)^{-1}\, dz.$$

Exercises

3A Define $T : C[0,1] \to C[0,1]$ by $Tf(t) := tf(t)$. Prove that T has no eigenvalues.

3B Let $S \in B(X)$ and suppose that $I \pm S$ are both invertible. Show that $I - S^2$ is invertible and

$$(I - S^2)^{-1} = \frac{(I-S)^{-1} + (I+S)^{-1}}{2}.$$

3C (Spectral mapping theorem) Let $T \in B(X)$ and let p be a polynomial. Show that

$$\sigma(p(T)) = p(\sigma(T)).$$

[*Hint:* factorize $p(z) - \lambda = c(z - z_1)\dots(z - z_n)$.]

3D Let $T \in B(X)$ and let $\lambda \in \partial\sigma(T)$. Show that λ is an approximate eigenvalue, i.e. there exist unit vectors $x_n \in X$ such that $\|(T - \lambda I)x_n\| \to 0$. [*Hint:* Use Exercise 2A.]

4. Hilbert spaces

4.1. Inner products.

DEFINITION. A map $\langle \cdot, \cdot \rangle : X \times X \to \mathbb{C}$ is an *inner product* if:

- $x \mapsto \langle x, y \rangle$ is linear, for each $y \in X$;
- $\langle y, x \rangle = \overline{\langle x, y \rangle}$, for all $x, y \in X$;
- $\langle x, x \rangle \geq 0$ for all $x \in X$, with equality iff $x = 0$.

Properties of inner products:

- $\|x\| := \langle x, x \rangle^{1/2}$ is a norm on X.
- $|\langle x, y \rangle| \leq \|x\|\|y\|$ (Cauchy–Schwarz inequality).
- If $\|x_n - x\| \to 0$ and $\|y_n - y\| \to 0$, then $\langle x_n, y_n \rangle \to \langle x, y \rangle$.
- $\|x + y\|^2 + \|x - y\|^2 = 2\|x\|^2 + 2\|y\|^2$ (parallelogram identity).
- $\langle x, y \rangle = (1/4)\sum_{k=0}^{3} i^k \|x + i^k y\|^2$ (polarization identity).

4.2. Hilbert spaces. A *Hilbert space* is a vector space H, equipped with an inner product, such that the resulting normed space is complete.

Examples:

- $H = \mathbb{C}^n$
  $\langle x, y \rangle := \sum_{k=1}^{n} x_k \overline{y}_k$
- $H = \ell^2 := \{x = (x_1, x_2, \dots) : \sum_1^\infty |x_k|^2 < \infty\}$
  $\langle x, y \rangle := \sum_{k=1}^{\infty} x_k \overline{y}_k$.
- $H = L^2[a,b] := \{f : [a,b] \to \mathbb{C} : \int_a^b |f(t)|^2\, dt < \infty\}$
  $\langle f, g \rangle := \int_a^b f(t)\overline{g(t)}\, dt$
- $H = L^2(S, \mu) := \{f : S \to \mathbb{C} : \int_S |f|^2\, d\mu < \infty\}$
  $\langle f, g \rangle := \int_S f\overline{g}\, d\mu$.

Henceforth, H designates a Hilbert space.

4.3. Orthonormal sequences.

DEFINITION. A (finite or infinite) sequence (e_k) in H is *orthonormal (o.n.)* if

$$\langle e_j, e_k \rangle = \begin{cases} 1, & j = k, \\ 0, & j \neq k. \end{cases}$$

Examples:

- $H = \mathbb{C}^n$; $e_1 = (1, 0, \ldots, 0), \ldots, e_n = (0, \ldots, 0, 1)$
- $H = \ell^2$; $e_1 = (1, 0, 0, \ldots), e_2 = (0, 1, 0, \ldots), \ldots$
- $H = L^2[0, 1]$; $e_k(t) = e^{2\pi i k t}$ $(k \in \mathbb{Z})$

4.4. Bessel's inequality.

THEOREM (Bessel's inequality). *Let (e_k) be an orthonormal sequence in H. Then, for all $x \in H$,*

$$\sum_k |\langle x, e_k\rangle|^2 \leq \|x\|^2.$$

PROOF. For $n \geq 1$, we have

$$0 \leq \Big\| x - \sum_{k=1}^n \langle x, e_k\rangle e_k \Big\|^2 = \|x\|^2 - \sum_{k=1}^n |\langle x, e_k\rangle|^2.$$

Let $n \to \infty$. □

4.5. Riesz–Fischer theorem.

THEOREM (Riesz–Fischer). *Let (e_k) be an o.n. sequence in H and let (c_k) be complex scalars. Then $\sum_{k=1}^\infty c_k e_k$ converges in H iff $\sum_{k=1}^\infty |c_k|^2 < \infty$.*

PROOF. Set $s_n := \sum_{k=1}^n c_k e_k$, and note that, for $n > m$,

$$\|s_n - s_m\|^2 = \Big\| \sum_{k=m+1}^n c_k e_k \Big\|^2 = \sum_{k=m+1}^n |c_k|^2.$$

Now use completeness of H. □

REMARKS. (1) $\sum_k c_k e_k$ is independent of the order of the terms.
(2) By Bessel and Riesz–Fischer, the series $\sum_k \langle x, e_k\rangle e_k$ converges.

4.6. Parseval's identity.

THEOREM. *Let (e_k) be an orthonormal sequence in H and let $x \in H$. TFAE:*

- *x lies in the closure of the span of the (e_k);*
- *$x = \sum_k \langle x, e_k\rangle e_k$;*
- *$\|x\|^2 = \sum_k |\langle x, e_k\rangle|^2$ (Parseval's identity).*

PROOF. The equivalence of the first two statements follows from

$$\Big\| x - \sum_{k=1}^n c_k e_k \Big\|^2 = \Big\| x - \sum_{k=1}^n \langle x, e_k\rangle e_k \Big\|^2 + \Big\| \sum_{k=1}^n (\langle x, e_k\rangle - c_k) e_k \Big\|^2.$$

The equivalence of the last two statements follows from

$$\Big\| x - \sum_{k=1}^n \langle x, e_k\rangle e_k \Big\|^2 = \|x\|^2 - \sum_{k=1}^n |\langle x, e_k\rangle|^2.$$ □

4.7. Orthonormal bases.

DEFINITION. An *orthonormal basis* of H is an o.n. sequence (e_k) that spans a dense subspace of H. By the preceding theorem, for all $x \in H$,

$$x = \sum_k \langle x, e_k \rangle e_k \quad \text{and} \quad \|x\|^2 = \sum_k |\langle x, e_k \rangle|^2.$$

A metric space is *separable* if it has a countable dense subset. Separable H include: $\mathbb{C}^n$, ℓ^2, $L^2[a,b]$.

THEOREM. *A Hilbert space has an orthonormal basis iff it is separable.*

PROOF. 'Only if' : consider $\{\sum_{k=1}^n c_k e_k : c_k \in \mathbb{Q} + i\mathbb{Q}, n \geq 1\}$.
'If': Choose (v_k) with dense span, then use Gram–Schmidt. □

4.8. Orthogonal complements. Given $S \subset H$, we define

$$S^\perp := \{x \in H : \langle x, s \rangle = 0 \text{ for all } x \in S\}.$$

Note that $S^\perp$ is a closed subspace of H (whether S is or not).

THEOREM. *Let M be a closed subspace of H. Then $H = M \oplus M^\perp$.*

PROOF. (For H separable.) Clearly $M \cap M^\perp = \{0\}$. We must show that $M + M^\perp = H$. Let $x \in H$. Let (e_k) be an o.n. basis of M and set

$$y := \sum_k \langle x, e_k \rangle e_k.$$

Then $y \in M$. Also $\langle x - y, e_k \rangle = 0$ for all k, so $x - y \in M^\perp$. Thus

$$x = y + (x - y) \in M + M^\perp. \qquad □$$

4.9. Linear functionals on a Hilbert space.

THEOREM (Riesz representation theorem). *Let $\phi : H \to \mathbb{C}$ be a continuous linear functional. Then there exists a unique $y \in H$ such that*

$$\phi(x) = \langle x, y \rangle \quad (x \in H).$$

PROOF. *Existence*: If $\phi = 0$, then we can take $y = 0$. Otherwise, $\ker \phi$ is a proper closed subspace of H. Since $H = \ker \phi \oplus (\ker \phi)^\perp$, there exists non-zero vector $y \in (\ker \phi)^\perp$. Then ϕ and $x \mapsto \langle x, y \rangle$ are linear functionals with the same kernel, so one is a multiple of the other. We can ensure that this multiple is 1 by replacing y by a suitable multiple of itself.

Uniqueness: If y_1, y_2 are two candidates, then $\langle x, y_1 - y_2 \rangle = 0$ for all x, in particular for $x = y_1 - y_2$. This implies $y_1 - y_2 = 0$. □

Exercises

4A Prove the properties of inner products listed in §4.1.
4B Prove that ℓ^2 is complete.
4C Let (e_k) be an o.n. basis of H and let $(c_k) \subset \ell^2$. Show that $\sum_k c_k e_k$ is independent of the order of the sum. [*Hint:* Show that $\sum_k \langle c_k e_k, y \rangle$ is absolutely convergent for each $y \in H$.]

4D (Gram–Schmidt algorithm) Let $(v_1, v_2, v_3, \dots)$ be a (finite or infinite) sequence of linearly independent vectors in H. Define inductively

$$f_k := v_k - \sum_{j=1}^{k-1} \langle v_k, e_j\rangle e_j \quad \text{and} \quad e_k := f_k/\|f_k\|.$$

Show that (e_k) is an orthonormal sequence and that $\operatorname{span}\{e_1, \dots, e_n\} = \operatorname{span}\{v_1, \dots, v_n\}$ for each n.

4E Show that every subset of a separable metric space is itself separable.

5. Operators on Hilbert spaces

5.1. Adjoint of an operator.

THEOREM. *Each $T \in B(H)$ has a unique* adjoint $T^* \in B(H)$ *such that*

$$\langle Tx, y\rangle = \langle x, T^*y\rangle \quad (x, y \in H).$$

PROOF. Fix $y \in H$. The map $x \mapsto \langle Tx, y\rangle$ is a continuous linear functional on H, so by the Riesz representation theorem, there exists a unique $z \in H$ such that

$$\langle Tx, y\rangle = \langle x, z\rangle \quad (x \in H).$$

Define $T^*y := z$. It is routine to check that T^* is linear. Lastly,

$$\|T^*x\|^2 = \langle T^*x, T^*x\rangle = \langle TT^*x, x\rangle \le \|T\|\|T^*x\|\|x\|,$$

whence $\|T^*x\| \le \|T\|\|x\|$. Thus $T^* \in B(H)$ and $\|T^*\| \le \|T\|$. □

5.2. Calculation of adjoints.

THEOREM. *Let (e_k) be an o.n. basis and suppose $Te_k = \sum_j a_{jk}e_j$ for all k. Then $T^*e_k = \sum_j \overline{a_{kj}}e_j$ for all k.*

PROOF. We have $T^*e_k = \sum_j \langle T^*e_k, e_j\rangle e_j$, and

$$\langle T^*e_k, e_j\rangle = \overline{\langle e_j, T^*e_k\rangle} = \overline{\langle Te_j, e_k\rangle} = \overline{\langle \sum_i a_{ij}e_i, e_k\rangle} = \overline{a_{kj}}.$$ □

Examples:

- If $T \in B(\ell^2)$ is the right shift, then T^* is the left shift.
- (Diagonal operators) If (e_k) is an o.n. basis for H and $Te_k = \lambda_k e_k$ for all k, then $T^*e_k = \overline{\lambda}_k e_k$ for all k.

5.3. Properties of adjoints.

Algebraic properties: $(S+T)^* = S^* + T^*$, $(\lambda T)^* = \overline{\lambda}T^*$, $(ST)^* = T^*S^*$, $T^{**} = T$.

Norm properties: $\|T^*\| = \|T\|$ and $\|T^*T\| = \|T\|^2$.

PROOF OF THE NORM PROPERTIES. We already know that $\|T^*\| \le \|T\|$. Also $\|T\| = \|T^{**}\| \le \|T^*\|$. Further $\|T^*T\| \le \|T^*\|\|T\| = \|T\|^2$. Finally,

$$\|Tx\|^2 = \langle Tx, Tx\rangle = \langle x, T^*Tx\rangle \le \|T^*T\|\|x\|^2,$$

whence $\|T\|^2 \le \|T^*T\|$. □

5.4. Self-adjoint operators.

DEFINITION. $T \in B(H)$ is *self-adjoint* if $T^* = T$.

Examples:

- For $T \in B(H)$, let $A := (T + T^*)/2$ and $B := (T - T^*)/2i$. Then A, B are self-adjoint and $T = A + iB$.
- If $T \in B(H)$, then T^*T is self-adjoint.
- Let $P : H \to M$ be the orthogonal projection onto a closed subspace M. Then P is self-adjoint.

5.5. Spectrum of self-adjoint operators.

THEOREM. *If T is a self-adjoint operator, then $\sigma(T) \subset \mathbb{R}$.*

PROOF. Let $\alpha + i\beta \in \sigma(T)$. Then $\alpha + i(\beta + t) \in \sigma(T + itI)$, so

$$|\alpha + i(\beta + t)|^2 \leq \|T + itI\|^2 = \|(T + itI)^*(T + itI)\| = \|T^2 + t^2 I\|.$$

Hence

$$\alpha^2 + \beta^2 + 2\beta t \leq \|T^2\|.$$

As this holds for all $t \in \mathbb{R}$, we must have $\beta = 0$. □

5.6. Normal operators.

DEFINITION. $T \in B(H)$ is *normal* if $TT^* = T^*T$.

Examples:

- Every self-adjoint operator is normal.
- Every diagonal operator is normal.
- Let A, B be self adjoint. Then $A + iB$ is normal iff $AB = BA$
- The right shift S on ℓ^2 is *not* normal: $S^*S = I \neq SS^*$.

5.7. Properties of normal operators.

Let $T \in B(H)$ be normal.

- $\|T^*x\| = \|Tx\|$ for all $x \in H$.
- If $Te = \lambda e$, then $T^*e = \overline{\lambda} e$.
- If $Te = \lambda e$ and $Tf = \mu f$ where $\lambda \neq \mu$, then $\langle e, f \rangle = 0$.
- $\|T^2\| = \|T\|^2$. Indeed:

$$\|T^2\|^2 = \|(T^2)^*(T^2)\| = \|(T^*T)^*(T^*T)\| = \|T^*T\|^2 = \|T\|^4.$$

- $\rho(T) = \|T\|$. Indeed: $\|T^{2^n}\| = \|T\|^{2^n}$ for all n, so

$$\rho(T) = \lim_{n\to\infty} \|T^{2^n}\|^{1/2^n} = \|T\|.$$

5.8. Spectral theorem in finite dimensions.

THEOREM. *Assume* $\dim H < \infty$. *Then every normal $T \in B(H)$ has an o.n. basis of eigenvectors.*

PROOF.

- Pick an o.n. basis of each eigenspace of T.
- Put them together to get a (finite) o.n. sequence (e_n).
- Let M be the (closed) subspace spanned by (e_n).
- Since T is normal, $T^*(M) \subset M$, whence $T(M^\perp) \subset M^\perp$.
- If $M^\perp \neq \{0\}$, then $T|M^\perp$ has an eigenvector e; then $e \in M \cap M^\perp$, so $e = 0$, contradiction.
- Conclusion: $M = H$ and (e_n) is an o.n. basis of H. □

5.9. Spectral theorem in infinite dimensions? If $\dim H = \infty$, then it is no longer true that every normal T has an o.n. basis of eigenvectors. Indeed, it may happen that T has no eigenvectors at all, even if T is self-adjoint.

EXAMPLE. $H = L^2[0,1]$ and $T : f(t) \mapsto tf(t)$.

We need a further ingredient...

Exercises

5A Prove that the adjoint T^* is a linear map.
5B Prove the algebraic properties of adjoints listed in §5.3.
5C Let M be a closed subspace of H and let $P : H \to M$ be the orthogonal projection of H onto M. Show that P is self-adjoint.
5D Let $T \in B(H)$. Show that T is a normal operator if and only if $\|Tx\| = \|T^*x\|$ for all $x \in H$. [*Hint for 'if':* Use the polarization identity (see §4.1).]
5E Let $T \in B(H)$. Show that $\|T\| = \rho(T^*T)^{1/2}$. Hence calculate $\left\|\begin{pmatrix} 1 & 1 \\ 0 & 1 \end{pmatrix}\right\|$.

6. Compact operators

6.1. Compact operators.

DEFINITION. $T \in B(H)$ is *compact* if, whenever (x_n) is a bounded sequence, (Tx_n) contains a convergent subsequence.

Examples:
- If $T \in B(H)$ and has finite rank (i.e. $\dim(T(H)) < \infty$), then it is compact.
- If $\dim H = \infty$, then I_H is not compact. [*Proof:* Let (e_n) be an o.n. sequence. Then (e_n) is bounded. But it contains no convergent subsequence since $\|e_n - e_m\| = \sqrt{2}$ for all $m \neq n$.]

The set of compact operators on H is denoted by $K(H)$.

6.2. Structure of $K(H)$.

THEOREM. *$K(H)$ is an ideal in $B(H)$.*

PROOF. Show:
- $S, T \in K(H) \Rightarrow (S + T) \in K(H)$.
- $S \in B(H), T \in K(H) \Rightarrow ST, TS \in K(H)$. □

COROLLARY. *If $\dim H = \infty$, then no invertible operator on H is compact.*

THEOREM. *$K(H)$ is closed in $B(H)$.*

PROOF. Let $(T_k) \in K(H)$ and let $T_k \to T$. Let (x_n) be bounded.
- There exists $N_1 \subset \mathbb{N}$ such that $(T_1 x_n)_{n \in N_1}$ converges.
- There exists $N_2 \subset N_1$ such that $(T_2 x_n)_{n \in N_2}$ converges. Etc.
- Take N to be the 1st element of N_1, the 2nd element of N_2, etc. Then $(T_k x_n)_{n \in N}$ converges for each k.
- As $\|T - T_k\| \to 0$, get $(Tx_n)_{n \in N}$ is Cauchy, so convergent.

Conclusion: T is compact. □

COROLLARY. *If T is a limit of bounded, finite-rank operators, then it is compact.*

THEOREM. *If $T \in K(H)$ then $T^* \in K(H)$.*

PROOF. Let (x_n) be a bounded sequence. As TT^* is compact, there exists $N \subset \mathbb{N}$ such that $(TT^*x_n)_{n\in N}$ converges. Since

$$\|T^*x_n - T^*x_m\|^2 \leq \|x_n - x_m\|\|TT^*(x_n - x_m)\|,$$

it follows that $(T^*x_n)_{n\in N}$ is Cauchy, and hence convergent. □

REMARKS. (1) To summarize: $K(H)$ is a closed $*$-ideal in $B(H)$.
(2) It can be shown that if H is infinite-dimensional and separable, then $K(H)$ is the unique non-trivial closed ideal in $B(H)$.

6.3. Compact diagonal operators.

THEOREM. *Let $T \in B(H)$, let (e_n) be an o.n. basis of H, and suppose that $Te_n = \lambda_n e_n$ for all n. Then T is compact iff $\lambda_n \to 0$.*

PROOF. Suppose that $\lambda_n \not\to 0$. Then there is a sequence $N \subset \mathbb{N}$ and $\delta > 0$ such that $|\lambda_n| \geq \delta$ for all $n \in N$. For $m, n \in N$ we then have

$$\|Te_n - Te_m\|^2 = |\lambda_n|^2 + |\lambda_m|^2 \geq 2\delta^2,$$

so $(Te_n)_{n\in N}$ has no convergent subsequence. So T is not compact.

Suppose conversely that $\lambda_n \to 0$. Let T_n be the diagonal operator with eigenvalues
$(\lambda_1, \ldots, \lambda_n, 0, 0, \ldots)$. Then T_n is finite rank and

$$\|T - T_n\| = \sup_{k>n} |\lambda_k| \to 0 \quad (n \to \infty).$$

So T is compact. □

6.4. Hilbert–Schmidt operators.

LEMMA. *Let $T \in B(H)$, and let (e_n) and (f_m) be o.n. bases of H. Then*

$$\sum_n \|Te_n\|^2 = \sum_m \|T^*f_m\|^2.$$

PROOF. By Parseval,

$$\sum_n \|Te_n\|^2 = \sum_n \sum_m |\langle Te_n, f_m\rangle|^2 = \sum_m \sum_n |\langle e_n, T^*f_m\rangle|^2 = \sum_m \|T^*f_m\|^2. \quad \square$$

DEFINITION. $T \in B(H)$ is a *Hilbert–Schmidt* operator if $\sum_n \|Te_n\|^2 < \infty$ for some (and hence for every) o.n. basis (e_n).

THEOREM. *Every Hilbert–Schmidt operator T is compact.*

PROOF. Let (e_n) be an o.n. basis. For $x \in H$, we have

$$Tx = T\Big(\sum_{k=1}^{\infty} \langle x, e_k\rangle e_k\Big) = \sum_{k=1}^{\infty} \langle x, e_k\rangle Te_k.$$

Set $T_n x := \sum_{k=1}^{n} \langle x, e_k\rangle Te_k$. By Cauchy–Schwarz and Parseval,

$$\|(T - T_n)x\| \leq \sum_{k=n+1}^{\infty} |\langle x, e_k\rangle| \|Te_k\| \leq \|x\| \Big(\sum_{k=n+1}^{\infty} \|Te_k\|^2\Big)^{1/2},$$

so $\|T - T_n\| \leq (\sum_{k=n+1}^{\infty} \|Te_k\|^2)^{1/2} \to 0$ as $n \to \infty$. As each T_n is of finite rank, it follows that T is compact. □

6.5. Integral operators.

THEOREM. *Let $K \in L^2([a,b] \times [a,b])$. For $f \in L^2[a,b]$, define*

$$T_K f(x) := \int_a^b K(x,y) f(y)\, dy.$$

(1) *$T_K \in B(L^2[a,b])$ with $\|T_K\| \le \|K\|_2$*
(2) *T_K is a Hilbert–Schmidt operator, hence compact.*

PROOF. (1) Write $K_x(y) := K(x,y)$.

- Since $K \in L^2([a,b]^2)$, we have $K_x \in L^2[a,b]$ for a.e. $x \in [a,b]$.
- For all such x, we have $T_K f(x) = \langle f, \overline{K_x} \rangle$.
- By Cauchy–Schwarz, $|T_K f(x)| \le \|K_x\|_2 \|f\|_2$ a.e.
- Hence $T_K f \in L^2([a,b])$ with $\|T_K f\|_2 \le \|K\|_2 \|f\|_2$.

(2) Let (e_n) be an o.n. basis of $L^2[a,b]$. Then

$$\sum_n \|T_K e_n\|^2 = \sum_n \int_a^b |(T_K e_n)(x)|^2\, dx = \sum_n \int_a^b |\langle e_n, \overline{K_x} \rangle|^2\, dx$$
$$= \int_a^b \sum_n |\langle e_n, \overline{K_x} \rangle|^2\, dx = \int_a^b \|\overline{K_x}\|_2^2\, dx = \|K\|_2^2 < \infty. \qquad \square$$

REMARK. All this works equally well in any countably generated, σ-finite measure space.

6.6. A compact operator with no eigenvalues. Define $T \in B(\ell^2)$ by $T := SD$, where S is the right shift and D is the diagonal operator with $De_n := e_n/n$ for all n.

- As S is bounded and D is compact, T is also compact.
- A computation gives $\|T^n\| = 1/n!$. By the spectral radius formula,

$$\rho(T) = \lim_{n\to\infty} \|T^n\|^{1/n} = \lim_{n\to\infty} (1/n!)^{1/n} = 0.$$

So the only possible eigenvalue of T is 0. But T is injective, so 0 is not an eigenvalue either.

Exercises

6A Prove that, if $S, T \in K(H)$, then $(S+T) \in K(H)$.
6B Prove that, if $S \in B(H)$ and $T \in K(H)$, then both $ST, TS \in K(H)$.
6C In the notation of §6.5, show that $(T_K)^* = T_L$, where $L(x,y) := \overline{K(y,x)}$.
6D In the example in §6.6, prove that $\|T^n\| = 1/n!$.

7. The spectral theorem

7.1. Spectral theorem.

THEOREM. *Let H be a separable Hilbert space and let $T \in B(H)$ be a compact normal operator. Then H has an o.n. basis of eigenvectors of T.*

The proof hinges on the following key lemma:

LEMMA. *A compact normal operator $T \in B(H)$ has an eigenvector.*

7.2. Proof of key lemma. Suppose first T is compact and *self-adjoint.* WLOG $\|T\| = 1$.

- Choose $x_n \in H$ with $\|x_n\| = 1$ such that $\|Tx_n\| \to 1$.
- $\|(T^2 - I)x_n\|^2 = \|T^2 x_n\|^2 + \|x_n\|^2 - 2\|Tx_n\|^2 \to 0$.
- By compactness of T, a subsequence $T^2 x_{n_j} \to$ some $y \in H$.
- As $(T^2 - I)x_{n_j} \to 0$, it follows that $x_{n_j} \to y$.
- Hence $\|y\| = 1$ and $(T^2 - I)y = 0$.
- If $(T - I)y = 0$, then y is an eigenvector of T.
- If $(T - I)y = z \neq 0$, then $(T + I)z = (T^2 - I)y = 0$, and so z is an eigenvector of T.

Conclusion: T has an eigenvector in this case.

Now suppose T is compact and *normal.*

- Set $A := (T + T^*)/2$ and $B := (T - T^*)/2i$.
- As A is compact and self-adjoint, it has an eigenvalue α.
- Set $K := \ker(A - \alpha I)$.
- As T is normal, $AB = BA$. It follows that $B(K) \subset K$.
- As $B|_K$ is compact and self-adjoint, it has an eigenvector e.
- So e is an eigenvector of both A and B.
- As $T = A + iB$, it follows that e is an eigenvector of T.

Conclusion: T has an eigenvector in this case too. □

7.3. Proof of spectral theorem. Let T be a compact normal operator on a separable H.

- Pick an o.n. basis of each eigenspace of T.
- Put them together to get a (countable) o.n. sequence (e_n).
- Let M be the closed subspace spanned by (e_n).
- Since T is normal, $T^*(M) \subset M$, whence $T(M^\perp) \subset M^\perp$.
- If $M^\perp \neq \{0\}$, then by the lemma $T|_{M^\perp}$ has an eigenvector e; then $e \in M \cap M^\perp$, so $e = 0$, contradiction.
- Conclusion: $M = H$ and (e_n) is an o.n. basis of H. □

7.4. Eigenvalue decay.

THEOREM. *Let $T \in B(H)$ with an o.n. basis (e_n) of eigenvectors, $Te_n = \lambda_n e_n$.*

(1) *If T is compact, then $\lambda_n \to 0$.*

(2) *If T is Hilbert–Schmidt, then $\sum_n |\lambda_n|^2 < \infty$.*

PROOF. (1) This is just the Theorem in §6.3.

(2) If T is Hilbert–Schmidt, then

$$\sum_n |\lambda_n|^2 = \sum_n \|Te_n\|^2 < \infty.$$ □

COROLLARY. *Every compact operator on H is the limit of a sequence of finite-rank operators.*

PROOF. If T is compact and self-adjoint, then it is a diagonal operator whose eigenvalues tend to zero, so it is the limit of finite-rank operators (see §6.3).

If T is compact but not self-adjoint, we write it as $T = A + iB$, where $A := (T + T^*)/2$ and $B := (T - T^*)/2i$ are both compact and self-adjoint. There exist finite-rank operators $A_n \to A$ and $B_n \to B$. Set $T_n := A_n + iB_n$. Then T_n is finite-rank and $T_n \to T$. □

7.5. Min-max principle. Let $\mathcal{M}_n$ denote the set of (closed) subspaces of H of dimension n.

THEOREM. *Let $T \in B(H)$ be a compact self-adjoint operator with positive eigenvalues, ordered so that $\lambda_1 \geq \lambda_2 \geq \cdots \geq 0$. Then, for each n,*

$$\lambda_n = \max_{M \in \mathcal{M}_n} \min_{\substack{x \in M \\ \|x\|=1}} \langle Tx, x\rangle = \min_{M \in \mathcal{M}_{n-1}} \max_{\substack{x \in M^\perp \\ \|x\|=1}} \langle Tx, x\rangle.$$

PROOF. We need two facts:

- For all x, we have $\langle Tx, x\rangle = \sum_k \lambda_k |\langle x, e_k\rangle|^2$.
- If $\dim M > \dim N$, then $M \cap N^\perp \neq \{0\}$ (as $P_N|_M : M \to N$ has a non-trivial kernel).

Let us write $M_n := \operatorname{span}\{e_1, \dots, e_n\}$.

- If $\dim M = n$, then $\exists x_0 \in M \cap M_{n-1}^\perp$ with $\|x_0\| = 1$, so
$$\min_{\substack{x \in M \\ \|x\|=1}} \langle Tx, x\rangle \leq \langle Tx_0, x_0\rangle = \sum_{k=n}^{\infty} \lambda_k |\langle x_0, e_k\rangle|^2 \leq \lambda_n.$$
- If $\dim M = n-1$, then then $\exists x_0 \in M_n \cap M^\perp$ with $\|x_0\| = 1$, so
$$\max_{\substack{x \in M^\perp \\ \|x\|=1}} \langle Tx, x\rangle \geq \langle Tx_0, x_0\rangle = \sum_{k=1}^{n} \lambda_k |\langle x_0, e_k\rangle|^2 \geq \lambda_n.$$
- Equality is attained in the first case if $M = M_n$ and in the second if $M = M_{n-1}$. □

7.6. Further developments. Let $T \in K(H)$. The *singular values* σ_k of T are the square roots of the eigenvalues of T^*T, listed in decreasing order.

- *Singular-value decomposition:* Given $T \in K(H)$, there are o.n. sequences (e_k) and (f_k) such that $Tx = \sum_k \sigma_k \langle x, e_k\rangle f_k$.
- *Distance formula:* If $T \in K(H)$ then, for each $n \geq 1$,
$$\sigma_n(T) = \min\{\|T - R\| : R \in B(H),\ \operatorname{rank} R < n\}.$$
Hence T is the limit of a sequence of finite-rank operators.
- *Weyl's inequality:* Let $T \in K(H)$ and let $\lambda_1, \dots, \lambda_n$ be n eigenvalues of T. Then $|\lambda_1\lambda_2 \dots \lambda_n| \leq \sigma_1\sigma_2 \dots \sigma_n$.
- *Schatten classes:* $S_p(H) := \{T : \sum_k |\sigma_k(T)|^p < \infty|\}$.
$S_2(H) =$ Hilbert–Schmidt operators.
$S_1(H) =$ trace-class operators.
- *Spectral theorem for general normal operators:* Every bounded normal operator on H is unitarily equivalent to multiplication by a bounded function on some L^2-space.

Exercises

7A (Singular-value decomposition) Let $T \in B(H)$ be a compact operator.

- Show that T^*T is a compact self-adjoint operator.
- Let (e_k) be an eigenbasis of T^*T. Show that the vectors (Te_k) are mutually orthogonal.
- Show that the eigenvalues of T^*T are positive, σ_k^2 say. Order the non-zero ones so that $\sigma_1 \geq \sigma_2 \geq \sigma_3 \geq \dots$.

- Set $f_k := Te_k/\sigma_k$ if $\sigma_k \neq 0$. Show that the resulting (e_k) and (f_k) are o.n. sequences and

$$Tx = \sum_k \sigma_k \langle x, e_k\rangle f_k \quad (x \in H).$$

7B Let $T \in B(H)$ be a compact operator, with singular-value decomposition $Tx = \sum_k \sigma_k \langle x, e_k\rangle f_k$.

- Show that, if $R \in B(H)$ and $\operatorname{rank}(R) < n$, then $\ker(R) \cap \operatorname{span}\{e_1, \dots, e_n\} \neq \{0\}$, and deduce that $\|T - R\| \geq \sigma_n$.
- Show that, if $Rx := \sum_{k=1}^{n-1} \sigma_k \langle x, e_k\rangle f_k$, then $\operatorname{rank}(R) < n$ and $\|T - R\| \leq \sigma_n$.
- Deduce the distance formula:

$$\sigma_n(T) = \min\{\|T - R\| : R \in B(H),\ \operatorname{rank}(R) < n\}.$$

7C Let H be a separable Hilbert space and E be a subset of H such that $\langle e, f\rangle = 0$ for all $e, f \in E$. Prove that E is at most countable. [This was used in proving the spectral theorem in §7.3.] [Hint: Show that the open balls $\{x \in H : \|x - e\| < \|e\|/2\}$ $(e \in E, e \neq 0)$ are disjoint.]

8. Sturm–Liouville equation

8.1. Formulation of the problem. Sturm–Liouville problem (SL): Find $y \in C^2[a, b]$ and $\lambda \in \mathbb{C}$ satisfying:

- **(SL1) differential equation:**

$$-(py')' + qy = \lambda y,$$

where $p \in C^1[a, b]$ and $p > 0$, and $q \in C[a, b]$ is real-valued, and
- **(SL2) boundary conditions:**

$$\begin{cases} \alpha_0 y(a) + \alpha_1 y'(a) = 0 \\ \beta_0 y(b) + \beta_1 y'(b) = 0, \end{cases}$$

where $(\alpha_0, \alpha_1), (\beta_0, \beta_1) \in \mathbb{R}^2 \setminus \{(0, 0)\}$.

8.2. Examples. The existence of non-trivial solutions y depends on the parameter λ.

EXAMPLE. Consider the Sturm–Liouville problem:

$$\begin{cases} -y'' = \lambda y \text{ on } [0, 1] \\ y(0) = 0,\ y(1) = 0. \end{cases}$$

This has a non-trivial solution y iff $\lambda = k^2\pi^2$ $(k = 1, 2, 3, \dots)$.
In this case $y(t) = c\sin(k\pi t)$.

REMARK. This particular SL-problem arises in the wave equation for a stretched string. Other applications include Schrödinger's equation and Legendre's equation.

8.3. Notation. In what follows:

- $Y := \{y \in C^2[a,b] : y \text{ satisfies (SL2)}\}$.
- $L : C^2[a,b] \to C[a,b]$ is defined by

$$Ly := -(py')' + qy.$$

Thus y is a solution to (SL) iff $y \in Y$ and $Ly = \lambda y$.
If $y \not\equiv 0$, we call it an *eigenfunction* of (SL) and λ its *eigenvalue.*

REMARK. This is an abuse of terminology, since $L(Y) \not\subset Y$.

8.4. The main theorem.

THEOREM. *With the above notation:*

- *The eigenvalues of (SL) form a real sequence* (λ_k) *such that*

$$\lambda_k \to \infty \quad \text{and} \quad \sum_k \lambda_k^{-2} < \infty.$$

- *The corresponding normalized eigenfunctions* (y_k) *form an o.n. basis of* $L^2[a,b]$.

Basic idea of the proof:

- Construct an integral operator T that is an inverse to the differential operator L.
- Apply the spectral theorem to T, and deduce result for L.

And now for the details ...

8.5. Proof of main theorem.

LEMMA 1 (Lagrange's identity). *If* $y_1, y_2 \in Y$, *then* $\langle Ly_1, y_2\rangle = \langle y_1, Ly_2\rangle$.

PROOF. Integrating by parts, we have

$$(*) \qquad \langle Ly_1, y_2\rangle = \int_a^b py_1'\overline{y_2'}\,dt + \int_a^b qy_1\overline{y_2}\,dt - \Big[py_1'\overline{y_2}\Big]_a^b.$$

Thus $\langle Ly_1, y_2\rangle - \langle y_1, Ly_2\rangle = -\Big[p.(y_1'\overline{y_2} - \overline{y_2'}y_1)\Big]_a^b = 0$, as $y_j \in Y$. □

Consequences:

- Every eigenvalue λ of (SL) is real.
- Eigenfunctions having distinct eigenvalues are orthogonal.

LEMMA 2. *There exists a constant* C *such that each eigenvalue* $\lambda > C$.

PROOF. Suppose not.

- Then there exist at least five eigenvalues $\lambda_1, \ldots, \lambda_5 < \min q$.
- Let $y_1, \ldots, y_5$ be the associated normalized eigenfunctions.
- By linear algebra, there exist $\gamma_1, \ldots, \gamma_5 \in \mathbb{C}$, not all zero, such that $y := \sum_1^5 \gamma_j y_j$ satisfies $y(a) = y'(a) = y(b) = y'(b) = 0$.
- By $(*)$, $\langle Ly, y\rangle = \int_a^b p|y'|^2 + \int_a^b q|y|^2 \ge (\min q)\sum_1^5 |\gamma_j|^2$.
- Also $\langle Ly, y\rangle = \sum_1^5 \lambda_j|\gamma_j|^2 < (\min q)\sum_1^5 |\gamma_j|^2$. Contradiction. □

Consequence: Replacing q by $q + C$ if necessary, WLOG we may suppose that all eigenvalues $\lambda > 0$. In particular, $L|_Y$ is injective.

LEMMA 3. *There exist real-valued* $u, v \in C^2[a,b]$ *such that*

- $Lu = Lv = 0$;
- $\alpha_0 u(a) + \alpha_1 u'(a) = 0$;
- $\beta_0 v(b) + \beta_1 v'(b) = 0$;
- $p.(u'v - v'u) \equiv 1$.

PROOF. We use the existence/uniqueness theorem for the initial-value problem (section 2).

- $\exists u \in C^2, u \not\equiv 0$ such that $Lu = 0$ and $\alpha_0 u(a) + \alpha_1 u'(a) = 0$, $\exists v \in C^2, v \not\equiv 0$ such that $Lv = 0$ and $\beta_0 v(b) + \beta_1 v'(b) = 0$.
- $p.(u'v - v'u)$ is a constant because
$$[p.(u'v - v'u)]' = (Lu)v + pu'v' - (Lv)u - pu'v' = 0.$$
- If the constant is 0, then $v \in Y$, contradicting $L|_Y$ injective.
- Replace u by a multiple of itself to make the constant 1. □

LEMMA 4. *With* u, v *as above, define* $G : [a,b] \times [a,b] \to \mathbb{R}$ *by*
$$G(s,t) := \begin{cases} u(s)v(t), & s \le t, \\ u(t)v(s), & t \le s, \end{cases}$$
and $T : L^2[a,b] \to L^2[a,b]$ *by*
$$Tf(s) := \int_a^b G(s,t)f(t)\,dt.$$
If $f \in C[a,b]$, *then* $Tf \in Y$ *and* $L(Tf) = f$. *Consequently*
$$T : C[a,b] \to Y$$
$$L : Y \to C[a,b]$$
are mutually inverse mappings.

PROOF. Fix $f \in C[a,b]$.

- We have
$$Tf(s) = v(s)\int_a^s uf + u(s)\int_s^b vf,$$
$$(Tf)'(s) = v'(s)\int_a^s uf + u'(s)\int_s^b vf.$$
It follows that $Tf \in C^2[a,b]$.
- Further
$$\alpha_0(Tf)(a) + \alpha_1(Tf)'(a) = (\alpha_0 u(a) + \alpha_1 u'(a))\int_a^b vf = 0.$$
Likewise $\beta_0(Tf)(b) + \beta_1(Tf)'(b) = 0$. So $Tf \in Y$.
- Lastly, writing $U(s) := \int_a^s uf$ and $V := \int_s^b vf$, we have
$$L(Tf) = -(p(v'U + u'V))' + q(vU + uV)$$
$$= (Lv)U + (Lu)V + p.(u'v - v'u)f = f.$$
□

CONCLUSION OF PROOF OF MAIN THEOREM.

- As $G \in L^2([a,b] \times [a,b])$, the operator T is Hilbert–Schmidt. Also, as G is real-valued and symmetric, T is self-adjoint.

- By the spectral theorem, $L^2[a,b]$ has an orthonormal basis of eigenvectors (y_k) of T and the eigenvalues satisfy $\sum_k |\mu_k|^2 < \infty$.
- The (y_k) automatically belong to Y. Indeed, since T maps $L^2[a,b]$ into $C[a,b]$ and $y_k = (1/\mu_k)Ty_k$, we have $y_k \in C[a,b]$. Likewise, since T maps $C[a,b]$ into Y, similar reasoning gives $y_k \in Y$.
- As T, L are mutually inverse, they have the same eigenvectors, and the eigenvalues of L are the reciprocals of those of T.
- Hence the (normalized) eigenfunctions of L form an o.n. basis of $L^2[a,b]$. Also the eigenvalues of L satisfy $\sum_k |\lambda_k|^{-2} < \infty$.
- Lastly, as already observed, the eigenvalues of L are real and bounded below. Consequently $\lambda_k \to \infty$ as $k \to \infty$. □

8.6. Further developments.

- The eigenvalues of L are simple (i.e. $\dim\ker(L - \lambda_k I) = 1$).
- The eigenfunctions y_k can be chosen real-valued.
- *Oscillation theorem:* y_k has exactly $k-1$ zeros in (a,b).
- SL with weights: $-(py')' + qy = \lambda ry$, where $r(t) > 0$.
- Singular Sturm–Liouville problems.
- Boundary-value problems in higher dimensions ...

Exercises

8A Let y_1, y_2 be two eigenfunctions of (SL) corresponding to the same eigenvalue λ. Show that there exist $\gamma_1, \gamma_2 \in \mathbb{C}$, not both zero, such that $z := \gamma_1 y_1 + \gamma_2 y_2$ satisfies $z(a) = z'(a) = 0$, and deduce that $z \equiv 0$. [*Hint for the last part:* Use the uniqueness of the solution to $(*)$ in §2.6.]

8B Let y be an eigenfunction of (SL). Show that $\overline{y}$ is also an eigenfunction with the same eigenvalue. Deduce that there exists a real-valued eigenfunction with the same eigenvalue.

Département de mathématiques et de statistique, Université Laval, Québec (QC), G1V0A6, Canada
Email address: ransford@mat.ulaval.ca

Contemporary Mathematics
Volume **720**, 2018
https://doi.org/10.1090/conm/720/14521

Spectral theory of partial differential equations

Richard S. Laugesen

Abstract. These lectures present highlights of spectral theory for selfadjoint partial differential operators, emphasizing problems with discrete spectrum. Spectral methods permeate the theory of partial differential equations. Linear PDEs are often solved by separation of variables, getting eigenvalues when the spectrum is discrete and continuous spectrum when it is not. Further, linearized stability of a steady state or traveling wave of a nonlinear PDE depends on the sign of the first eigenvalue, or more generally on the location of the spectrum in the complex plane.

We define eigenvalues in terms of quadratic forms on a general Hilbert space. Particular applications include the eigenvalues of the Laplacian under Dirichlet and Neumann boundary conditions. Rayleigh-type principles characterize the first and higher eigenvalues, and lead to a number of comparison and domain monotonicity properties. Lastly, the role of eigenvalues in stability analysis is investigated for a reaction-diffusion equation in one spatial dimension.

Computable examples are presented before the general theory. Some ideas are used before being properly defined, but overall students gain a better understanding of the purpose of the theory by gaining first a solid grounding in specific examples.

Contents

2010 *Mathematics Subject Classification.* 35F15.

1. Resources, prerequisites and notation

Online resources

Expanded lecture notes covering more topics
`https://wiki.cites.illinois.edu/wiki/display/MATH595STP/Math+595+STP`

Java applets for vibrating membranes (and more)
`http://www.falstad.com/mathphysics.html`

Sage applets for vibrating membranes
`http://www.math.illinois.edu/~laugesen/applets.html`

Prerequisites

We assume familiarity with elementary Hilbert space theory: inner product, norm, Cauchy–Schwarz, orthonormal basis (ONB), bounded operators, and compact operators. Our treatment of discrete spectra builds on the spectral theorem for compact, selfadjoint operators.

Function spaces

All functions are assumed measurable. Readers unfamiliar with Sobolev spaces need only know that:

$$\begin{aligned} L^2 &= \{\text{square integrable functions}\}, \\ H^1 = W^{1,2} &= \{L^2\text{-functions with 1 derivative in } L^2\}, \\ H^1_0 = W^{1,2}_0 &= \{H^1\text{-functions that equal zero on the boundary}\}. \end{aligned}$$

These characterizations are not mathematically precise, but suffice for our purposes. Later we will recall the standard inner products that make these Sobolev spaces into Hilbert spaces.

For more on Sobolev space theory, and related concepts of weak solutions and elliptic regularity, see [**Evans**].

Laplacian

$$\Delta = \nabla \cdot \nabla = \Big(\frac{\partial}{\partial x_1}\Big)^2 + \cdots + \Big(\frac{\partial}{\partial x_d}\Big)^2$$

Eigenfunctions satisfy $\boxed{-\Delta u = \lambda u}$, and we write the eigenvalues in increasing order as

$$\lambda_1 \leq \lambda_2 \leq \lambda_3 \leq \cdots \to \infty.$$

Divergence theorem

Given a bounded domain $\Omega \subset \mathbb{R}^d$ with smooth enough boundary, and a vector field F on the closure of the domain, one has

$$\int_\Omega \nabla \cdot F \, dx = \int_{\partial\Omega} F \cdot n \, dS$$

where n denotes the outward unit normal vector on $\partial\Omega$. In 1-dimension, the Divergence Theorem is simply the Fundamental Theorem of Calculus:

$$\int_{(a,b)} F'(x) \, dx = -F(a) + F(b)$$

where the negative sign indicates the leftward orientation of n at $x = a$.

Integration by parts

$$\int_\Omega \frac{\partial u}{\partial x_j} v\, dx = -\int_\Omega u \frac{\partial v}{\partial x_j}\, dx + \int_{\partial\Omega} uv\, n_j\, dS$$

where n_j is the jth component of the normal vector.

Proof. Apply the Divergence Theorem to $F = (0, \dots, 0, uv, 0, \dots, 0)$.

Green's formulas

$$\int_\Omega (\nabla u \cdot \nabla v + v\Delta u)\, dx = \int_{\partial\Omega} v \frac{\partial u}{\partial n}\, dS \tag{1.1}$$

$$\int_\Omega (u\Delta v - v\Delta u)\, dx = \int_{\partial\Omega} \Big(u\frac{\partial v}{\partial n} - v\frac{\partial u}{\partial n}\Big)\, dS \tag{1.2}$$

where the normal derivative is defined as the normal component of the gradient vector:

$$\frac{\partial u}{\partial n} = \nabla u \cdot n.$$

Proof of Green's formulas. Apply the Divergence theorem to $F = v\nabla u$. Interchange u and v, and subtract.

Boundary conditions

Ω: domain in $\mathbb{R}^d, d \geq 1$, $u(x)$: function defined on Ω
BC means boundary condition

Dirichlet BC: $u = 0$ on $\partial\Omega$

Neumann BC: $\dfrac{\partial u}{\partial n} = 0$ on $\partial\Omega$

2. Computable spectra and qualitative properties—Laplacian

Goal. To develop a library of explicitly computable spectra (which will motivate the later general theory) and to extract from these examples some qualitative properties such as scaling and asymptotic growth rates.

These computable spectra are classical, and so proofs are left to the reader or omitted, although the Weyl asymptotic law is proved in detail for rectangles.

Notation. Let Ω be a domain in $\mathbb{R}^d, d \geq 1$. Fix $L > 0$.

- *Dirichlet* BC means $u = 0$ on $\partial\Omega$,
- *Neumann* BC means $\frac{\partial u}{\partial n} = 0$ on $\partial\Omega$.

For a vibrating drum, Dirichlet BCs mean the drum is fixed at the boundary, while Neumann BCs mean the drum is free (attached to a frictionless vertical support). For the diffusion or heat equation, Dirichlet BCs mean the boundary is refrigerated to maintain the temperature at zero, while Neumann BCs mean the boundary is perfectly insulated and so has heat flux zero.

Spectra of the Laplacian under these BCs. Recall the Laplacian is

$$\Delta = \nabla \cdot \nabla = \Big(\frac{\partial}{\partial x_1}\Big)^2 + \cdots + \Big(\frac{\partial}{\partial x_d}\Big)^2.$$

Its eigenfunctions satisfy $\boxed{-\Delta u = \lambda u}$, and we put the eigenvalues in increasing order as

$$\lambda_1 \leq \lambda_2 \leq \lambda_3 \leq \cdots \to \infty.$$

Section 5 proves existence of the spectrum under various BCs. Explicit formulas for these eigenvalues can be computed on just a handful of special domains. This section summarizes the most important computable cases.

Note. One usually normalizes the eigenfunctions in L^2, in order to get an ONB, but for simplicity we do not normalize the following examples.

One dimension $\boxed{-u'' = \lambda u}$

1. *Circle* $\mathbb{T} = \mathbb{R}/2\pi\mathbb{Z}$, periodic BC: $u(-\pi) = u(\pi), u'(-\pi) = u'(\pi)$.
   Eigenfunctions e^{ijx} for $j \in \mathbb{Z}$, or equivalently $1, \cos(jx), \sin(jx)$ for $j \geq 1$.
   Eigenvalues $\lambda_j = j^2$ for $j \in \mathbb{Z}$, or $\lambda = 0^2, 1^2, 1^2, 2^2, 2^2, \ldots$

2. *Interval* $(0, L)$

(a) Dirichlet BC: $u(0) = u(L) = 0$.
   Eigenfunctions $u_j(x) = \sin(j\pi x/L)$ for $j \geq 1$.
   Eigenvalues $\lambda_j = (j\pi/L)^2$ for $j \geq 1$, *e.g.* $L = \pi \Rightarrow \lambda = 1^2, 2^2, 3^2, \ldots$

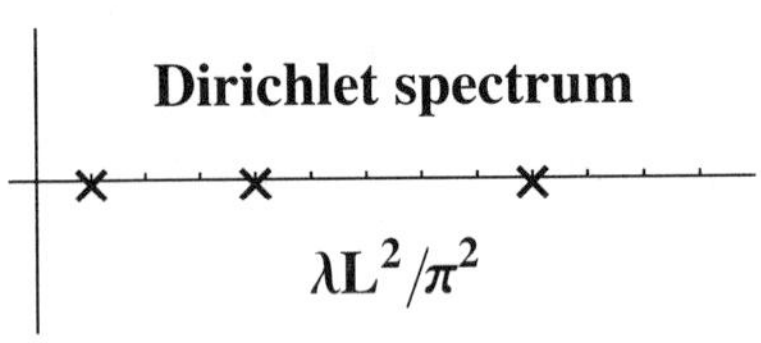

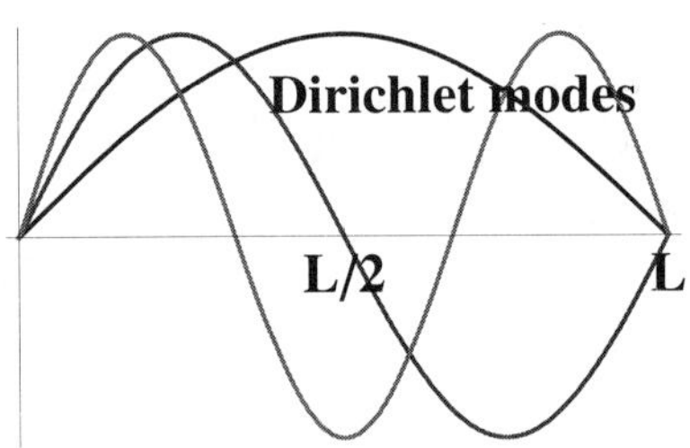

(b) Neumann BC: $u'(0) = u'(L) = 0$.
   Eigenfunctions $u_j(x) = \cos\big((j-1)\pi x/L\big)$ for $j \geq 1$ (note $u_1 \equiv 1$).
   Eigenvalues $\mu_j = \big((j-1)\pi/L\big)^2$ for $j \geq 1$, *e.g.* $L = \pi \Rightarrow \mu = 0^2, 1^2, 2^2, 3^2, \ldots$

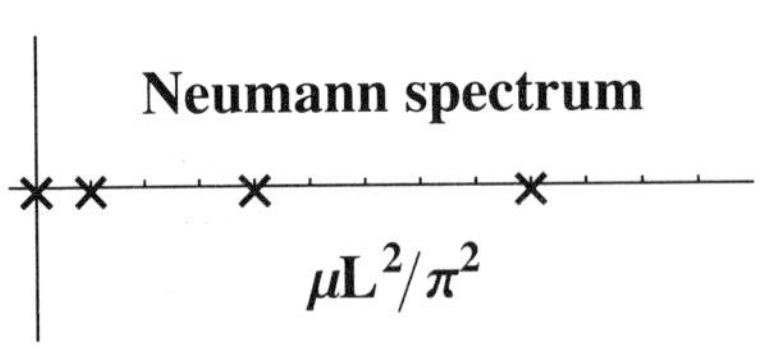

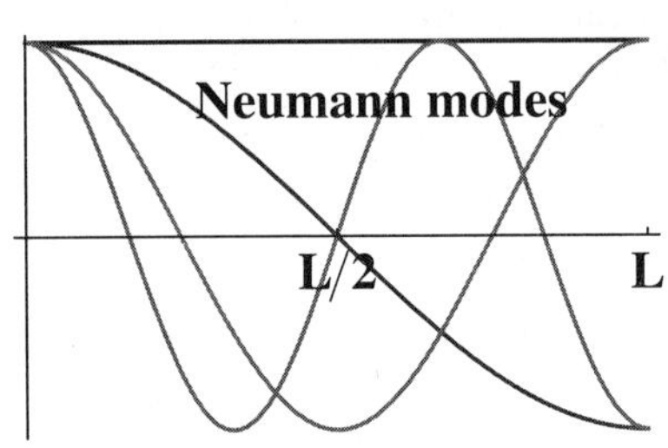

Spectral features in 1 dim: $-u'' = \lambda u$

i. Spatial frequency increases as j increases. Thus the temporal and spatial frequencies increase together.

ii. Symmetry: eigenfunctions are either even or odd with respect to the midpoint of the interval, that is, under the substitution $x \mapsto L - x$.

iii. Concavity: eigenfunctions are concave up wherever u is negative ($u < 0$ implies $u'' > 0$), and concave down wherever u is positive ($u > 0$ implies $u'' < 0$). Physically, this occurs because the tension in the vibrating string wants to pull it back toward the rest state.

iv. Scaling: the eigenvalue λ must balance d^2/dx^2 in the eigenvalue equation, and so $\lambda \sim (\text{length scale})^{-2}$. Intuitively, long strings produce low tones. More precisely, the eigenvalues scale like the reciprocal square of the length of the interval:

$$\lambda_j\big((0,L)\big) = \frac{\lambda_j\big((0,1)\big)}{L^2}$$

and similarly for the Neumann eigenvalue μ_j.

v. Growth rate: eigenvalues grow at a regular rate, $\lambda_j \sim (\text{const.})j^2$

Two dimensions $\boxed{-\Delta u = \lambda u}$

Write $\Delta = \partial^2/\partial x^2 + \partial^2/\partial y^2$.

1. *Rectangle* $\Omega = (0,L) \times (0,M)$ (product of intervals).

We may assume the sides of the rectangle are parallel to the coordinate axes because the Laplacian, and hence its spectrum, is rotationally and translationally invariant; see the Exercises at the end of the section.

One finds eigenvalues by separating variables in rectangular coordinates x, y. See the formulas below and the figures at the end of the section.

(a) Dirichlet BC: $u = 0$
Eigenfunctions $u(x) = \sin(j\pi x/L)\sin(k\pi y/M)$ for $j, k \geq 1$.
Eigenvalues $\lambda = (j\pi/L)^2 + (k\pi/M)^2$ for $j, k \geq 1$.
e.g. $L = M = \pi \Rightarrow \lambda = j^2 + k^2$ for $j, k \geq 1$; that is, $\lambda = 2, 5, 5, 8, 10, 10, \ldots$
These eigenvalues are the squares of distances from the origin to the positive integer lattice points.

(b) Neumann BC: $\frac{\partial u}{\partial n} = 0$
Eigenfunctions $u(x) = \cos\big((j-1)\pi x/L\big)\cos\big((k-1)\pi y/M\big)$ for $j, k \geq 1$.
Eigenvalues $\mu = \big((j-1)\pi/L\big)^2 + \big((k-1)\pi/M\big)^2$ for $j, k \geq 1$.
e.g. $L = M = \pi \Rightarrow \mu = 0, 1, 1, 2, 4, 4, \ldots$

2. *Disk* $\Omega = \{x \in \mathbb{R}^2 : |x| < R\}$
Separate variables using polar coordinates r, θ.

(a) Dirichlet BC: $u = 0$
Eigenfunctions

$$J_0(rj_{0,m}/R) \text{ for } m \geq 1,$$
$$J_n(rj_{n,m}/R)\cos(n\theta) \text{ and } J_n(rj_{n,m}/R)\sin(n\theta) \text{ for } n \geq 1, m \geq 1.$$

Eigenvalues $\lambda = (j_{n,m}/R)^2$ for $n \geq 0, m \geq 1$, where

$$J_n = \text{Bessel function of order } n,$$
$$j_{n,m} = m\text{-th positive root of } J_n(r) = 0.$$

When $n = 0$, the modes are purely radial. When $n \geq 1$, the modes have angular dependence (both cosine and sine modes), and $\lambda_{n,m}$ has multiplicity 2. The Bessel functions satisfy

$$r^2 J_n''(r) + rJ_n'(r) + (r^2 - n^2)J_n(r) = 0$$

and behave like r^n near the origin: $J_n(r) \simeq (\text{const.})r^n$ when $r \simeq 0$.

From the graphs of the Bessel functions J_0, J_1, J_2 we can read off the first four roots:

$$j_{0,1} \simeq 2.40, \quad j_{1,1} \simeq 3.83, \quad j_{2,1} \simeq 5.13, \quad j_{0,2} \simeq 5.52.$$

These roots generate the first six eigenvalues (remembering eigenvalues are double when $n \geq 1$).

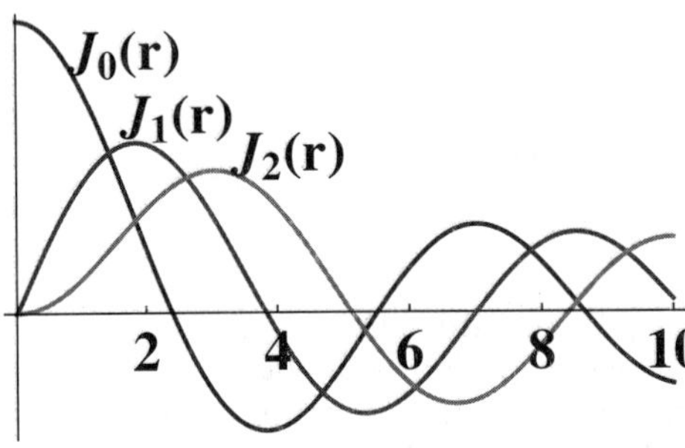

(b) Neumann BC: $\frac{\partial u}{\partial n} = 0$. Use roots of $J_n'(r) = 0$; see [**Bandle**, Chapter III].

3. *Equilateral triangle* of sidelength L.

Separation of variables fails, but one may reflect repeatedly to a hexagonal lattice whose eigenfunctions are trigonometric. See [**Mathews and Walker, McCartin**].

Dirichlet eigenvalues:

$$\lambda = \tfrac{16\pi^2}{9L^2}(j^2 + jk + k^2) \qquad \text{for } j, k \geq 1.$$

Neumann eigenvalues:

$$\mu = \tfrac{16\pi^2}{9L^2}(j^2 + jk + k^2) \qquad \text{for } j, k \geq 0.$$

Spectral features in 2 dim: $-\Delta u = \lambda u$

i. Scaling: the eigenvalue λ must balance Δ, and so $\lambda \sim (\text{length scale})^{-2}$. Intuitively, big drums produce low tones. More precisely, Dirichlet and Neumann eigenvalues scale like the reciprocal square of the scale of the domain:

$$\lambda_n(t\Omega) = \frac{\lambda_n(\Omega)}{t^2}$$

and similarly for the Neumann eigenvalue μ_n. This scaling relation can be verified explicitly in the examples above, and holds also for general domains as explained in the exercises.

ii. Sub/superharmonicity: eigenfunctions are subharmonic wherever u is negative (since $u < 0$ implies $\Delta u > 0$), and superharmonic wherever u is positive (since $u > 0$ implies $\Delta u < 0$). Physically, the tension in the membrane wants to pull it back toward the rest state.

iii. Degnerate domains: Dirichlet and Neumann spectra behave quite differently when the domain degenerates. Consider the rectangle, for example. Fix one side length L, and let the other side length M tend to 0. Then the first positive Dirichlet eigenvalue blows up, since taking $j = k = 1$ gives eigenvalue $(\pi/L)^2 + (\pi/M)^2 \to \infty$. Meanwhile the first positive Neumann eigenvalue is constant (independent of M) because taking $j = 1, k = 0$, gives eigenvalue $(\pi/L)^2$.

iv. Growth rate: eigenvalues of the rectangle grow at a regular rate:

PROPOSITION 2.1. *(Weyl's law for rectangles) The rectangle* $(0, L) \times (0, M)$ *satisfies*

$$\lambda_n \sim \mu_n \sim \frac{4\pi n}{Area} \qquad as\ n \to \infty,$$

where $Area = LM$ *is the area of the rectangle and* $\lambda_1, \lambda_2, \lambda_3, \dots$ *and* $\mu_1, \mu_2, \mu_3, \dots$ *are the Dirichlet and Neumann eigenvalues respectively, in increasing order.*

PROOF. We give a proof for Dirichlet eigenvalues; the Neumann case is similar. Define for $\alpha > 0$ the eigenvalue counting function

$$\begin{aligned} N(\alpha) &= \#\{\text{eigenvalues} \leq \alpha\} \\ &= \#\{(j,k) \in \mathbb{N} \times \mathbb{N} : \frac{j^2}{\alpha L^2/\pi^2} + \frac{k^2}{\alpha M^2/\pi^2} \leq 1\} \\ &= \#\{(j,k) \in \mathbb{N} \times \mathbb{N} : (j,k) \in E\} \end{aligned}$$

where E is the ellipse $(x/a)^2 + (y/b)^2 \leq 1$ with $a = \sqrt{\alpha}L/\pi, b = \sqrt{\alpha}M/\pi$.

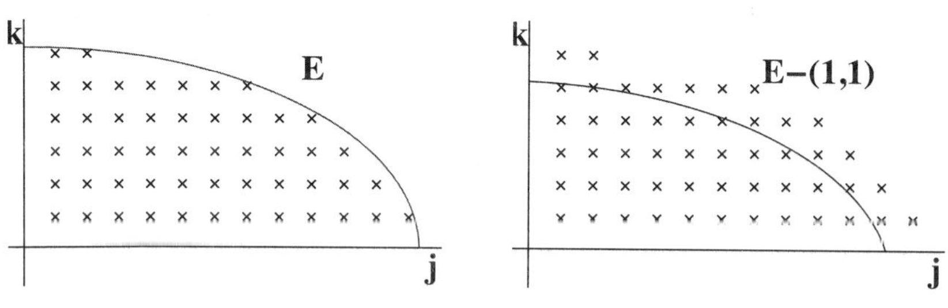

We associate each lattice point $(j, k) \in E$ with the square

$$S(j,k) = [j-1, j] \times [k-1, k]$$

whose upper right corner lies at (j, k). These squares all lie within E, and so by comparing areas we find

$$N(\alpha) \leq (\text{area of } E \text{ in first quadrant}) = \frac{1}{4}\pi ab = \frac{\text{Area}}{4\pi}\alpha.$$

In the reverse direction, the union of the squares having $(j, k) \in E$ covers a copy of E shifted down and left by one unit, as a little thought shows:

$$\cup S(j,k) \supset \big(E - (1,1)\big) \cap (\text{first quadrant}).$$

Comparing areas implies

$$\begin{aligned} N(\alpha) &\geq \frac{1}{4}\pi ab - a - b \\ &= \frac{LM}{4\pi}\alpha - \frac{L+M}{\pi}\sqrt{\alpha} \\ &= \frac{\text{Area}}{4\pi}\alpha - \frac{\text{Perimeter}}{2\pi}\sqrt{\alpha}. \end{aligned}$$

Combining our upper and lower estimates shows that

$$N(\alpha) \sim \frac{\text{Area}}{4\pi}\alpha$$

as $\alpha \to \infty$. To complete the proof we simply invert this last asymptotic, with the help of the lemma below. □

LEMMA 2.2. *(Inversion of asymptotics) Fix $c > 0$. Then:*

$$N(\alpha) \sim \frac{\alpha}{c} \quad \text{as } \alpha \to \infty \qquad \Longrightarrow \qquad \lambda_n \sim cn \quad \text{as } n \to \infty.$$

PROOF. Formally substituting $\alpha = \lambda_n$ and $N(\alpha) = n$ takes us from the first asymptotic to the second. The difficulty with making this substitution rigorous is that if λ_n is a multiple eigenvalue, then $N(\lambda_n)$ can exceed n. In other words, $N(\lambda_n) \geq n$ but equality need not hold.

To circumvent the problem, we argue as follows. Given $\varepsilon > 0$, we know from $N(\alpha) \sim \alpha/c$ that

$$(1-\varepsilon)\frac{\alpha}{c} < N(\alpha) < (1+\varepsilon)\frac{\alpha}{c}$$

for all large α. Substituting $\alpha = \lambda_n$ into the right inequality and recalling $N(\lambda_n) \geq n$ implies that

$$n < (1+\varepsilon)\frac{\lambda_n}{c}$$

for all large n. Letting $0 < \delta < 1$ and substituting $\alpha = \lambda_n - \delta$ into the left inequality implies, since $N(\lambda_n - \delta) < n$, that

$$(1-\varepsilon)\frac{\lambda_n - \delta}{c} < n$$

for each large n. Hence by letting $\delta \to 0$,

$$(1-\varepsilon)\frac{\lambda_n}{c} \leq n.$$

We deduce that

$$\frac{1}{1+\varepsilon} < \frac{\lambda_n}{cn} \leq \frac{1}{1-\varepsilon}$$

for all sufficiently large n (depending on ε), and since ε is arbitrary we conclude

$$\lim_{n\to\infty} \frac{\lambda_n}{cn} = 1$$

as desired. □

Remark. More generally, one has **Weyl's Asymptotic Law**:

$$\lambda_n \sim 4\pi n/\text{Area}$$

for all bounded domains in 2 dimensions, regardless of shape or boundary conditions.

Exercises. *Max/min problems for eigenvalues of rectangles and domains in the plane*

2.1 — Find the shape of rectangle that minimizes $\lambda_1 A$ among all rectangles, where A denotes the area.
Note. The eigenvalue is multiplied by area in this problem in order to obtain a scale-invariant quantity: notice $\lambda_1 A$ does not change value when the side lengths of the rectangle are multiplied by a constant factor (replacing L and M with tL and tM, for example).

2.2 — Guess the shape of domain that minimizes $\lambda_1 A$ among all planar domains. (Faber–Krahn theorem.)

2.3 — Find the shape of rectangle that maximizes $\mu_2 A$ among all rectangles.

2.4 — Guess the shape of domain that maximizes $\mu_2 A$ among all planar domains. (Szegő–Weinberger theorem.)

The moral is the same for each of the preceding problems — namely that nature prefers symmetric optimizers.

Inverse spectral problems

2.5 — Show that a square is determined by its fundamental tone, that is, by its first Dirichlet eigenvalue. (In other words, given a square domain and its first eigenvalue, one can determine the sidelength of the square.)

2.6 — How many Dirichlet eigenvalues are needed to determine a rectangle?

2.7 — How many Dirichlet eigenvalues do you expect would determine a triangle? (Open problem! [**Antunes and Freitas**] investigated numerically. More about eigenvalues on triangles can be found in [**Henrot et al.**])

2.8 — How many Neumann eigenvalues are needed to determine a rectangle?

Eigenfunctions of a right isosceles triangle

2.9 — Consider the 45–45–90 degree triangle $\{(x,y) : 0 < x < y < \pi\}$, which is the region of the square $(0,\pi) \times (0,\pi)$ lying above the diagonal line $x = y$.

(i) Find an eigenfunction of the Laplacian on this triangle that satisfies the Dirichlet boundary condition on the left and top sides, and the Neumann condition on the diagonal side.

(ii) Find an eigenfunction of the Laplacian on this triangle that satisfies the Dirichlet boundary condition on all three sides.

(iii) Find infinitely many Dirichlet eigenfunctions on the triangle.

Invariance under reflections, translations and rotations in the plane

2.10 — Show the Laplacian is invariant under reflections across the y-axis:

$$\Delta\big(u(-x,y)\big) = (\Delta u)(-x,y)$$

whenever u is a smooth function of two variables. One may write the result more formally as

$$\Delta(u \circ F) = (\Delta u) \circ F$$

where

$$F\left(\begin{smallmatrix} x \\ y \end{smallmatrix}\right) = \begin{pmatrix} -x \\ y \end{pmatrix}$$

is the reflection of $\mathbb{R}^2$ across the y-axis.

2.11 — Show the Laplacian is invariant under translations:

$$\Delta(u \circ T) = (\Delta u) \circ T$$

where u is a smooth function of two variables and

$$T\left(\begin{smallmatrix} x \\ y \end{smallmatrix}\right) = \begin{pmatrix} x+a \\ y+b \end{pmatrix}$$

is a translation of $\mathbb{R}^2$ by the constant vector $\left(\begin{smallmatrix} a \\ b \end{smallmatrix}\right)$.

2.12 — Show the Laplacian is invariant under rotations:

$$\Delta(u \circ R) = (\Delta u) \circ R$$

where u is a smooth function of two variables and

$$R\left(\begin{smallmatrix} x \\ y \end{smallmatrix}\right) = \begin{pmatrix} \cos\alpha & -\sin\alpha \\ \sin\alpha & \cos\alpha \end{pmatrix} \begin{pmatrix} x \\ y \end{pmatrix}$$

is a rotation of $\mathbb{R}^2$ by angle α.

Remark. Rotational invariance of the Laplacian might seem surprising, since the Laplacian is defined in terms of rectangular coordinates.

2.13 — Show the Dirichlet eigenvalues of a domain remain unchanged if the domain is reflected, translated or rotated. The same holds true for Neumann eigenvalues. (Physically, this invariance means vibrating membranes are unaffected by the coordinate systems we place upon them.)

These invariance properties of the spectrum hold in higher dimensions also.

Scaling invariance

2.14 — Show the Laplacian is invariant under dilations:

$$\Delta\big(u(tx,ty)\big) = t^2(\Delta u)(tx,ty)$$

whenever u is a smooth function of two variables and $t > 0$. That is,

$$\Delta(u \circ D) = t^2(\Delta u) \circ D$$

where

$$D\left(\begin{smallmatrix} x \\ y \end{smallmatrix}\right) = t\begin{pmatrix} x \\ y \end{pmatrix}$$

is the dilation or scaling of $\mathbb{R}^2$ by the factor t.

2.15 — Show the Dirichlet eigenvalues of a domain Ω rescale as follows:

$$\lambda_n(t\Omega) = \frac{\lambda_n(\Omega)}{t^2}$$

whenever $t > 0$, and similarly for the Neumann eigenvalues μ_n. (Physically, this scaling formula means large drums produce low tones.)

These rescaling formulas hold in higher dimensions also.

Weyl asymptotic and higher eigenvalues

2.14 — Prove that for rectangles, the Weyl asymptotic expression provides a lower bound on each Dirichlet eigenvalue:

$$\lambda_n A \geq 4\pi n, \qquad n = 1, 2, 3 \ldots$$

Pólya's Conjecture claims that this inequality holds on every (bounded) domain in 2-dimensions. (Open problem! Known for tiling domains, but for general domains the conjecture remains open for all $n \geq 3$.)

Notes and comments. For the computable spectra stated in this section, see [**Strauss**] Chapters 4 and 10, and [**Farlow**] Lesson 30, among many other good sources.

Question. What does a "typical" eigenfunction look like, in each of the examples above? See the figures below...

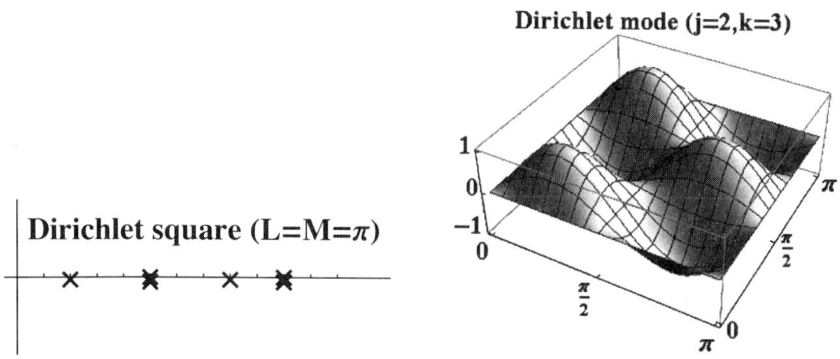

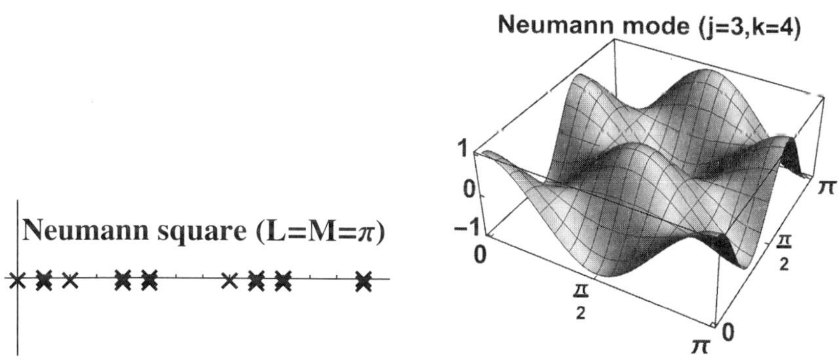

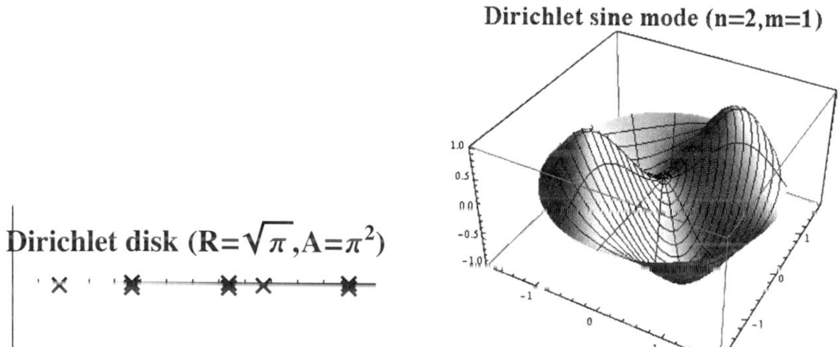

3. Discrete spectral theorem for sesquilinear forms

Goal. To state the spectral theorem for a coercive sesquilinear form on a dense, compactly imbedded Hilbert space, and to prove it using the spectral theorem for compact selfadjoint operators.

Later sections will apply this spectral theorem to unify the explicit examples of Section 2.

Preview 1 — weak eigenvectors of matrices. Consider a Hermitian matrix A (so that $A^T = \overline{A}$) and suppose u is an eigenvector with eigenvalue γ, meaning $Au = \gamma u$. Take the dot product with an arbitrary vector v to obtain

$$Au \cdot \overline{v} = \gamma u \cdot \overline{v}, \qquad \forall v \in \mathbb{C}^d.$$

Call this condition the "weak form" of the eigenvector equation. Clearly it implies the original "strong" form, because $(Au - \gamma u) \cdot \overline{v} = 0$ for all v implies $Au - \gamma u = 0$.

Notice the Hermitian property of the matrix guarantees the left side of the weak equation is conjugate-symmetric: $Au \cdot \overline{v} = \overline{Av \cdot \overline{u}}$. Symmetry ensures that all eigenvalues are real ($\overline{\gamma} = \gamma$), by choosing $v = u$ in the weak form.

The weak form of the eigenvector equation indicates the correct approach for generalizing from the matrix case to Hilbert spaces, in this section.

Preview 2 — weak eigenfunctions of the Laplacian. Consider an eigenfunction of the Laplacian in a domain Ω, satisfying $-\Delta u = \lambda u$. Multiply by a function $v \in H_0^1(\Omega)$ and integrate to obtain

$$-\int_\Omega v \Delta u \, dx = \lambda \int_\Omega uv \, dx.$$

Assume u and v are real valued for simplicity. Green's formula (1.1) and the assumption that $v = 0$ on $\partial\Omega$ imply that

$$\int_\Omega \nabla u \cdot \nabla v \, dx = \lambda \langle u, v \rangle_{L^2(\Omega)}, \qquad \forall v \in H_0^1(\Omega).$$

Call this condition the "weak form" of the eigenfunction equation. Our plan is to construct an ONB of such weak eigenfunctions using the Hilbert space methods of this section, and later invoke PDE regularity theory to conclude that weak eigenfunctions are in fact smooth functions satisfying the equation $-\Delta u = \lambda u$. Section 5 will implement this plan.

Now we construct the spectral theorem for sesquilinear forms in Hilbert space.

Hypotheses. Consider two infinite dimensional Hilbert spaces $\mathcal{H}$ and $\mathcal{K}$ over $\mathbb{R}$ (or $\mathbb{C}$):

$\mathcal{H}$: inner product $\langle u, v \rangle_\mathcal{H}$, norm $\|u\|_\mathcal{H}$,
$\mathcal{K}$: inner product $\langle u, v \rangle_\mathcal{K}$, norm $\|u\|_\mathcal{K}$.

Assume $\mathcal{H}$ is separable (has a countable dense subset) and:

1. $\mathcal{K}$ is continuously and densely imbedded in $\mathcal{H}$, meaning a continuous linear injection

$$\iota : \mathcal{K} \to \mathcal{H}$$

exists such that $\iota(\mathcal{K})$ dense in $\mathcal{H}$. Thus we may regard $\mathcal{K}$ as a subset of $\mathcal{H}$.

2. The imbedding $\mathcal{K} \hookrightarrow \mathcal{H}$ is **compact**, meaning if E is a bounded subset of $\mathcal{K}$ then E is precompact when considered as a subset of $\mathcal{H}$. Equivalently, every bounded sequence in $\mathcal{K}$ has a subsequence that converges in $\mathcal{H}$.

3. A map $a : \mathcal{K} \times \mathcal{K} \to \mathbb{R}$ (or $\mathbb{C}$) exists that is **sesquilinear**:

$$u \mapsto a(u,v) \text{ is linear, for each fixed } v,$$
$$v \mapsto a(u,v) \text{ is linear (or conjugate linear), for each fixed } u,$$

and **continuous**:

$$|a(u,v)| \leq (\text{const.})\|u\|_{\mathcal{K}}\|v\|_{\mathcal{K}},$$

and **symmetric**:

$$a(v,u) = a(u,v) \quad (\text{or } \overline{a(u,v)}).$$

4. a is **coercive** on $\mathcal{K}$, meaning

$$a(u,u) \geq c\|u\|_{\mathcal{K}}^2 \qquad \forall u \in \mathcal{K},$$

for some $c > 0$. Hence $a(u,u) \asymp \|u\|_{\mathcal{K}}^2$.

Symmetry and coercivity imply that:

> $a(u,v)$ defines an inner product whose norm is equivalent to the $\|\cdot\|_{\mathcal{K}}$-norm.

Spectral theorem.

THEOREM 3.1. *Under the hypotheses above, there exist vectors* $u_1, u_2, u_3, \ldots \in \mathcal{K}$ *and numbers*

$$0 < \gamma_1 \leq \gamma_2 \leq \gamma_3 \leq \cdots \to \infty$$

such that:

- u_j *is an eigenvector of* $a(\cdot,\cdot)$ *with eigenvalue* γ_j, *meaning*

$$a(u_j, v) = \gamma_j \langle u_j, v\rangle_{\mathcal{H}} \qquad \forall v \in \mathcal{K}, \tag{3.1}$$

- $\{u_j\}$ *is an ONB for* $\mathcal{H}$,
- $\{u_j/\sqrt{\gamma_j}\}$ *is an ONB for* $\mathcal{K}$ *with respect to the* a-*inner product.*

The decomposition

$$\boxed{f = \sum_j \langle f, u_j\rangle_{\mathcal{H}}\, u_j} \tag{3.2}$$

holds with convergence in $\mathcal{H}$ *for each* $f \in \mathcal{H}$, *and holds with convergence in* $\mathcal{K}$ *for each* $f \subset \mathcal{K}$.

Formula (3.1) is known as the (weak) eigenvalue equation for a.

The idea of the proof is to show that a certain "inverse" operator associated with the sesquilinear form is compact and selfadjoint on $\mathcal{H}$, and then apply the spectral theorem for compact selfadjoint operators. In terms of differential equations, a corresponds to a differential operator such as $-\Delta$, which is unbounded, and the inverse corresponds to an integral operator $(-\Delta)^{-1}$, which is bounded and in fact compact. Indeed, we will begin the proof by solving the analogue of Poisson's equation $-\Delta u = f$ weakly in our Hilbert space setting, with the help of the Riesz Representation Theorem.

Proof of Theorem 3.1. [Omitted during the lectures.]

Step 1. We claim for each $f \in \mathcal{H}$ there exists a unique $u \in \mathcal{K}$ such that

$$a(u,v) = \langle f, v\rangle_{\mathcal{H}} \qquad \forall v \in \mathcal{K}, \tag{3.3}$$

and that the map

$$\begin{aligned} B : \mathcal{H} &\to \mathcal{K} \\ f &\mapsto u \end{aligned}$$

is linear and bounded. To prove this claim, fix $f \in \mathcal{H}$ and define a bounded linear functional $F(v) = \langle v, f \rangle_{\mathcal{H}}$ on $\mathcal{K}$, noting for the boundedness that

$$\begin{aligned} |F(v)| &\leq \|v\|_{\mathcal{H}} \|f\|_{\mathcal{H}} && \text{by Cauchy–Schwarz} \\ &\leq (\text{const.}) \|v\|_{\mathcal{K}} \|f\|_{\mathcal{H}} && \text{since } \mathcal{K} \text{ is imbedded in } \mathcal{H} \\ &\leq (\text{const.}) a(v,v)^{1/2} \|f\|_{\mathcal{H}} \end{aligned}$$

by coercivity. Hence by the Riesz Representation Theorem on $\mathcal{K}$ (with respect to the a-inner product on $\mathcal{K}$), a unique $u \in \mathcal{K}$ exists such that $F(v) = a(v,u)$ for all $v \in \mathcal{K}$. That is,

$$\langle v, f \rangle_{\mathcal{H}} = a(v,u) \qquad \forall v \in \mathcal{K},$$

as desired for (3.3). Thus the map $B : f \mapsto u$ is well defined. Clearly it is linear. And

$$a(u,u) = |F(u)| \leq (\text{const.}) a(u,u)^{1/2} \|f\|_{\mathcal{H}}$$

implies $a(u,u)^{1/2} \leq (\text{const.}) \|f\|_{\mathcal{H}}$, so that B is bounded from $\mathcal{H}$ to $\mathcal{K}$ (using the a-norm on $\mathcal{K}$), which proves our claim in Step 1.

Step 2. $\iota \circ B : \mathcal{H} \to \mathcal{K} \to \mathcal{H}$ is compact, since ι imbeds $\mathcal{K}$ compactly into $\mathcal{H}$. Thus we may regard B as a compact bounded linear operator from $\mathcal{H}$ to itself. Observe B is selfadjoint on $\mathcal{H}$, since for all $f, g \in \mathcal{H}$ one has

$$\begin{aligned} \langle Bf, g \rangle_{\mathcal{H}} &= \overline{\langle g, Bf \rangle_{\mathcal{H}}} && \text{by symmetry of the inner product} \\ &= \overline{a(Bg, Bf)} && \text{by definition of } B \text{ in (3.3),} \\ &= a(Bf, Bg) && \text{by symmetry of } a, \\ &= \langle f, Bg \rangle_{\mathcal{H}} && \text{by definition of } B \text{ in (3.3),} \end{aligned}$$

which implies $B^* = B$.

Further, B is injective, because if $Bf = 0$ then (3.3) implies $\langle f, v \rangle_{\mathcal{H}} = 0$ for all $v \in \mathcal{K}$ and hence for all $v \in \mathcal{H}$ (by density of $\mathcal{K}$ in $\mathcal{H}$), from which we conclude $f = 0$. Thus the kernel of B consists of just the zero vector.

Therefore the spectral theorem for compact selfadjoint operators (stated at the end of this section) provides an ONB for $\mathcal{H}$ consisting of eigenvectors of B, with

$$Bu_j = \beta_j u_j$$

for some real eigenvalues $\beta_j \to 0$. Note $\beta_j \neq 0$ because the kernel of B contains only the zero vector. The decomposition (3.2) holds for $f \in \mathcal{H}$ because $\{u_j\}$ forms an ONB for $\mathcal{H}$.

Step 3. Dividing by the eigenvalue shows $u_j = B(u_j/\beta_j)$, which guarantees that u_j belongs to the range of B and therefore $u_j \in \mathcal{K}$. Next observe the eigenvalues are all positive, because

$$\beta_j a(u_j, v) = a(Bu_j, v) = \langle u_j, v \rangle_{\mathcal{H}} \qquad \forall v \in \mathcal{K},$$

and choosing $v = u_j \in \mathcal{K}$ and using coercivity shows that $\beta_j > 0$. Thus the reciprocals

$$\gamma_j = \frac{1}{\beta_j}$$

are positive and tend to infinity, and satisfy

$$a(u_j, v) = \gamma_j \langle u_j, v\rangle_{\mathcal{H}} \qquad \forall v \in \mathcal{K},$$

which is the desired eigenvalue formula (3.1). After reordering, we may further assume the γ_j are increasing: $0 < \gamma_1 \leq \gamma_2 \leq \gamma_3 \leq \cdots \to \infty$.

Step 4. Finally, the set $\{u_j/\sqrt{\gamma_j}\}$ is a-orthonormal because

$$\begin{aligned} a(u_j, u_k) &= \gamma_j \langle u_j, u_k\rangle_{\mathcal{H}} \\ &= \gamma_j \delta_{jk} \\ &= \sqrt{\gamma_j}\sqrt{\gamma_k}\, \delta_{jk}. \end{aligned}$$

This orthonormal set is complete in $\mathcal{K}$, because if $a(u_j, v) = 0$ for all j then $\langle u_j, v\rangle_{\mathcal{H}} = 0$ for all j by (3.1), so that $v = 0$. Therefore each $f \in \mathcal{K}$ can be decomposed as

$$f = \sum_j a(f, u_j/\sqrt{\gamma_j})\, u_j/\sqrt{\gamma_j}$$

with the series converging in $\mathcal{K}$, and this decomposition reduces to (3.2) because $a(f, u_j) = \gamma_j \langle f, u_j\rangle_{\mathcal{H}}$. □

Remark. Eigenvectors corresponding to distinct eigenvalues are automatically orthogonal, since

$$\begin{aligned} (\gamma_j - \gamma_k)\langle u_j, u_k\rangle_{\mathcal{H}} &= \gamma_j \langle u_j, u_k\rangle_{\mathcal{H}} - \overline{\gamma_k \langle u_k, u_j\rangle_{\mathcal{H}}} \\ &= a(u_j, u_k) - \overline{a(u_k, u_j)} \\ &= 0 \end{aligned}$$

by symmetry of a.

The proof above used:

THEOREM 3.2 (Spectral theorem for compact selfadjoint operators). *Assume $\mathcal{H}$ is a separable, infinite dimensional Hilbert space, and $B : \mathcal{H} \to \mathcal{H}$ is a linear, compact, selfadjoint operator.*

Then $\mathcal{H}$ has a countable ONB $\{u_k\}_{k=1}^{\infty}$ consisting of eigenvectors of B, with

$$Bu_k = \beta_k u_k,$$

for some real eigenvalues β_k.

Further, if infinitely many of the β_k are nonzero then they can be arranged as a sequence converging to 0. In particular, each nonzero eigenvalue has finite multiplicity (the corresponding eigenspace is finite dimensional).

The finite dimensional version of the theorem simply says that a selfadjoint matrix (either real symmetric or complex Hermitian) possesses an ONB of eigenvectors.

Notes and comments. The discrete spectral Theorem 3.1 can be found in various textbooks, such as [**Blanchard and Brüning**] Section 6.3. It does not seem to have a standard name.

4. Variational characterizations of eigenvalues

Goal. To obtain minimax and maximin characterizations of the eigenvalues of the sesquilinear form in Section 3.

Motivation and hypotheses. How can one estimate the eigenvalues if the spectrum cannot be computed explicitly? We develop a variational characterization of eigenvalues. The intuition comes from the special case of eigenvalues of a Hermitian (or real symmetric) matrix A, for which the sesquilinear form is $a(u,v) = Au \cdot \overline{v}$ and the first eigenvalue is

$$\gamma_1 = \min_{v \neq 0} \frac{Av \cdot \overline{v}}{v \cdot \overline{v}}.$$

Poincaré's minimax characterization of the eigenvalues. We work under the assumptions of the discrete spectral theorem in Section 3, for the sesquilinear form a. Recall the ordering

$$\gamma_1 \leq \gamma_2 \leq \gamma_3 \leq \cdots \to \infty.$$

Define the *Rayleigh quotient* of u to be

$$\frac{a(u,u)}{\langle u,u\rangle_{\mathcal{H}}}.$$

Rayleigh principle. The first eigenvalue γ_1 equals the minimum value of the Rayleigh quotient:

$$\boxed{\gamma_1 = \min_{f \in \mathcal{K} \setminus \{0\}} \frac{a(f,f)}{\langle f,f\rangle_{\mathcal{H}}}.} \tag{4.1}$$

The minimum is attained if and only if f is an eigenfunction with eigenvalue γ_1.

Proof of Rayleigh principle. The arbitrary vector $f \in \mathcal{K}$ can be expanded in terms of the ONB of eigenvectors as

$$f = \sum_j c_j u_j$$

where $c_j = \langle f, u_j\rangle_{\mathcal{H}}$. This series converges in both $\mathcal{H}$ and $\mathcal{K}$, as proved in Section 3. Hence we may substitute the series into the Rayleigh quotient to obtain

$$\begin{aligned} \frac{a(f,f)}{\langle f,f\rangle_{\mathcal{H}}} &= \frac{\sum_{j,k} c_j \overline{c_k} a(u_j,u_k)}{\sum_{j,k} c_j \overline{c_k} \langle u_j,u_k\rangle_{\mathcal{H}}} \\ &= \frac{\sum_j |c_j|^2 \gamma_j}{\sum_j |c_j|^2} \end{aligned} \tag{4.2}$$

since the eigenvectors $\{u_j\}$ are orthonormal in $\mathcal{H}$ and the collection $\{u_j/\sqrt{\gamma_j}\}$ is a-orthonormal in $\mathcal{K}$ (that is, $a(u_j,u_k) = \gamma_j \delta_{jk}$). The expression (4.2) is obviously greater than or equal to γ_1.

Equality holds if and only if f is a first eigenfunction, that is, if and only if $f = \sum_{j=1}^J c_j u_j$ where $\gamma_1 = \cdots = \gamma_J < \gamma_{J+1}$. (In most applications the first eigenvalue is simple, in which case $J = 1$ and equality holds in the Rayleigh quotient if and only if f is a multiple of u_1.)

Poincaré principle. The higher eigenvalues are given by the following minimax formula: for $j \geq 1$,

$$\boxed{\gamma_j = \min_S \max_{f \in S \setminus \{0\}} \frac{a(f,f)}{\langle f, f \rangle_{\mathcal{H}}}} \tag{4.3}$$

where S ranges over all j-dimensional subspaces of $\mathcal{K}$.

Remark. The Rayleigh and Poincaré principles provide *upper bounds* on eigenvalues, since they express γ_1 and γ_j as minima. That is, an upper bound on γ_1 follows from substituting any vector f into the Rayleigh quotient in (4.1), and an upper bound on γ_j is obtained by choosing S to be any j-dimensional subspace and evaluating the maximum over $f \in S$ of the Rayleigh quotient in (4.3).

Proof of Poincaré principle. We prove the minimax formula (4.3) for $j = 2$, and leave the case of higher j-values as an exercise.

"$\geq$" direction of the proof. Choose $S = \{c_1 u_1 + c_2 u_2 : c_1, c_2 \text{ scalars}\}$ to be the span of the first two eigenvectors. Then

$$\max_{f \in S \setminus \{0\}} \frac{a(f,f)}{\langle f, f \rangle_{\mathcal{H}}} = \max_{(c_1,c_2) \neq (0,0)} \frac{\sum_{j=1}^2 |c_j|^2 \gamma_j}{\sum_{j=1}^2 |c_j|^2} = \gamma_2.$$

Hence $\gamma_2 \geq$ right side of (4.3).

"$\leq$" direction of the proof. To prove the opposite inequality, consider an arbitrary 2-dimensional subspace $S \subset \mathcal{K}$. Note this subspace is arbitrary, and need not equal the span of two of the eigenvectors.

The subspace contains a nonzero vector g that is orthogonal to u_1. (Proof: given a basis $\{v_1, v_2\}$ for the subspace, there exist scalars d_1, d_2 not both zero such that $d_1 \langle v_1, u_1 \rangle_{\mathcal{H}} + d_2 \langle v_2, u_1 \rangle_{\mathcal{H}} = 0$. Hence the vector $g = d_1 v_1 + d_2 v_2$ satisfies $\langle g, u_1 \rangle_{\mathcal{H}} = 0$.) Thus $c_1 = 0$ in the orthonormal expansion for g, and so by (4.2),

$$\frac{a(g,g)}{\langle g, g \rangle_{\mathcal{H}}} = \frac{\sum_{j=2}^\infty |c_j|^2 \gamma_j}{\sum_{j=2}^\infty |c_j|^2} \geq \gamma_2.$$

Hence

$$\max_{f \in S \setminus \{0\}} \frac{a(f,f)}{\langle f, f \rangle_{\mathcal{H}}} \geq \frac{a(g,g)}{\langle g, g \rangle_{\mathcal{H}}} \geq \gamma_2,$$

which implies that $\gamma_2 \leq$ right side of (4.3). □

Rayleigh principle for second eigenvalue. A variant of the Rayleigh principle is sometimes used for the second eigenvalue:

$$\gamma_2 = \min_{0 \neq f \perp u_1} \frac{a(f,f)}{\langle f, f \rangle_{\mathcal{H}}}$$

where the trial vector satisfies $f \in \mathcal{K}, f \neq 0$ and $\langle f, u_1 \rangle_{\mathcal{H}} = 0$. The proof mimics that of the Rayleigh principle, except now with $c_1 = 0$ since f is orthogonal to the first eigenvector

Notes and comments. Variational principles can be developed not just for individual eigenvalues, but also for combinations such as the sum of the first n eigenvalues, and the sum of the first n reciprocal eigenvalues. For example, the sum of the first n eigenvalues has a "minimum" characterization similar to the Rayleigh principle, but now involving pairwise orthogonal trial vectors:

$$\begin{aligned}&\gamma_1 + \cdots + \gamma_n \\ &= \min\Big\{\frac{a(f_1,f_1)}{\langle f_1,f_1\rangle_{\mathcal{H}}} + \cdots + \frac{a(f_n,f_n)}{\langle f_n,f_n\rangle_{\mathcal{H}}} : f_j \in \mathcal{K}\setminus\{0\}, \langle f_j,f_k\rangle_{\mathcal{H}} = 0 \text{ when } j \neq k\Big\}.\end{aligned}$$

See [**Bandle**] Section III.1.2.

5. Application: Discrete spectrum for the Laplacian

Goal. To apply the spectral theorem from Section 3 to the Dirichlet and Neumann Laplacians.

Dirichlet Laplacian. We want an ONB of eigenfunctions satisfying

$$\begin{aligned} -\Delta u &= \lambda u & &\text{in } \Omega \\ u &= 0 & &\text{on } \partial\Omega \end{aligned}$$

where Ω is a domain in $\mathbb{R}^d$ having finite volume.

To verify the hypotheses of the discrete spectral Theorem 3.1 we let
$\mathcal{H} = L^2(\Omega)$, inner product $\langle u,v\rangle_{L^2} = \int_\Omega uv\,dx$,
$\mathcal{K} = H_0^1(\Omega) =$ Sobolev space, which is the completion of $C_0^\infty(\Omega)$ (the space of smooth functions having compact support in Ω) under the inner product

$$\langle u,v\rangle_{H^1} = \int_\Omega (\nabla u\cdot\nabla v + uv)\,dx.$$

All functions are assumed real valued, for simplicity. Note $L^2(\Omega)$ is separable [**Ciarlet**, Theorem 2.5-4].
Density: $C_0^\infty \subset H_0^1 \subset L^2$ and C_0^∞ is dense in L^2, so H_0^1 is dense in L^2.
Continuous imbedding $H_0^1 \hookrightarrow L^2$:

$$\begin{aligned}\|u\|_{L^2} &= \Big(\int_\Omega u^2\,dx\Big)^{1/2} \\ &\leq \Big(\int_\Omega (|\nabla u|^2 + u^2)\,dx\Big)^{1/2} \\ &= \|u\|_{H^1}.\end{aligned}$$

Compact imbedding $H_0^1 \hookrightarrow L^2$: see Rellich's Theorem A.4 in the Appendix.

Sesquilinear form: Define

$$a(u,v) = \int_\Omega \nabla u \cdot \nabla v \, dx + \int_\Omega uv \, dx = \langle u, v\rangle_{H^1}, \qquad u, v \in H_0^1(\Omega).$$

Clearly a is linear in each variable, and symmetric and continuous on $H_0^1(\Omega)$.

Coercivity: $a(u,u) = \|u\|_{H^1}^2$.

Hence the discrete spectral Theorem 3.1 gives an ONB $\{u_j\}$ for $L^2(\Omega)$ (with $u_j \in H_0^1(\Omega)$ also) and eigenvalues which we denote $\gamma_j = \lambda_j + 1$ satisfying

$$\langle u_j, v\rangle_{H^1} = (\lambda_j + 1)\langle u_j, v\rangle_{L^2} \qquad \forall v \in H_0^1(\Omega).$$

That is,

$$\int_\Omega \nabla u_j \cdot \nabla v \, dx = \lambda_j \int_\Omega u_j v \, dx \qquad \forall v \in H_0^1(\Omega), \tag{5.1}$$

which means

$$-\Delta u_j = \lambda_j u_j \quad \text{weakly.}$$

Thus u_j is a weak eigenfunction of the Laplacian with eigenvalue λ_j. Elliptic regularity theory says u_j is C^∞-smooth in Ω [**Gilbarg and Trudinger**, Corollary 8.11].

Next we show u_j satisfies the eigenfunction equation classically. Formula (5.1) and Green's formula (1.1) imply

$$\int_\Omega (-\Delta u_j) v \, dx = \int_\Omega (\lambda_j u_j) v \, dx \qquad \forall v \in C_0^\infty(\Omega), \tag{5.2}$$

where for simplicity we work with trial functions in $C_0^\infty(\Omega) \subset H_0^1(\Omega)$; note the boundary term vanished in Green's formula because $v = 0$ on $\partial\Omega$. Thus

$$\int_\Omega (\Delta u_j + \lambda_j u_j) v \, dx = 0 \qquad \forall v \in C_0^\infty(\Omega).$$

If $\Delta u_j + \lambda_j u_j > 0$ on some open set (or is < 0 there) then we may choose v to be a nonnegative smooth function with compact support in that set such that $\int_\Omega (\Delta u_j + \lambda_j u_j) v \, dx > 0$ (or < 0), which contradicts the last displayed equation. Hence $\Delta u_j + \lambda_j u_j = 0$ at every point, or

$$-\Delta u_j = \lambda_j u_j \qquad \text{in } \Omega,$$

which means u_j is an eigenfunction in the classical sense.

Dirichlet boundary condition: $u_j = 0$ on $\partial\Omega$ in the sense of Sobolev spaces (the trace theorem) since H_0^1 is the closure of C_0^∞. This boundary condition holds classically on any smooth portion of $\partial\Omega$, by elliptic regularity results.

Positivity of the Dirichlet eigenvalues: We have

$$\lambda_j = \frac{\int_\Omega |\nabla u_j|^2 \, dx}{\int_\Omega u_j^2 \, dx} \geq 0 \tag{5.3}$$

by choosing $v = u_j$ in the weak formulation (5.1). In fact $\lambda_j > 0$, as follows. If $\lambda_j = 0$ then $|\nabla u_j| \equiv 0$ by the last formula and so u_j is constant. Since $u_j = 0$ on the boundary we conclude $u_j \equiv 0$ in the domain. But u_j cannot vanish identically because it has L^2 norm equal to 1. Hence the Dirichlet eigenvalues are positive, with

$$0 < \lambda_1 \leq \lambda_2 \leq \lambda_3 \leq \cdots \to \infty.$$

Aside. Alternatively one may use a Sobolev inequality for H_0^1 to conclude the Dirichlet eigenvalues are positive. A suitable such inequality can be proved directly for $u \in C_0^\infty(\Omega)$, when the domain is bounded:

$$\begin{aligned}
\|u\|_{L^2}^2 &= \int_\Omega u^2\,dx \\
&= -\int_\Omega 2u\frac{\partial u}{\partial x_1}x_1\,dx \\
&\qquad\qquad \text{by integration by parts in the } x_1\text{-direction} \\
&\leq \big(2\max_{x\in\overline{\Omega}}|x_1|\big)\,\|u\|_{L^2}\|\partial u/\partial x_1\|_{L^2} \\
&\leq (\text{const.})\|u\|_{L^2}\|\nabla u\|_{L^2},
\end{aligned}$$

so that

$$\|u\|_{L^2} \leq (\text{const.})\|\nabla u\|_{L^2}$$

where the constant depends on the bounded domain Ω. Hence this Sobolev inequality holds also for $u \in H_0^1(\Omega)$. Consequently the left side of (5.3) is bounded below by a positive constant, giving $\lambda_j > 0$.

Neumann Laplacian. One seeks an ONB of eigenfunctions such that

$$\begin{aligned}
-\Delta u &= \mu u \qquad &&\text{in } \Omega \\
\frac{\partial u}{\partial n} &= 0 &&\text{on } \partial\Omega
\end{aligned}$$

where Ω is a bounded domain in $\mathbb{R}^d$ with Lipschitz boundary.

To verify the hypotheses of the discrete spectral Theorem 3.1 we let

$\mathcal{H} = L^2(\Omega)$,

$\mathcal{K} = H^1(\Omega) =$ Sobolev space, which is the completion of $C^\infty(\overline{\Omega})$ under the inner product $\langle u, v\rangle_{H^1}$; see [**Gilbarg and Trudinger**, p. 174].

Now argue as above for the Dirichlet Laplacian. The compact imbedding is provided by Rellich's Theorem A.5, which relies on Lipschitz smoothness of the boundary.

Writing the eigenvalues in the discrete spectral Theorem 3.1 as $\gamma_j = \mu_j + 1$, one finds

$$\int_\Omega \nabla u_j \cdot \nabla v\,dx = \mu_j \int_\Omega u_j v\,dx \qquad \forall v \in H^1(\Omega). \tag{5.4}$$

In particular, restricting to $v \in H_0^1(\Omega)$ implies

$$-\Delta u_j = \mu_j u_j \quad \text{weakly.}$$

Hence u_j is smooth, by elliptic regularity. Arguing as in the Dirichlet case using $v \in C_0^\infty(\Omega)$, one shows u_j is an eigenfunction in the classical sense, satisfying $-\Delta u_j = \mu_j u_j$ in Ω.

Nonnegativity of the Neumann eigenvalues: choosing $v = u_j$ in (5.4) proves $\mu_j \geq 0$. The first Neumann eigenvalue is zero: $\mu_1 = 0$ with a constant eigenfunction $u_1 \equiv \text{const.} \neq 0$. (Note this constant function belongs to $H^1(\Omega)$, though not to $H_0^1(\Omega)$.) Hence

$$0 = \mu_1 \leq \mu_2 \leq \mu_3 \leq \cdots \to \infty.$$

Neumann "natural" boundary condition: formula (5.4) asserts more than its Dirichlet counterpart (5.1) does, because (5.4) holds for all $v \in H^1(\Omega)$

rather than just $v \in H_0^1(\Omega)$. We will use this additional information to show our weak eigenfunctions in (5.4) automatically satisfy the Neumann boundary condition $\partial u_j/\partial n = 0$. This Neumann boundary condition holds even though it is **not** imposed in the function space $\mathcal{K}$.

(Functions in $H^1(\Omega)$ do not generally have vanishing normal derivative. In contrast, for the Dirichlet eigenfunctions considered earlier, the boundary condition is imposed directly by the choice of function space $H_0^1(\Omega)$: every function in that space equals zero on the boundary.)

The weak form of the eigenfunction equation (5.4) for an eigenfunction u with eigenvalue μ says

$$\int_\Omega \nabla u \cdot \nabla v \, dx = \mu \int_\Omega uv \, dx \qquad \forall v \in H^1(\Omega). \tag{5.5}$$

Recall u is smooth in Ω, by elliptic regularity theory. From (5.5) and Green's formula (1.1) we find

$$\int_\Omega (-\Delta u) v \, dx + \int_{\partial\Omega} \frac{\partial u}{\partial n} v \, dS = \int_\Omega (\mu u) v \, dx \qquad \forall v \in C^\infty(\overline{\Omega}), \tag{5.6}$$

where for simplicity we work with trial functions in $C^\infty(\overline{\Omega}) \subset H^1(\Omega)$. If $v \in C_0^\infty(\Omega)$ then the boundary term vanishes in (5.6), and so

$$\int_\Omega (\Delta u + \mu u) v \, dx = 0 \qquad \forall v \in C_0^\infty(\Omega).$$

If $\Delta u + \mu u > 0$ on some open set (or is < 0 there) then we may choose v to be a nonnegative smooth function with compact support in that set such that $\int_\Omega (\Delta u + \mu u) v \, dx > 0$ (or < 0), which is impossible. Hence $\Delta u + \mu u = 0$. Thus by using trial functions in $C_0^\infty(\Omega)$ we have obtained the eigenfunction equation

$$-\Delta u = \mu u \qquad \text{in } \Omega.$$

Next, to get the Neumann boundary condition, we will use trial functions v that do not vanish identically on the boundary. Assume for simplicity that the boundary is smooth, so that u extends smoothly to $\overline{\Omega}$ by elliptic regularity. Substituting $-\Delta u = \mu u$ into (5.6) reduces that formula to

$$\int_{\partial\Omega} \frac{\partial u}{\partial n} v \, dS = 0 \qquad \forall v \in C^\infty(\overline{\Omega}).$$

If $\partial u/\partial n > 0$ on some relatively open subset of the boundary (or if it is < 0 there), then we may choose a nonnegative smooth function v on $\mathbb{R}^d$ whose restriction to $\partial\Omega$ is supported in the relatively open set and for which $\int_{\partial\Omega} (\partial u/\partial n) v \, dS > 0$ (or < 0), which is a contradiction. Hence

$$\frac{\partial u}{\partial n} = 0 \qquad \text{on } \partial\Omega,$$

which is the Neumann boundary condition.

Note. If the boundary is only piecewise smooth, then one may prove the Neumann condition on the smooth portions of the boundary, by the above reasoning.

Eigenfunction expansions in the L^2 and H^1 norms. The L^2-ONB of eigenfunctions $\{u_j\}$ of the Laplacian gives the decomposition

$$f = \sum_j \langle f, u_j \rangle_{L^2} u_j \tag{5.7}$$

with convergence in the L^2 and H^1 norms, whenever f belongs to the following spaces:

$$f \in \begin{cases} H_0^1(\Omega) & \text{for Dirichlet,} \\ H^1(\Omega) & \text{for Neumann.} \end{cases}$$

These claims follow immediately from the discrete spectral Theorem 3.1, in view of our work above.

Exercises. *Regular Sturm–Liouville problem*

5.1 — Consider the eigenvalue problem

$$\begin{aligned} -(pu')' + qu &= \lambda wu \quad \text{on the interval } (0, L), \\ u(0) = 0, &\quad u(L) = 0, \end{aligned}$$

where the coefficient functions p, q, w are smooth and real valued on $[0, L]$, and p and w are positive on $[0, L]$.

(i) Formulate this regular Sturm–Liouville eigenvalue problem in terms of appropriate Hilbert spaces $\mathcal{H}$ and $\mathcal{K}$ and a sesquilinear form $a(u, v)$.

(ii) Verify the hypotheses of the discrete spectral Theorem 3.1.

(iii) Find a lower bound on the first eigenvalue λ_1 in terms of the coefficient functions. For simplicity, assume in this part that q is nonnegative.

Elliptic operator

5.2 — Consider the second order elliptic eigenvalue problem

$$-\nabla \cdot \big(A(x)\nabla u\big) = \lambda u$$

on a bounded domain $\Omega \subset \mathbb{R}^d$, where the real symmetric $d \times d$ matrix $A(x)$ depends smoothly on $x \in \overline{\Omega}$ and is uniformly elliptic, meaning

$$y^T A(x) y \geq c|y|^2, \qquad x, y \in \mathbb{R}^d,$$

for some constant $c > 0$. (In other words, the lowest eigenvalue of the matrix is bounded below away from 0.)

Formulate the Dirichlet eigenvalue problem for this elliptic operator in terms of the function spaces H_0^1 and L^2, for a suitable sesquilinear form $a(u, v)$, and verify the hypotheses of the discrete spectral Theorem 3.1.

6. Application: Monotonicity properties of eigenvalues

Goal. To establish monotonicity results for Dirichlet and Neumann eigenvalues of the Laplacian.

Rayleigh quotients. The spectral problems in Section 5 have Rayleigh quotients and function spaces as follows.

$$\text{Dirichlet Laplacian:} \quad \frac{\int_\Omega |\nabla f|^2\,dx}{\int_\Omega f^2\,dx}, \qquad f \in H_0^1(\Omega).$$

$$\text{Neumann Laplacian:} \quad \frac{\int_\Omega |\nabla f|^2\,dx}{\int_\Omega f^2\,dx}, \qquad f \in H^1(\Omega).$$

The Poincaré principle (4.3) gives a minimax formula for the eigenvalues in terms of these Rayleigh quotients:

$$\lambda_j = \min_S \max_{f \in S\setminus\{0\}} \frac{\int_\Omega |\nabla f|^2\,dx}{\int_\Omega f^2\,dx},$$

$$\mu_j = \min_U \max_{f \in U\setminus\{0\}} \frac{\int_\Omega |\nabla f|^2\,dx}{\int_\Omega f^2\,dx},$$

where S ranges over j-dimensional subspaces of $H_0^1(\Omega)$, and U ranges over j-dimensional subspaces of $H^1(\Omega)$.

Neumann $\leq$ Dirichlet. Free membranes give lower tones than fixed membranes:

THEOREM 6.1 (Neumann–Dirichlet comparison). *Let Ω be a bounded domain in $\mathbb{R}^d$ with Lipschitz boundary.*

Then the Neumann eigenvalues of the Laplacian lie below their Dirichlet counterparts:

$$\mu_j \leq \lambda_j \qquad \forall j \geq 1.$$

PROOF. Every subspace S is also a valid U, since $H_0^1 \subset H^1$ by definition. Thus the minimum for the Neumann eigenvalue μ_j is taken over a larger class of subspaces than the minimum for the Dirichlet eigenvalue λ_j. Hence $\mu_j \leq \lambda_j$. $\square$

Incidentally, remember the first Neumann eigenvalue is zero: $\mu_1 = 0$.

Domain monotonicity for Dirichlet spectrum. Making a drum smaller increases its frequencies of vibration:

THEOREM 6.2. *Let Ω and $\widetilde{\Omega}$ be domains of finite volume in $\mathbb{R}^d$ with eigenvalues λ_j and $\widetilde{\lambda}_j$ for the Dirichlet Laplacian, respectively. If $\Omega \supset \widetilde{\Omega}$ then*

$$\lambda_j \leq \widetilde{\lambda}_j \qquad \forall j \geq 1.$$

PROOF. Poincaré's minimax principle gives

$$\lambda_j = \min_S \max_{f \in S\setminus\{0\}} \frac{\int_\Omega |\nabla f|^2\,dx}{\int_\Omega f^2\,dx}, \qquad \widetilde{\lambda}_j = \min_{\widetilde{S}} \max_{f \in \widetilde{S}\setminus\{0\}} \frac{\int_{\widetilde{\Omega}} |\nabla f|^2\,dx}{\int_{\widetilde{\Omega}} f^2\,dx},$$

where S ranges over j-dimensional subspaces of $H_0^1(\Omega)$ and $\widetilde{S}$ ranges over j-dimensional subspaces of $H_0^1(\widetilde{\Omega})$.

Every subspace $\widetilde{S}$ is also a valid S since $H_0^1(\widetilde{\Omega}) \subset H_0^1(\Omega)$, by extending $f \in H_0^1(\widetilde{\Omega})$ to equal 0 outside $\widetilde{\Omega}$. This extension by 0 does not change the value of the Rayleigh quotient, and hence does not change the value of the maximum over $f \in \widetilde{S}$. Therefore $\lambda_j \leq \widetilde{\lambda}_j$. □

Counterexample to Neumann domain monotonicity. A smaller domain can have smaller Neumann eigenvalues, as the figure below shows for a rectangle contained in a unit square. Recall from Section 2 that $\mu_2 = \pi^2/L^2$ for a rectangle with longer side length L. The square has both sides of length 1 and hence $\mu_2 = \pi^2$, while the rectangle has longer side length $(0.9)\sqrt{2}$ (just a little shorter than the diagonal length $\sqrt{2}$) and so $\widetilde{\mu}_2 = \pi^2/(1.62)$. Thus the second Neumann eigenvalue is smaller for the rectangle even though the rectangle is smaller than the square. Thus domain monotonicity can fail for Neumann eigenvalues.

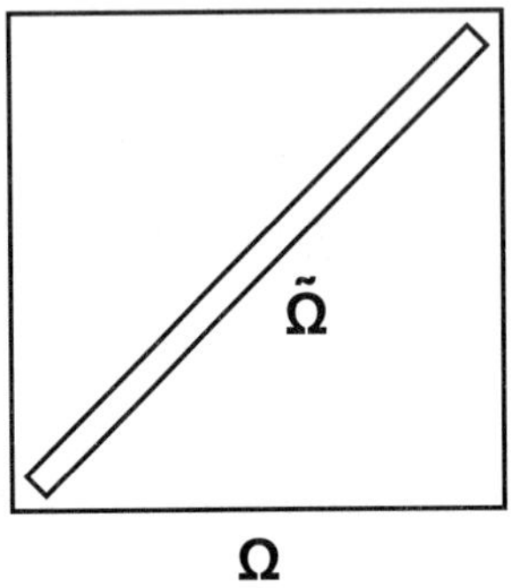

To see where the Dirichlet domain monotonicity proof breaks down for Neumann eigenvalues, note that although one can extend a function in $H^1(\widetilde{\Omega})$ to belong to $H^1(\Omega)$, the extended function must generally be nonzero outside $\widetilde{\Omega}$ and so the L^2-norm of the function and its gradient will differ from those of the original function; hence the Rayleigh quotient will differ too.

Rescaling gives a kind of monotonicity that is sometimes useful for Neumann eigenvalues:

$$\mu_j(t\Omega) = \frac{\mu_j(\Omega)}{t^2},$$

as shown in the Section 2 exercises.

Restricted reverse monotonicity for Neumann spectrum. Neumann monotonicity does hold in a certain restricted situation, with the inequality reversed from the Dirichlet case — the smaller drum has smaller tones.

THEOREM 6.3. *Let Ω and $\widetilde{\Omega}$ be bounded Lipschitz domains in $\mathbb{R}^d$ with eigenvalues μ_j and $\widetilde{\mu}_j$ for the Neumann Laplacian, respectively.*

If $\Omega \supset \widetilde{\Omega}$ and $\Omega \setminus \widetilde{\Omega}$ has measure zero, then

$$\mu_j \geq \widetilde{\mu}_j \qquad \forall j \geq 1.$$

One might imagine the smaller domain $\widetilde{\Omega}$ as being constructed by removing a hypersurface of measure zero from Ω, thus introducing an additional boundary surface that behaves like a "tear" in the fabric of the membrane. Reverse monotonicity then makes sense, because an eigenfunction can take different values on the two sides of this additional piece of boundary, enabling the eigenfunction to "relax" and the eigenvalue (frequency) to decrease.

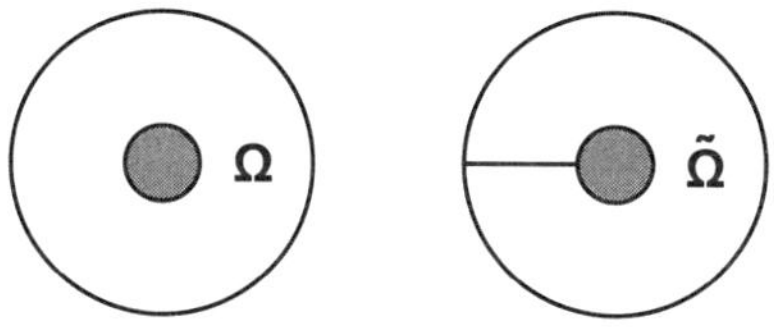

In contrast, introducing additional boundary surfaces to a Dirichlet problem would have the opposite effect: the eigenfunction would face additional constraints, and hence the eigenvalue (frequency) would increase.

PROOF OF THEOREM 6.3. Poincaré's minimax principle gives

$$\mu_j = \min_S \max_{f \in S \setminus \{0\}} \frac{\int_{\widetilde{\Omega}} |\nabla f|^2 \, dx}{\int_{\widetilde{\Omega}} f^2 \, dx}, \qquad \widetilde{\mu}_j = \min_{\widetilde{S}} \max_{f \in \widetilde{S} \setminus \{0\}} \frac{\int_{\widetilde{\Omega}} |\nabla f|^2 \, dx}{\int_{\widetilde{\Omega}} f^2 \, dx},$$

where S ranges over j-dimensional subspaces of $H^1(\Omega)$ and $\widetilde{S}$ ranges over j-dimensional subspaces of $H^1(\widetilde{\Omega})$. In the Rayleigh quotient for μ_j we should integrate over Ω instead of $\widetilde{\Omega}$, but it makes no difference because $\Omega \setminus \widetilde{\Omega}$ has measure zero.

Every subspace S is also a valid $\widetilde{S}$, since functions in $H^1(\Omega)$ obviously restrict to functions in $H^1(\widetilde{\Omega})$. Therefore $\mu_j \geq \widetilde{\mu}_j$. □

7. Case study: Stability of steady states for reaction–diffusion PDEs

Goal. To linearize a nonlinear reaction–diffusion PDE around a steady state, and determine stability using the spectral theory of the linearized operator.

Reaction–diffusion PDEs. Assume throughout this section that $f : \mathbb{R} \to \mathbb{R}$ is a smooth function with f' bounded above. Let $X > 0$. We study the reaction–diffusion PDE

$$u_t = u_{xx} + f(u) \tag{7.1}$$

on the interval $(0, X)$ with Dirichlet boundary conditions $u(0) = u(X) = 0$. Physical interpretations include:

(i) $u =$ temperature and $f =$ rate of heat generation,
(ii) $u =$ chemical concentration and $f =$ reaction rate of chemical creation.

Intuitively, the 2nd order diffusion term in the PDE is **stabilizing** (since $u_t = u_{xx}$ is the usual diffusion equation), whereas the 0th order reaction term can be **destabilizing** (since solutions to $u_t = f(u)$ will grow, when f is positive). Thus the reaction–diffusion PDE features a competition between stabilizing and destabilizing effects. This competition leads to nonconstant steady states, and interesting stability behavior.

Next we characterize the steady states of the reaction–diffusion equation, and then investigate their stability.

Steady states. $U(x)$ is a steady state if

$$U'' + f(U) = 0, \qquad 0 < x < X. \tag{7.2}$$

More than one steady state can exist. For example if $f(y) = y$ and $X = \pi$, then $U(x) \equiv 0$ is a steady state and so is $U(x) = \sin x$.

The general **nonlinear oscillator equation** (7.2) can in principle be solved by multiplying by U' and integrating to obtain

$$\frac{1}{2}(U')^2 + F(U) = c,$$

where F is an antiderivative of f and c is constant. Then one rearranges to the form $dx/dU = \pm 1/\sqrt{2(c - F(U))}$ which can be integrated to obtain x as a function of U.

Linearizing the PDE. Perturb a steady state by considering

$$u = U + \varepsilon\phi,$$

where the perturbation $\phi(x,t)$ is assumed to satisfy the Dirichlet BC $\phi = 0$ at $x = 0$ and $x = X$, for each t. Substituting u into the reaction–diffusion equation (7.1) gives

$$\begin{aligned} u_t &= u_{xx} + f(u) \\ 0 + \varepsilon\phi_t &= (U_{xx} + \varepsilon\phi_{xx}) + f(U + \varepsilon\phi) \\ &= U_{xx} + \varepsilon\phi_{xx} + f(U) + f'(U)\varepsilon\phi + O(\varepsilon^2). \end{aligned}$$

The terms of order ε^0 on the right side equal zero by the steady state equation $U_{xx} + f(U) = 0$. We discard terms of order ε^2 and higher. (This approximation seems reasonable when ε is small and ϕ is not too large. To justify the approximation rigorously, one would need to prove a *linearization theorem* for the reaction diffusion PDE.)

The remaining terms, which have order ε^1, give the linearized equation

$$\phi_t = \phi_{xx} + f'(U)\phi.$$

That is,

$$\phi_t + L\phi = 0$$

where L is the second order linear operator

$$Lw = -w_{xx} - f'(U)w.$$

Separation of variables gives (formally) solutions of the form

$$\phi(x,t) = \sum_j c_j e^{-\tau_j t} w_j(x),$$

where the eigenvalues τ_j and Dirichlet eigenfunctions w_j satisfy

$$Lw_j = \tau_j w_j$$

with $w_j(0) = w_j(X) = 0$.

The steady state U of the reaction–diffusion PDE is called

- **linearly stable** if $\tau_1 \geq 0$, because in that case all coefficients in ϕ remain bounded as t increases,
- **linearly unstable** if $\tau_1 < 0$, because in that case the first coefficient $c_1 e^{-\tau_1 t}$ grows as t increases.

Thus the task is to investigate the spectrum of L and the sign of its first eigenvalue τ_1.

Spectrum of linearized operator L. We take:
$\Omega = (0, X)$,
$\mathcal{H} = L^2(0, X)$, inner product $\langle u, v\rangle_{L^2} = \int_0^X uv\,dx$,
$\mathcal{K} = H_0^1(0, X)$, inner product

$$\langle u, v\rangle_{H^1} = \int_0^X (u'v' + uv)\,dx.$$

Compact imbedding $H_0^1 \hookrightarrow L^2$: by Rellich's Theorem A.4.
Symmetric sesquilinear form:

$$a(u, v) = \int_0^X \big(u'v' - f'(U)uv + Cuv\big)\,dx$$

where $C = 1 + \sup f'$.

Coercivity: the definition of C insures

$$a(u, u) \geq \int_0^X \big((u')^2 + u^2\big)\,dx = \|u\|_{H^1}^2.$$

The discrete spectral Theorem 3.1 now yields an ONB of eigenfunctions $\{w_j\}$ with eigenvalues γ_j such that

$$a(w_j, v) = \gamma_j\langle w_j, v\rangle_{L^2} \qquad \forall v \in H_0^1(0, X).$$

Writing $\gamma_j = \tau_j + C$ we get

$$\int_0^X \big(w_j'v' - f'(U)w_jv\big)\,dx = \tau_j \int_0^X w_jv\,dx \qquad \forall v \in H_0^1(0, X).$$

That is, the eigenfunctions satisfy

$$Lw_j = \tau_j w_j \qquad \text{weakly}$$

and hence also classically.

Stability of the zero steady state. Assume $f(0) = 0$, so that $U \equiv 0$ is a steady state. Its stability is easily determined, as follows.

The linearized operator is $Lw = -w'' - f'(0)w$, which has Dirichlet eigenvalues

$$\tau_j = \Big(\frac{j\pi}{X}\Big)^2 - f'(0)$$

on the interval $(0, X)$, with eigenfunctions $\sin(j\pi x/X), j \geq 1$. From the formula for $\tau_1 = (\pi/X)^2 - f'(0)$, one immediately deduces stability criteria:

PROPOSITION 7.1 (Stability of the zero steady state).
(a) If $f(0) = 0$ and $f'(0) \leq 0$ then the zero steady state is linearly stable.
(b) If $f(0) = 0$ and $f'(0) > 0$ then the zero steady state is linearly stable on short intervals ($X \leq \pi/\sqrt{f'(0)}$), but is linearly unstable when the interval is long ($X > \pi/\sqrt{f'(0)}$).

The reaction–diffusion equation is called **long-wave unstable** in case (b).

Instability of sign-changing steady states. The next criterion depends on a sign-changing property of the steady state.

THEOREM 7.2 (Linear instability [**Schaaf2**, Proposition 4.1.2]). *If the steady state U changes sign on $(0, X)$ then $\tau_1 < 0$.*

For example, suppose $f(y) = y$ so that the steady state equation is $U'' + U = 0$. If $X = 2\pi$ then the steady state $U = \sin x$ is linearly unstable, by the theorem. In this example we can compute the spectrum of L exactly: the lowest eigenfunction is $w = \sin(x/2)$ with eigenvalue $\tau_1 = \left(\frac{1}{2}\right)^2 - 1 < 0$.

PROOF. If U changes sign then it has a positive local maximum and a negative local minimum in $(0, X)$, recalling that $U = 0$ at the endpoints. Obviously U' must be nonzero at some point between these local extrema, and so there exist points $0 < x_1 < x_2 < X$ such that

$$U'(x_1) = U'(x_2) = 0$$

and $U' \neq 0$ on (x_1, x_2). Define a trial function

$$w = \begin{cases} U' & \text{on } (x_1, x_2), \\ 0 & \text{elsewhere.} \end{cases}$$

(We motivate this choice of trial function at the end of the proof.) Then w is piecewise smooth, and is continuous since $w = U' = 0$ at x_1 and x_2. Therefore $w \in H_0^1(0, X)$, and $w \not\equiv 0$ since $U' \neq 0$ on (x_1, x_2).

The numerator of the Rayleigh quotient for w is

$$\begin{aligned} \int_0^X \left((w')^2 - f'(U)w^2\right) dx &= \int_{x_1}^{x_2} \left(-w'' - f'(U)w\right)w \, dx \qquad \text{by parts} \\ &= 0 \end{aligned}$$

since

$$-w'' = -U''' = \big(f(U)\big)' = f'(U)U' = f'(U)w. \tag{7.3}$$

Hence $\tau_1 \leq 0$, by using w as a trial function in the Rayleigh principle for the first eigenvalue.

Suppose $\tau_1 = 0$. Then w is an eigenfunction with eigenvalue 0, by the condition for equality in the Rayleigh quotient (4.1). Since eigenfunctions are smooth, one has $w'(x_2) = 0$ by taking the derivative from the right. Then $w(x_2) = w'(x_2) = 0$, and so $w \equiv 0$ by uniqueness for the second order linear ODE (7.3). On the other hand, $w \not\equiv 0$ by construction. Hence $\tau_1 < 0$. □

Motivation for the choice of trial function. The derivative U' lies in the nullspace of L since

$$LU' = -(U')'' - f'(U)U' = -\big(U'' + f(U)\big)' = 0.$$

In other words, U' is an eigenfunction with eigenvalue 0, which almost proves instability, since instability would correspond to a negative eigenvalue. Of course, U' does not satisfy the Dirichlet boundary conditions at the endpoints, which is why we restrict to the subinterval (x_1, x_2) in the proof above in order to obtain a valid trial function.

Structural conditions for linearized instability. Our next instability criterion is *structural*, meaning it depends on properties of the reaction function f rather than on properties of the particular steady state U.

THEOREM 7.3 (Linear instability). *Assume the steady state U is nonconstant, and that*

$$f(0) = 0, \quad f''(0) = 0, \quad f''' > 0.$$

Then $\tau_1 < 0$.

For example, every nonconstant steady state is unstable if $f(y) = y^3 - y$. (The derivative $f'(y) = 3y^2 - 1$ admittedly has no upper bound, but one may make it bounded by modifying f outside a neighborhood of U; the instability proof below is unaffected.)

PROOF. First we collect facts about boundary values, to be used later when integrating by parts:

$$\begin{aligned} U &= 0 \text{ at } x = 0, X && \text{by the Dirichlet BC,} \\ f(U) &= 0 \text{ at } x = 0, X && \text{since } f(0) = 0, \\ U'' &= 0 \text{ at } x = 0, X && \text{because } U'' = -f(U), \\ f''(U) &= 0 \text{ at } x = 0, X && \text{since } f''(0) = 0. \end{aligned}$$

The Rayleigh principle for L says that

$$\tau_1 = \min \left\{ \frac{\int_0^X \left((w')^2 - f'(U)w^2\right) dx}{\int_0^X w^2 \, dx} : w \in H_0^1(0, X) \right\}.$$

Choose a trial function

$$w = U''$$

for reasons explained after the proof. Notice w is not the zero function, since U is not linear. The numerator of the Rayleigh quotient for w is

$$\begin{aligned} &\int_0^X \left((U''')^2 - f'(U)(U'')^2\right) dx \\ &= \int_0^X \left(-U'''' - f'(U)U''\right)U'' \, dx && \text{by parts on the first term} \\ &= \int_0^X f''(U)(U')^2 U'' \, dx && \text{by substituting for } U'''' \text{ from the second} \\ & && \text{derivative of steady state equation (7.2)} \\ &= \frac{1}{3} \int_0^X f''(U)\left[(U')^3\right]' dx \\ &= -\frac{1}{3} \int_0^X f'''(U)(U')^4 \, dx && \text{by parts} \\ &< 0 \end{aligned}$$

since $f''' > 0$ and U is nonconstant. Hence $\tau_1 < 0$, by the Rayleigh principle. □

Motivation for the choice of trial function. Our trial function $w = U''$ corresponds to a perturbation $u = U + \varepsilon\phi \approx U + \varepsilon e^{-\tau_1 t}U''$, which will want to evolve (assuming $\varepsilon > 0$ and the steady state U looks something like a sine function) from the steady state towards the constant function.

Notes and comments. Monotonicity criteria for the time map (or "period map") of the family of steady states lead to rather precise structural stability results for reaction–diffusion equations. See the overview and extensions in [**Laugesen and Pugh 1**, Section 7].

Exercises.

7.1 — Show that Theorem 7.3 (linear instability for nonconstant steady states) continues to hold under the alternative assumptions

$$f(0) = 0, \qquad f(y)f''(y) > 0 \quad \text{for all } y \neq 0.$$

Verify those assumptions when

$$f(y) = 2y - \int_0^y e^{-z^2}\, dz,$$

and show f''' changes sign in this example, so that Theorem 7.3 does not apply.

7.2 — Prove linear instability of every nonconstant steady state of the reaction-diffusion equation under the Neumann boundary conditions $u'(0) = u'(X) = 0$, assuming either

$$f''' > 0 \qquad \text{or} \qquad f(y)f''(y) > 0 \quad \text{for all } y \neq 0.$$

Appendix A. Compact imbeddings of Sobolev space into L^2

Goal. Develop compact imbedding theorems of Rellich type for $H_0^1 \hookrightarrow L^2$ on domains of finite volume, and $H^1 \hookrightarrow L^2$ on bounded Lipschitz domains, as used in Section 5.

Fourier multiplier operators. Boundedness and compactness properties of Fourier multipliers are established in the next lemmas, to be used later when proving compact imbeddings of Sobolev space.

All functions in this appendix are complex valued unless specified otherwise, and $L^2(\Omega)$ means $L^2(\Omega;\mathbb{C})$ and so on. The Fourier transform of f is

$$\widehat{f}(\xi) = (\mathcal{F}f)(\xi) = \int_{\mathbb{R}^d} f(x)e^{-2\pi i\xi\cdot x}\, dx.$$

Write M_w for the Fourier multiplier operator

$$M_w f = \mathcal{F}^{-1}(w\mathcal{F}f).$$

LEMMA A.1 (Boundedness). *If $g, w \in L^\infty(\mathbb{R}^d)$ then*

$$gM_w : L^2(\mathbb{R}^d) \to L^2(\mathbb{R}^d)$$

is linear and bounded.

PROOF. Since the Fourier transform acts isometrically on L^2, it is obvious from the definition of $M_w f = \mathcal{F}^{-1}(w\mathcal{F}f)$ and boundedness of the functions g and w that gM_w is a bounded operator, with norm at most $\|g\|_{L^\infty}\|w\|_{L^\infty}$. □

LEMMA A.2 (Compactness). *If $g, w \in L^2(\mathbb{R}^d)$ then*

$$gM_w : L^2(\mathbb{R}^d) \to L^2(\mathbb{R}^d)$$

is linear and compact.

PROOF. First we show the operator is well defined and bounded. Since $w \in L^2$, the multiplier can be written as a convolution, namely

$$M_w f = \mathcal{F}^{-1}(w\mathcal{F}f) = (\mathcal{F}^{-1}w) * f, \qquad f \in L^2(\mathbb{R}^d),$$

and so

$$\|gM_w f\|_{L^2} \leq \|g\|_{L^2}\|M_w f\|_{L^\infty} \leq \|g\|_{L^2}\|w\|_{L^2}\|f\|_{L^2}.$$

Thus gM_w is bounded on L^2. Further, it is Hilbert–Schmidt and therefore compact, because the operator

$$(gM_w f)(x) = g(x)\int_{\mathbb{R}^d}(\mathcal{F}^{-1}w)(x-y)f(y)\,dy$$

has integral kernel $K(x,y) = g(x)(\mathcal{F}^{-1}w)(x-y)$, which is square integrable:

$$\int_{\mathbb{R}^d}\int_{\mathbb{R}^d} K(x,y)^2\,dxdy = \|g\|_{L^2}^2\|w\|_{L^2}^2.$$

□

LEMMA A.3. *If $g \in L^2 \cap L^\infty(\mathbb{R}^d), w \in L^\infty(\mathbb{R}^d)$ and $w(\xi) \to 0$ as $|\xi| \to \infty$, then*

$$gM_w : L^2(\mathbb{R}^d) \to L^2(\mathbb{R}^d)$$

is linear and compact.

PROOF. First, gM_w is bounded on L^2 by Lemma A.1, since g and w are bounded. For compactness, write

$$w_n = w1_{B(n)}$$

where $B(n)$ is the ball of radius n in $\mathbb{R}^d$. Then w_n is a bounded function with compact support, and so belongs to $L^2(\mathbb{R}^d)$. Since $g \in L^2$ by hypothesis, Lemma A.2 now yields compactness of gM_{w_n}.

We will show $gM_{w_n} \to gM_w$ in the operator norm, so that gM_w is compact. Indeed,

$$\begin{aligned}\|(gM_{w_n} - gM_w)f\|_{L^2} &\leq \|g\|_{L^\infty}\|(M_{w_n} - M_w)f\|_{L^2}\\ &= \|g\|_{L^\infty}\|(w_n - w)\mathcal{F}f\|_{L^2}\\ &\leq \|g\|_{L^\infty}\sup_{|\xi|\geq n}|w(\xi)|\,\|\mathcal{F}f\|_{L^2}\end{aligned}$$

and so as $n \to \infty$,

$$\|gM_{w_n} - gM_w\|_{L^2\to L^2} \leq \|g\|_{L^\infty}\sup_{|\xi|\geq n}|w(\xi)| \to 0.$$

□

Compactness of $H_0^1 \hookrightarrow L^2$ on finite volume domains. Now comes the main result of the appendix.

THEOREM A.4 (Rellich for H_0^1). *If $\Omega \subset \mathbb{R}^d$ is open with finite volume then the imbedding*

$$H_0^1(\Omega) \hookrightarrow L^2(\Omega)$$

is compact.

This appendix treats complex valued functions. The compactness of the imbedding follows immediately also for real valued functions.

PROOF. Take a bounded sequence $\{f_k\}$ in $H_0^1(\Omega)$, say with $\|f_k\|_{H^1(\Omega)} \le C$. Extend f_k to equal 0 outside Ω, so that the extended function belongs to $H^1(\mathbb{R}^d)$. Define $F_k \in L^2(\mathbb{R}^d)$ by

$$\mathcal{F}F_k = (1+4\pi^2|\xi|^2)^{1/2}\mathcal{F}f_k,$$

where we observe

$$\|F_k\|_{L^2(\mathbb{R}^d)} = \|f_k\|_{H^1(\Omega)} \le C,$$

so that the sequence $\{F_k\}$ is bounded in $L^2(\mathbb{R}^d)$.

Let $g = 1_\Omega$ and notice $g \in L^1 \cap L^\infty(\mathbb{R}^d)$ since Ω has finite volume, so that in particular $g \in L^2(\mathbb{R}^d)$. Next let $w(\xi) = (1+4\pi^2|\xi|^2)^{-1/2}$, so that $w \in L^\infty(\mathbb{R}^d)$ and $w(\xi) \to 0$ as $|\xi| \to \infty$. Hence gM_w is compact on $L^2(\mathbb{R}^d)$ by Lemma A.3.

Applying this compact operator to F_k yields existence of a subsequence of $\{gM_wF_k\}$ that converges in $L^2(\mathbb{R}^d)$. Since

$$gM_wF_k = 1_\Omega f_k,$$

we have a subsequence of $\{f_k\}$ that converges in $L^2(\Omega)$, which proves that the imbedding of $H_0^1(\Omega)$ into $L^2(\Omega)$ is compact. □

Compactness of $H^1 \hookrightarrow L^2$ on Lipschitz domains. The imbedding of H_0^1 implies an imbedding of H^1.

THEOREM A.5 (Rellich for H^1). *If $\Omega \subset \mathbb{R}^d$ is a bounded domain with Lipschitz boundary then the imbedding*

$$H^1(\Omega) \hookrightarrow L^2(\Omega)$$

is compact.

PROOF. Choose a bounded domain Ω_0 that contains the closure of Ω, and write

$$E : H^1(\Omega) \to H_0^1(\Omega_0)$$

for the extension operator [**Evans and Gariepy**]. A bounded sequence $\{f_k\}$ in $H^1(\Omega)$ gives a bounded sequence $\{Ef_k\}$ in $H_0^1(\Omega_0)$. Some subsequence of the Ef_k converges in $L^2(\Omega_0)$ by Rellich's Theorem A.4, and restricting to Ω (where $Ef_k = f_k$) shows convergence of the subsequence in $L^2(\Omega)$. □

Notes and comments. The proof by Fourier transform and Hilbert–Schmidt operators of the compact imbedding $H_0^1 \hookrightarrow L^2$ (Rellich's Theorem for finite volume domains) was shown to me by Dirk Hundertmark.

The proof extends straightforwardly to give compactness of the imbedding for fractional order Sobolev spaces too: $H_0^s(\Omega) \hookrightarrow L^2(\Omega)$ compactly for all $s > 0$, whenever Ω has finite volume. Hence one obtains discrete spectrum for the fractional Laplacian on domains of finite volume. For more on spectral theory of the fractional Laplacian, see the review article [**Fra18**].

Further, the proof of Theorem A.4 shows that $H^s(\mathbb{R}^d)$ restricts compactly into $L^2(\Omega)$, whenever the domain Ω has finite volume.

References

[Antunes and Freitas] P. R. S. Antunes and P. Freitas, *On the inverse spectral problem for Euclidean triangles*, Proc. R. Soc. Lond. Ser. A Math. Phys. Eng. Sci. **467** (2011), no. 2130, 1546–1562, DOI 10.1098/rspa.2010.0540. MR2795790

[Bandle] C. Bandle, *Isoperimetric inequalities and applications*, Monographs and Studies in Mathematics, vol. 7, Pitman (Advanced Publishing Program), Boston, Mass.-London, 1980. MR572958

[Blanchard and Brüning] P. Blanchard and E. Brüning, *Variational methods in mathematical physics*, Texts and Monographs in Physics, Springer-Verlag, Berlin, 1992. A unified approach; Translated from the German by Gillian M. Hayes. MR1230382

[Ciarlet] P. G. Ciarlet, *Linear and nonlinear functional analysis with applications*, Society for Industrial and Applied Mathematics, Philadelphia, PA, 2013. MR3136903

[Evans] L. C. Evans, *Partial differential equations*, 2nd ed., Graduate Studies in Mathematics, vol. 19, American Mathematical Society, Providence, RI, 2010. MR2597943

[Evans and Gariepy] L. C. Evans and R. F. Gariepy, *Measure theory and fine properties of functions*, Studies in Advanced Mathematics, CRC Press, Boca Raton, FL, 1992. MR1158660

[Farlow] S. J. Farlow, *Partial differential equations for scientists and engineers*, Dover Publications, Inc., New York, 1993. Revised reprint of the 1982 original. MR1243874

[Fra18] R. L. Frank, *Eigenvalue bounds for the fractional Laplacian: a review*, Recent developments in nonlocal theory, De Gruyter, Berlin, 2018, pp. 210–235. MR3824213

[Gilbarg and Trudinger] D. Gilbarg and N. S. Trudinger, *Elliptic partial differential equations of second order*, Classics in Mathematics, Springer-Verlag, Berlin, 2001. Reprint of the 1998 edition. MR1814364

[Henrot et al.] A. Henrot (ed.), *Shape optimization and spectral theory*, De Gruyter Open, Warsaw, 2017. MR3681143

[Laugesen and Pugh 1] R. S. Laugesen and M. C. Pugh, *Properties of steady states for thin film equations*, European J. Appl. Math. **11** (2000), no. 3, 293–351, DOI 10.1017/S0956792599003794. MR1844589

[Mathews and Walker] J. Mathews and R. L. Walker, *Mathematical methods of physics*, Second edition, W. A. Benjamin, Inc., New York, 1970. MR0261817

[McCartin] B. J. McCartin, *Eigenstructure of the equilateral triangle. I. The Dirichlet problem*, SIAM Rev. **45** (2003), no. 2, 267–287, DOI 10.1137/S003614450238720. MR2010379

[Schaaf2] R. Schaaf, *Global solution branches of two-point boundary value problems*, Lecture Notes in Mathematics, vol. 1458, Springer-Verlag, Berlin, 1990. MR1090827

[Strauss] W. A. Strauss, *Partial differential equations*, 2nd ed., John Wiley & Sons, Ltd., Chichester, 2008. An introduction. MR2398759

Department of Mathematics, University of Illinois, Urbana, Illinois 61801
Email address: laugesen@illinois.edu

Contemporary Mathematics
Volume **720**, 2018
https://doi.org/10.1090/conm/720/14522

From classical mechanics to quantum mechanics

Richard Froese

ABSTRACT. These are the notes for five one hour lectures delivered at the 2016 CRM summer school in Spectral Theory at Laval University. They were intended to provide some general background on classical and quantum mechanics to advanced undergraduate and beginning graduate students in mathematics. The first four lectures contain an outline of these two theories from a mathematical point view and a comparison of the classical and quantum descriptions of some simple systems. The last lecture is about hidden variables and is meant to illustrate the essential strangeness of the quantum description of nature. I have tried to make these notes more accessible by including an outline of some basic ideas in measure theory and operator theory. A similar outline of the basics of manifolds, tangent and cotangent spaces and differential forms would have been useful for understanding the sections on Lagrangian submanifolds and Hamilton-Jacobi equations. Unfortunately I did not have time to include this. I am grateful to Cyrus Bhiladvala and the referee for making suggestions for improvement.

CONTENTS

1. Classical mechanics

1.1. Newton's equations and Hamilton's equations. We begin by showing how classical mechanics describes the motion of a single particle of mass m moving in configuration space $\mathbb{R}^n$ under the influence of an external conservative force field. The force felt by the particle at the point $x = (x_1, \ldots, x_n) \in \mathbb{R}^n$ is $F(x) = -\nabla V(x)$, where the potential V is a real valued function. The motion is described by the trajectory $x(t)$, where $t \in \mathbb{R}$ is time, and $x(t)$ is the position at time t. According to Newton's law $F = m\mathbf{a}$, the trajectory satisfies the equation

$$m\ddot{x} = -\nabla V(x). \tag{1}$$

2010 *Mathematics Subject Classification.* Primary 70-01, 81-01, 70H05, 70H20, 81Q20, 81P10.

This is a system of n second order ODEs so under suitable conditions on V, there is a unique solution $x(t)$ if we impose initial conditions

$$\begin{aligned} x(0) &= x_0 \\ \dot{x}(0) &= v_0. \end{aligned} \tag{2}$$

There is a standard trick for turning a system of n second order equations into an equivalent system of $2n$ first order equations. If $x(t)$ solves Newton's equation (1) with initial conditions (2), then, defining the momentum as $p(t) = m\dot{x}(t)$, we find that $(x(t), p(t)) \in \mathbb{R}^n \times \mathbb{R}^n$ satisfies Hamilton's equations

$$\begin{aligned} \dot{x} &= \frac{1}{m}p \\ \dot{p} &= m\ddot{x} = -\nabla V(x), \end{aligned} \tag{3}$$

with initial conditions

$$\begin{aligned} x(0) &= x_0 \\ p(0) &= mv_0 = p_0. \end{aligned} \tag{4}$$

On the other hand, if $(x(t), p(t))$ solves Hamilton's equations (3) with initial condition (4), then $x(t)$ solves Newton's equation (1) with initial conditions (2).

Hamilton's equations are a first order system in phase space $\mathbb{R}^n \times \mathbb{R}^n$. The right side of (3) defines a vector field on phase space, namely

$$(x, p) \mapsto \begin{bmatrix} \frac{1}{m}p \\ \\ -\nabla V(x) \end{bmatrix}.$$

The trajectories $(x(t), p(t))$ are the corresponding flow.

Although it is not obvious, the vector field on the right of (3) is very special. Let $H(x, p)$ be the real valued function on phase space defined by

$$H(x, p) = \frac{p^2}{2m} + V(x).$$

Here $p^2 = \langle p, p \rangle$ is the square of the standard Euclidean norm of p in $\mathbb{R}^n$. The function $H(x, p)$, called the Hamiltonian, represents the total energy (kinetic plus potential) of the particle. Written in terms of H, Hamilton's equations have the form

$$\begin{aligned} \dot{x} &= \frac{\partial H}{\partial p}(x, p) \\ \dot{p} &= -\frac{\partial H}{\partial x}(x, p), \end{aligned} \tag{5}$$

where

$$\frac{\partial H}{\partial x} = \begin{bmatrix} \frac{\partial H}{\partial x_1} \\ \vdots \\ \frac{\partial H}{\partial x_n} \end{bmatrix}, \quad \frac{\partial H}{\partial p} = \begin{bmatrix} \frac{\partial H}{\partial p_1} \\ \vdots \\ \frac{\partial H}{\partial x_p} \end{bmatrix}$$

denote the gradients with respect to the position variables $x_1, \ldots, x_n$ and momentum variables $p_1, \ldots, p_n$ respectively.

To get a feeling for Hamilton's equations let's look at three simple examples. For a Hamiltonian of the form $H(x,p) = \frac{p^2}{2m} + V(x)$ we can get a mental image of the motion by imagining a particle sliding without friction on a hill whose height at the point x is $V(x)$.

In these examples we will pay attention to the presence of bound states and scattering states. Bound states are trajectories for which $x(t)$ remains bounded for all time, while scattering states are those where $|x(t)|$ tends to infinity when $t \to \pm\infty$.

1.2. Example 1: Free motion. If the force $F = 0$ then the motion is free. This happens if $V(x) = V_0$ is constant. In this case Hamilton's equations are

$$\begin{aligned} \dot{x} &= \frac{1}{m} p \\ \dot{p} &= 0. \end{aligned}$$

The second equation says that the momentum $p(t) = p_0$ is constant. Then $x(t) = x_0 + \frac{1}{m} p_0 t$. So the trajectories in configuration space are straight lines in the direction of the momentum.

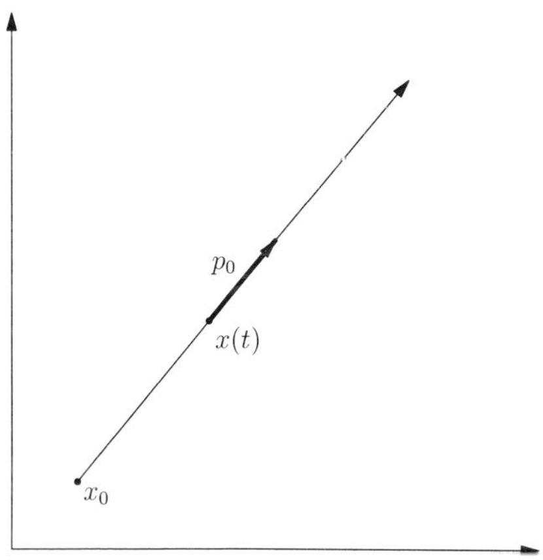

Trajectory in position space $n = 2$

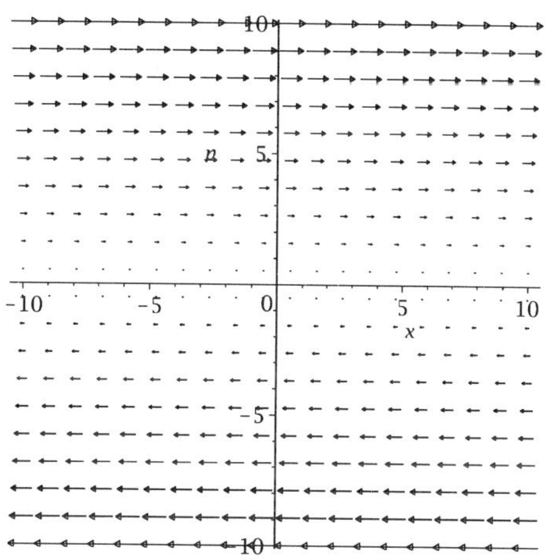

Vector field in phase space $n = 1$

In this example, states with $p_0 \neq 0$ move with uniform velocity. Clearly $|x(t)| \to \infty$ for these states, so they are scattering states. States with $p_0 = 0$ are bound states that stay fixed at x_0 for all time.

1.3. Example 2: Harmonic oscillator. The harmonic oscillator potential in n dimensions is given by

$$V(x) = \frac{m}{2}\left(\omega_1^2 x_1^2 + \cdots + \omega_n^2 x_n^2\right)$$

where $\omega_1, \ldots, \omega_n$ are positive constants. Hamilton's equations for the harmonic oscillator are

$$\dot{x}_i = \frac{1}{m} p_i$$
$$\dot{p}_i = -m\omega_i^2 x_i.$$

for $i = 1, \ldots, n$. For simplicity let's take $m = \omega_i = 1$. Then we have n independent 2×2 linear systems

$$\begin{bmatrix} \dot{x}_i \\ \dot{p}_i \end{bmatrix} = \begin{bmatrix} 0 & 1 \\ -1 & 0 \end{bmatrix} \begin{bmatrix} x_i \\ p_i \end{bmatrix}$$

whose solutions are given by clockwise rotations in phase space

$$\begin{bmatrix} x_i \\ p_i \end{bmatrix} = \exp\left(t \begin{bmatrix} 0 & 1 \\ -1 & 0 \end{bmatrix} \right) \begin{bmatrix} x_{i,0} \\ p_{i,0} \end{bmatrix} = \begin{bmatrix} \cos(t) & \sin(t) \\ -\sin(t) & \cos(t) \end{bmatrix} \begin{bmatrix} x_{i,0} \\ p_{i,0} \end{bmatrix}.$$

Another way to solve this system is to consider the complex functions

$$a_k = \frac{1}{\sqrt{2}} (x_k + i p_k)$$

for $k = 1, \ldots, n$. Then

$$\dot{a}_k = \frac{1}{\sqrt{2}} (\dot{x}_k + i \dot{p}_k) = \frac{1}{\sqrt{2}} (p_k - i x_k) = -i a_k.$$

Thus $a_k(t) = e^{-it} a_{k,0}$ for some initial conditions $a_{k,0}$, from which we can recover $x_k(t)$ and $p_k(t)$ by taking the real and imaginary parts. So if $a_{k,0} = A_k e^{i\alpha_k}$ then

$$x_k(t) = A_k \cos(t - \alpha_k)$$
$$p_k(t) = -A_k \sin(t - \alpha_k).$$

Solutions are circles in phase space traversed counter-clockwise. All trajectories are bound states.

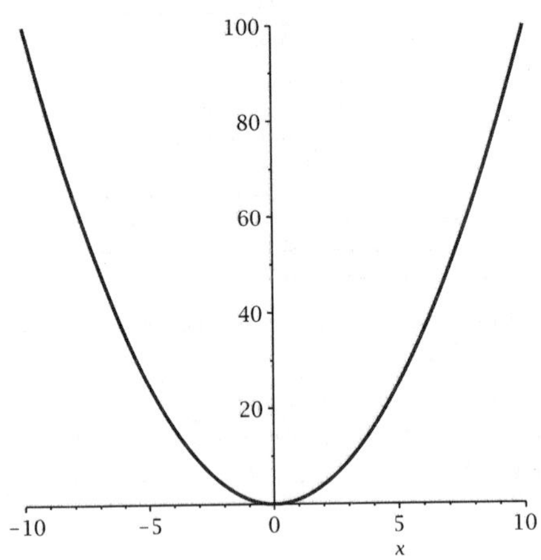

Harmonic oscillator potential $n = 1$

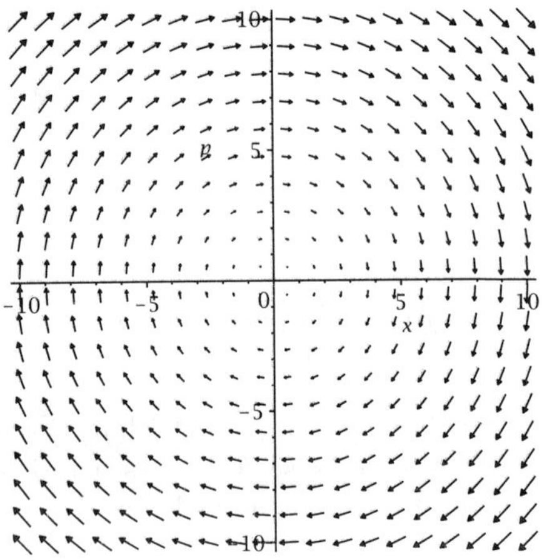

Harmonic oscillator vector field in phase space

PROBLEM 1. What are the solutions for general m and ω_i?

1.4. Example 3: Two bump potential $n = 1$. Now consider a potential $V(x)$ that looks like this:

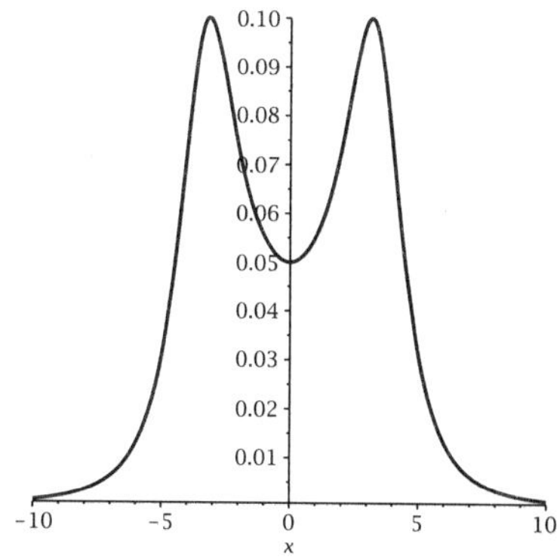

Two bump potential

Hamilton's equations are

$$\dot{x} = \frac{1}{m}p$$
$$\dot{p} = -V'(x).$$

Although we cannot write down the exact solution, we can get a qualitative idea of the motion by sketching the Hamiltonian vector field. Here is a plot, with some exceptional solution curves

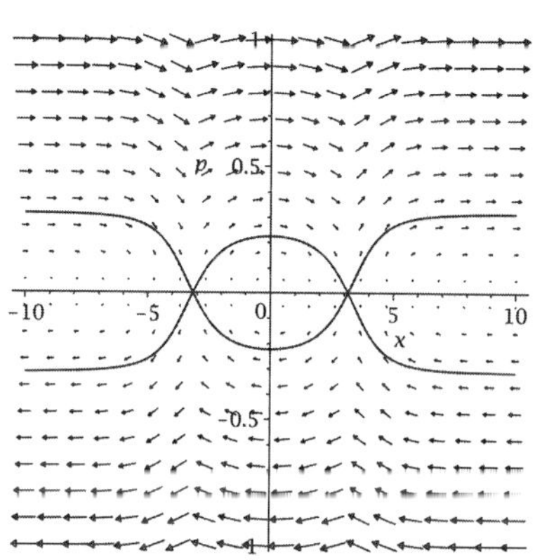

Hamiltonian vector field for two bump potential

PROBLEM 2. Which states are bound states, which are scattering states, and which are neither?

1.5. Symplectic form, Poisson bracket and time evolution of observables. The vector field X_H on the right side of Hamilton's equations (5) has the form

$$X_H = \begin{bmatrix} \dfrac{\partial H}{\partial p} \\ \\ -\dfrac{\partial H}{\partial x} \end{bmatrix} = J\nabla H$$

where

$$J = \begin{bmatrix} 0 & I \\ -I & 0 \end{bmatrix}, \quad \nabla H = \begin{bmatrix} \dfrac{\partial H}{\partial x} \\ \dfrac{\partial H}{\partial p} \end{bmatrix}.$$

Replacing H in this construction with any smooth function f on phase space, we can produce vector fields X_f. These will define a corresponding flow. But the Hamiltonian function H is special because the flow it generates is the time evolution.

In classical mechanics, (sufficiently nice) functions $f(x,p)$ on phase space are called *observables*. They correspond to measurable quantities. Here are some examples.

Observable	$f(x,p)$
position	x_i
momentum	p_i
angular momentum in 2-d	$x_1 p_1 - x_2 p_1$
energy	$H(x,p)$

To determine how the value of an observable changes in time as the particle moves along a trajectory we use the chain rule to compute

$$\frac{d}{dt} f(x(t),p(t)) = \left\langle \nabla f, \begin{bmatrix} \dot{x}(t) \\ \dot{p}(t) \end{bmatrix} \right\rangle = \langle \nabla f, J \nabla H \rangle \tag{6}$$

where the gradients are evaluated at $(x(t), p(t))$. We now introduce some notation to rewrite this equation. Define ω to be the antisymmetric bilinear form whose value on vectors X and Y (thought of as tangent vectors to phase space) is given by

$$\omega[X,Y] = \langle X, JY \rangle . \tag{7}$$

This ω is called the *symplectic form*. Then the *Poisson bracket* of two observables f and g is defined as

$$\{f,g\} = \omega[\nabla f, \nabla g] = \sum_i \frac{\partial f}{\partial x_i}\frac{\partial g}{\partial p_i} - \frac{\partial f}{\partial p_i}\frac{\partial g}{\partial x_i}.$$

The Poisson bracket is an antisymmetric bilinear form on (differentiable) observables. It is a derivation, which means that it satisfies the product rule

$$\{f, gh\} = g\{f,h\} + \{f,g\}h.$$

It also satisfies the Jacobi identity

$$\{\{f,g\},h\} + \{\{h,f\},g\} + \{\{g,h\},f\} = 0.$$

Using this notation, equation (6) can be written

$$\frac{d}{dt} f = \{f, H\}$$

with all functions evaluated at $(x(t), p(t))$. Since $\{f, f\} = 0$ for any f we see that energy is conserved:

$$\frac{d}{dt}H = \{H, H\} = 0.$$

Other observables f will be constants of motion provided $\{f, H\} = 0$. Here is an example. For the harmonic oscillator in n dimensions, we can define n observables $I_1, \ldots, I_n$, called action variables, as

$$I_k(x, p) = \frac{1}{2m}p_k^2 + \frac{m}{2}\omega_k^2 x_k^2.$$

Clearly $H = I_1 + \cdots + I_n$ and for $j \neq k$, $\{I_k, I_j\} = 0$ since they depend on disjoint sets of variables. Thus $\{I_k, H\} = \{I_k, I_1 + \cdots + I_n\} = \{I_k, I_k\} = 0$. So the action variables are constants of the motion.

When $n = 1$ the constant energy surfaces are curves in $\mathbb{R}^2$. In this case we can see the shape of the orbits simply by plotting them. For the harmonic oscillator with $n = 1$ the curves are circles $\{x^2 + p^2 = 2E\}$ when $E > 0$ or the point $(0, 0)$ when $E = 0$. For the two-bump potential a plot of the constant energy curves looks like this.

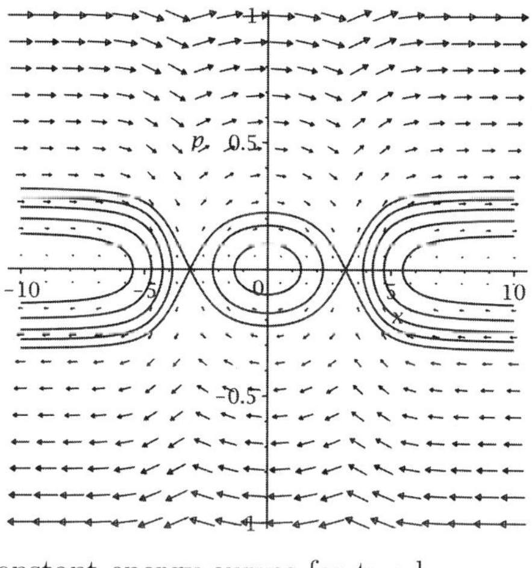

Some constant energy curves for two bump potential

So we can get a pretty good picture of the flow without much work.

When there are n constants of the motion $f_1, \ldots, f_n$, as for the harmonic oscillator, the orbits lie on the intersection of all the level surfaces. When $\{f_i, f_j\} = 0$, such systems (under some regularity conditions) are called completely integrable. The intersections of the level surfaces turn out to be n dimensional tori and the motion is straight line motion.

The other extreme is if the orbit is spread out evenly on the $n-1$ dimensional constant energy surface. In this case the system is called ergodic.

PROBLEM 3. Verify that action functions for the harmonic oscillator verify $\{I_j, I_k\} = 0$ for every j and k.

PROBLEM 4. What happens to the position x_j under the flow generated by momentum p_k. What about the other way around?

1.6. Symplectic flow and Poincaré recurrence. The Hamiltonian flow Φ_t associated with X_H is the map $\Phi_t : (x,p) \mapsto (x(t), p(t))$, where $(x(t), p(t))$ solves Hamilton's equations with intial condition (x,p). In other words

$$\frac{d}{dt}\Phi_t(x,p) = X_H(\Phi_t((x,p))), \qquad \Phi_0((x,p)) = (x,p)$$

We want to show that the Hamiltonian flow preserves the symplectic form. To formulate this result recall that for fixed t the differentiable map $\Phi_t : \mathbb{R}^n \times \mathbb{R}^n \to \mathbb{R}^n \times \mathbb{R}^n$ induces a map on tangent vectors, sending X in the tangent space at (x,p) to $\Phi_t'(x,p)X$ in the tangent space at $\Phi_t((x,p))$, where $\Phi_t'(x,p)$ is the Jacobian matrix (also denoted $d\Phi_t(x,p)$).

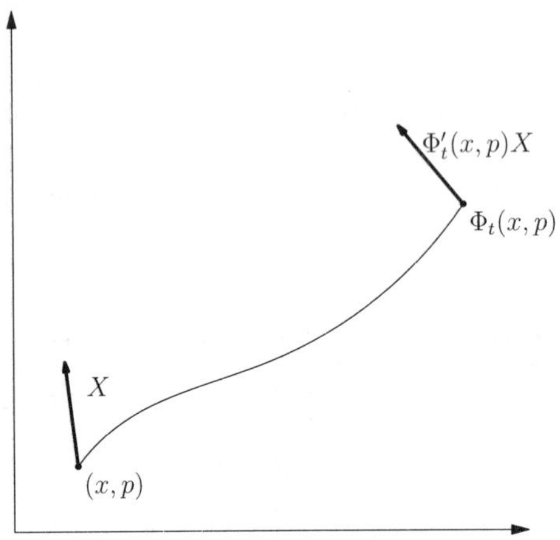

The flow Φ_t and the map $\Phi_t'(x,p)$

If we write

$$\Phi_t((x,p)) = (f(t,x,p), g(t,x,p)).$$

then

$$\Phi_t'((x,p)) = \begin{bmatrix} \frac{\partial f}{\partial x}(t,x,p) & \frac{\partial f}{\partial p}(t,x,p) \\ \frac{\partial g}{\partial x}(t,x,p) & \frac{\partial g}{\partial p}(t,x,p) \end{bmatrix}$$

Saying that the flow preserves the symplectic form means

(8) $$\omega[X,Y] = \omega[\Phi_t'((x,p))X, \Phi_t'((x,p))Y]$$

for every t.

PROPOSITION 1. *The Hamiltonian flow Φ_t preserves the symplectic form*

PROOF. We will show that Φ_t' is a symplectic matrix, which means that

(9) $$\Phi_t'^T J \Phi_t' = J$$

If (9) holds then $\omega[\Phi_t' X, \Phi_t' Y] = \langle \Phi_t' X, J\Phi_t' Y\rangle = \langle X, \Phi_t'^T J \Phi_t' Y\rangle = \langle X, JY\rangle = \omega[X,Y]$. Thus (9) implies (8).

Writing out the equation for the flow, we have

$$\frac{d}{dt}\Phi_t(x,p) = J\nabla H(\Phi_t(x,p)).$$

Computing the Jacobian of both sides, and exchanging the order of the partial differentiation with respect to x and p with the time derivative gives

$$\frac{d}{dt}\Phi_t'(x,p) = JH''(\Phi_t(x,p))\Phi_t'(x,p)$$

where H'' is the Hessian of H. Now

$$\begin{aligned}\frac{d}{dt}(\Phi_t'^T J\Phi_t') &= (JH''\Phi_t')^T J\Phi_t' + \Phi'^T J(JH''\Phi_t') \\ &= \Phi_t'^T H''^T J^T J\Phi_t' + \Phi_t'^T JJH''\Phi_t' \\ &= \Phi_t'^T H''\Phi_t' - \Phi_t'^T H''\Phi_t' \\ &= 0\end{aligned}$$

Here we used $J^T J = I$, $J^2 = -I$ and $H''^T = H''$. This implies that $\Phi_t'^T J\Phi_t'$ is constant.When $t = 0$ we have $\Phi_0' = I$ so the constant value is J. □

COROLLARY 1. *(Liouville's theorem) The Hamiltonian flow preserves the phase space volume.*

PROOF. Taking the determinant of both sides of (9) yields $\det(\Phi_t')^2 = 1$ so $|\det(\Phi_t')| = 1$. Now if A is a (measurable) set in phase space then

$$\mathrm{Vol}(\Phi_t(A)) = \int_A |\det(\Phi_t')(x,p)|\, dx^n dp^n = \int_A dx^n dp^n = \mathrm{Vol}(A)$$

□

COROLLARY 2. *(Poincaré's recurrence theorem) Suppose that $\Phi_t(V) \subset V$ for all times t and for some set V in phase space with Vol(V) $< \infty$. Let $v \in V$ and $T > 0$. Then any neighbourhood $U \subseteq V$ of v contains a point that returns to U after time T.*

PROOF. Let $F = \Phi_T$ and denote the k fold iterate $F \circ F \circ \cdots \circ F$ by F^k. Then $U, F(U), F^2(U), \ldots$ are contained in V and have the same non-zero volume. If they were all disjoint then V would have infinite volume. Thus $F^k(U) \cap F^j(U) \neq \emptyset$ for some $k > j$ which implies $F^{k-j}(U) \cap U \neq \emptyset$. □

PROBLEM 5. Show that if M is a symplectic matrix $M^T JM = J$ then $\det(M) = 1$.

1.7. Summary of classical mechanics (so far). A *pure state* of the system is a point (x,p) in phase space. We call these pure states to distinguish them from mixed states, introduced later. A pure state is meant to describe the system completely with no uncertainty.

Observables are smooth real valued functions f on phase space. There is a pairing between states and observables given by

$$\langle f|(x,p)\rangle = f(x,p).$$

This is a real number representing the result of making a measurement of the observable f when the system is in the state (x,p).

There is a distinguished observable $H(x,p)$ representing the total energy of the system. The time evolution of states is given by Hamiltonian flow Φ_t. This flow preserves the symplectic form. The time evolution on observables is defined via the pairing. Explicitly, $f(x,p,t) = \Phi_t^* f$, where $\langle \Phi_t^* f | (x,p) \rangle = \langle f | \Phi_t(x,p) \rangle$. The time evolution on observables obeys

$$f' = \{f, H\},$$

where $\{\cdot,\cdot\}$ is the Poisson bracket.

1.8. The Hamilton Jacobi equation. One of the points of contact between classical and quantum mechanics is the Hamilton Jacobi equation.

$$H(x, \nabla S(x,t)) + \frac{\partial S}{\partial t}(x,t) = 0 \tag{10}$$

and its variant

$$H(x, \nabla s(x,E)) - E = 0 \tag{11}$$

We want to explain how these equation can be solved using the Hamiltonian flow for H. Since the solution to these equations can be used to construct approximate solutions to the Schrödinger equation in various situations, the equations provide a link between classical and quantum mechanics. We will consider an example of this later in the course.

Although it is not really important for us, it is remarkable that sometimes solutions to (11) can be used to determine the Hamiltonian flow for H.

It is maybe not surprising that first order equations can be solved using flows. But there is some pretty geometry connected with these equations that I'll try to describe. A reference for this material (in fact for everything in this section) is Arnold's classic book on classical mechanics [A]. I have tried to give an informal account of the main ideas. This section uses the language of differential forms. In the context of classical mechanics, these are described in Arnold [A] and in Marsden and Ratiu [MR].

Our phase space is $\mathcal{P} = \mathbb{R}^n \times \mathbb{R}^n$ with co-ordinates $(x_1, \ldots, x_n, p_1, \ldots, p_n)$. In the language of differential forms the symplectic form we introduced above is

$$\omega = \sum_{i=1}^{n} dp_1 \wedge dx_i \tag{12}$$

A more general setting for classical mechanics is a symplectic manifold. This is an even dimensional manifold, typically the co-tangent bundle of configuration space, equipped with a closed non-degenerate 2-form ω. (Closed means that $d\omega = 0$ and nondegenerate means that if $\omega[X,Y] = 0$ for every Y, then $X = 0$.) Darboux's theorem states that near every point one can find a co-ordinate system so that ω is given by (12). Actually there are always many such co-ordinate systems. They are called symplectic co-ordinates, or canonical co-ordinates.

Then the connection between (12) and (7) is this. Let $(x_1, \ldots, x_n, p_1, \ldots, p_n)$ be a symplectic co-ordinate system. Any vector ξ in the tangent space of $\mathcal{P}$ at q can be associated with the derivative at $t = 0$ of a curve $q(t)$ in $\mathcal{P}$ with $q(0) = q$.

If $(x_1(t),\ldots,x_n(t),p_1(t),\ldots,p_n(t))$ are the co-ordinates of $q(t)$ then ξ has the co-ordinate representation

$$X = \begin{bmatrix} \dot{x}_1(0) \\ \vdots \\ \dot{x}_n(0) \\ \dot{p}_1(0) \\ \vdots \\ \dot{p}_n(0) \end{bmatrix}.$$

Given two such tangent vectors ξ and η with co-ordinate representations X and Y we find that $\sum_{i=1}^{n} dp_1 \wedge dx_i[\xi,\eta] = \langle X, JY\rangle$.

1.9. Lagrangian submanifolds. We now introduce a special class of submanifolds of phase space. They have the important property that often they can be written as the graph of a gradient. Let's give the definition and explain this.

A submanifold $\mathcal{L}$ of $2n$ dimensional phase space $\mathcal{P}$ with symplectic form ω is *Lagrangian* if

(i) $\dim(\mathcal{L}) = n$
(ii) ω vanishes on $\mathcal{L}$.

Condition (ii) means that for any 2 vectors X, Y tangent to $\mathcal{L}$ at some point, $\omega[X,Y] = 0$. We can check this as follows. Take any two curves $(x(t),p(t))$ and $(\tilde{x}(t),\tilde{p}(t))$ that stay in $\mathcal{L}$ and both pass through $q \in \mathcal{L}$ when $t=0$. Let $X = \begin{bmatrix} \dot{x}(0) \\ \dot{p}(0) \end{bmatrix}$ and $Y = \begin{bmatrix} \dot{\tilde{x}}(0) \\ \dot{\tilde{p}}(0) \end{bmatrix}$. Then we must have $\langle X, JY\rangle = 0$. If $\mathcal{L}$ satisfies (ii) but not (i) it is called *isotropic*.

Here are some examples. When $n = 1$ then a n-dimensional submanifold of phase space $\mathbb{R}\times\mathbb{R}$ is a curve. There is only one tangent direction, so all tangent vectors are a multiple of a single vector X. By the antisymmetry of $\omega[X,X] = 0$. Another example is the submanifold

$$\mathcal{L}_1 = \{(x_0,p) : p \in \mathbb{R}^n\}.$$

containing all points in phase space lying above a fixed x_0 in configuration space. Tangent vectors to $\mathcal{L}_1$ have the form $\begin{bmatrix} 0 \\ A \end{bmatrix}$ and $\left\langle \begin{bmatrix} 0 \\ A \end{bmatrix}, J\begin{bmatrix} 0 \\ B \end{bmatrix} \right\rangle = 0$, for any $A, B \in \mathbb{R}^n$. A final simple example is

$$\mathcal{L}_2 = \{(x,p) \in \mathbb{R}^n \times \mathbb{R}^n : x = p\}.$$

In this case tangent vectors have the form $\begin{bmatrix} A \\ A \end{bmatrix}$ for $A \in \mathbb{R}^n$ and $\left\langle \begin{bmatrix} A \\ A \end{bmatrix}, J\begin{bmatrix} B \\ B \end{bmatrix} \right\rangle = 0$, for any $A, B \in \mathbb{R}^n$.

An n-dimensional submanifold $\mathcal{L}$ of phase space is a *graph over* $x = (x_1,\ldots,x_n)$ if there are functions $p(x) = (p_1(x),\ldots,p_n(x))$ so that $(x,p) \in \mathcal{L} \Leftrightarrow p = p(x)$. In this case $x = (x_1,\ldots,x_n)$ are co-ordinates for $\mathcal{L}$. In the examples above $\mathcal{L}_1$ is a graph over $p_1,\ldots,p_n$ but not over $x_1,\ldots,x_n$. On the other hand $\mathcal{L}_2$ is a graph over both $x_1,\ldots,x_n$ and $p_1,\ldots,p_n$.

Here is the main point of this section.

PROPOSITION 2. *Let $(x_1, \ldots, x_n, p_1, \ldots, p_n)$ be symplectic co-ordinates. Suppose that $\mathcal{L}$ is a graph over $(x_1, \ldots, x_n)$. Then $\mathcal{L}$ is Lagrangian $\Leftrightarrow$ there is a function $S(x)$ such that $p(x) = \nabla S(x)$.*

Remark: This proposition holds more generally for co-ordinates $(\tilde{x}_1, \ldots, \tilde{x}_n, \tilde{p}_1, \ldots, \tilde{p}_n)$ where $\sum d\tilde{p}_i \wedge d\tilde{x}_i = f\omega$ is a multiple of ω. So $(\tilde{x}_1, \ldots, \tilde{x}_n, \tilde{p}_1, \ldots, \tilde{p}_n)$ could be $(\tilde{x}_1, \ldots, \tilde{x}_n, \tilde{p}_1, \ldots, \tilde{p}_n) = (p_1, \ldots, p_n, x_1, \ldots, x_n)$ in which case $f = -1$. Sometimes this extra flexibility is useful.

PROOF. If $p(x) = \nabla S(x)$, then tangent vectors to $\mathcal{L}$ at the point $(x, p(x))$ have the form $\begin{bmatrix} A \\ S''(x)A \end{bmatrix}$ where $A \in \mathbb{R}^n$ and $S''(x) = \begin{bmatrix} \dfrac{\partial^2 S(x)}{\partial x_i \partial x_j} \end{bmatrix}$ is the Hessian of S. Then

$$\left\langle \begin{bmatrix} A \\ S''(x)A \end{bmatrix}, J \begin{bmatrix} B \\ S''(x)B \end{bmatrix} \right\rangle = \langle A, S''(x)B \rangle - \langle S''(x)A, B \rangle = 0$$

follows from the symmetry $S'' = S''^T$ of S''. Thus $\mathcal{L}$ is Lagrangian.

Conversely, if $\mathcal{L}$ is Lagrangian then the one form $\alpha = \sum p_i dx_i$ satisfies $d\alpha = \sum dp_i \wedge dx_i = \omega$. Thus $d\alpha = 0$ when restricted to $\mathcal{L}$. Since we are in a situation where every closed loop is spanned by a two dimensional surface (topologically $\mathcal{L} = \mathbb{R}^n$), this implies that $\alpha = dS$ for some function S on $\mathcal{L}$. We can think of S as function of the co-ordinates $(x_1, \ldots, x_n)$. Then

$$\alpha = \sum p_i(x) dx_i = dS = \sum \frac{\partial S}{\partial x_i}(x) dx_i$$

which implies $p_i(x) = \dfrac{\partial S}{\partial x_i}(x)$. □

The function $S(x)$ is called a generating function.

If we know the functions $p_i(x)$ that determine $\mathcal{L}$, we can compute $S(x)$ (which is only determined up to a constant $S(x_0)$) by choosing a base point x_0 and curve $\gamma = ((x(t), p(x(t)))$ in $\mathcal{L}$ from $(x_0, p(x_0))$ to $(x, p(x))$. Then

$$\begin{aligned} S(x) &= S(x_0) + \int_\gamma dS \\ &= S(x_0) + \int_\gamma \sum p_i dx_i \\ &= S(x_0) + \int_0^1 \sum p_i(x(t)) \dot{x}(t) dt \end{aligned}$$

1.10. Legendre transforms. Suppose that the Lagrangian manifold $\mathcal{L}$ is simultaneously a graph over $(x_1, \ldots, x_n)$ and $(p_1, \ldots, p_n)$. Then there are functions $S(x)$ and $s(p)$ such that

$$(x, p) \in \mathcal{L} \Leftrightarrow p = \nabla S(x) \Leftrightarrow x = \nabla s(p).$$

In this situation s is called a Legendre transform of S. Notice that if γ is a path connecting (x_0, p_0) to (x, p) in $\mathcal{L}$ then

$$\begin{aligned} S(x) - S(x_0) &= \int_\gamma \sum p_i dx_i \\ &= \int_\gamma \sum d(p_i x_i) - x_i dp_i \\ &= \langle p, x\rangle - \langle p_0, x_0\rangle - s(p) + s(p_0) \end{aligned}$$

If we choose constants and co-ordinates such that $S(x_0) + s(p_0) - \langle p_0, x_0\rangle = 0$ then

$$S(x) = \langle p, x\rangle - s(p)$$

with p determined by $x = \nabla s(p)$. This is the classical formula for the Legendre transform (at least for differentiable functions).

We can also consider partial Legendre transforms. If $\mathcal{L}$ is a graph over say $p_1, x_2, \cdots, x_n$ then we can apply our proposition to the co-ordinate system $p_1, x_2, \ldots, x_n, -x_1, p_2, \ldots, p_n$. Then there is a generating function $s(p_1, x_2, \ldots, x_n)$ such that $ds = -x_1 dp_1 + \sum_{i=2}^n p_i dx_i = -d(x_1 p_1) + \sum_{i=1}^n p_i dx_i$ so that (up to constants)

$$s(p_1, x_2, \ldots, x_n) = -x_1 p_1 + S(x_1, \ldots, x_n).$$

In this situation s is a partial Legendre transform of S.

1.11. Flow outs. In this section we show how to use the Hamiltonian flow to enlarge a constant energy isotropic manifold to a constant energy Lagrangian manifold.

PROPOSITION 3. *Let $H(x, p)$ be a Hamiltonian and suppose $\mathcal{L}_0$ is an $(n-1)$ dimensional submanifold of phase space such that*

(i) *$\mathcal{L}_0$ is isotropic, i.e., $\omega[X, Y] = 0$ for X, Y tangent to $\mathcal{L}_0$,*
(ii) *$\mathcal{L}_0$ is contained in a constant energy surface $\{(x, p) : H(x, p) = E\}$*
(iii) *The Hamiltonian vector field X_H is nowhere tangent to $\mathcal{L}_0$.*

Let $\mathcal{L}_0^+$ be the n-dimensional submanifold in phase space swept out by the Hamiltonian flow for H with initial conditions on $\mathcal{L}_0$. Then $\mathcal{L}_0^+$ is is Lagrangian and also contained in the constant energy surface $\{(x, p) : H(x, p) = E\}$.

PROOF. Since Hamiltonian flow preserves ω we need only check $\omega[X, Y] = 0$ for vectors X, Y tangent to $\mathcal{L}_0^+$ at the initial points in $\mathcal{L}_0$. Such tangent vectors have the form $X = X_1 + aX_H$, $Y = Y_1 + bX_H$ where X_1, Y_1 are tangent to $\mathcal{L}_0$. Since $\mathcal{L}_0$ is isotropic, $\omega[X_1, Y_1] = 0$ and $\omega[X_H, X_H] = 0$ by antisymmetry. Finally, $\omega[X_1, X_H] = \langle X_1, \nabla H\rangle = 0$ because X_1 points in a direction in the constant energy surface. Similarly $\omega[Y_1, X_H] = 0$. This implies $\omega[X, Y] = 0$.

Since the Hamiltonian flow preserves energy, and all the initial points are contained in $\{(x, p) : H(x, p) = E\}$, the flow out $\mathcal{L}_0^+$ is contained in $\{(x, p) : H(x, p) = E\}$ too. □

1.12. Extended phase space and the geometry of the Hamilton Jacobi equations. Consider a two dimensional phase space $\mathcal{P}_1 = \mathbb{R} \times \mathbb{R}$ with variables (τ, E) and symplectic form $\omega_1 = dE \wedge d\tau$ and Hamiltonian $H_1(\tau, E) = E$. Then we can form extended phase space $\mathcal{P}'$ as the product

$$\mathcal{P}' = \mathcal{P} \times \mathcal{P}_1 = \mathbb{R}^n \times \mathbb{R}^n \times \mathbb{R} \times \mathbb{R}$$

with variables (x, p, τ, E) and symplectic form

$$\omega' = \omega - \omega_1 = \sum_{i=1}^{n} dp_i \wedge dx_i - dE \wedge d\tau.$$

We define a new Hamiltonian on $\mathcal{P}'$ as

$$H'(x, p, \tau, E) = H(x, p) - H_1(\tau, E) = H(x, p) - E$$

Here is the geometric meaning of solving the Hamilton Jacobi equations (10) and (11):

> Finding solutions amounts to finding a Lagrangian submanifold of $\mathcal{L}' \subset \mathcal{P}'$ contained in the constant energy surface $H'(x, p, \tau, E) = 0$.

To see this, suppose we have found such an $\mathcal{L}'$ that is a graph over (x, τ). Then $\mathcal{L}' = \{(x, p, \tau, E : p = \nabla_x S, E = \partial S/\partial\tau\}$ for some generating function $S(x, \tau)$. Since $\mathcal{L}'$ lies in the surface $H' = 0$, the generating function $S(x, \tau)$ will satisfy $H'(x, \nabla_x S, \tau, \partial S/\partial\tau) = 0$. But this is the Hamilton-Jacobi equation (10) (with $t = \tau$). Similarly, if $\mathcal{L}'$ is a graph over (x, E) then the generating function $s(x, E)$ will satisfy (11). Once we know $\mathcal{L}'$, the generating functions $S(x, \tau)$ and $s(x, E)$ can be found as described above. They will be partial Legendre transforms.

How can we find a Lagrangian submanifold of extended phase space that does the job? One way of constructing $\mathcal{L}'$ is to start with a Lagrangian submanifold $\mathcal{L}$ of $\mathcal{P}$ and imbed it as an isotropic submanifold in $\mathcal{P}'$ contained in the constant energy surface $H' = 0$ via $(x, p) \mapsto (x, p, 0, H(x, p))$. We can then flow out these points under the Hamiltonian flow for H', to obtain a Lagrangian submanifold in $\mathcal{P}'$ in the constant enery surface $H' = 0$.

PROBLEM 6. Show that the Hamiltonian vector field $X_{H'}$ is never tangent to this imbedded isotropic submanifold.

To perform the flow out, we can use the following proposition, which we state for more general products.

PROPOSITION 4. *Consider the extended phase space*

$$\mathcal{P}' = \mathbb{R}^n \times \mathbb{R}^n \times \mathbb{R}^k \times \mathbb{R}^k$$

with variables $(x_1, \ldots, x_n, p_1, \ldots, p_n, X_1, \ldots, X_k, P_1, \ldots, P_k)$ *and symplectic form*

$$\omega' = \sum_{i=1}^{n} dp_i \wedge dx_i - \sum_{j=1}^{k} dP_j \wedge dX_j.$$

Suppose we have a Hamiltonian of the form $H'(x, p, X, P) = H(x, p) - H_1(X, P)$. *Let* $\mathcal{L}'$ *be a Lagrangian submanifold in* $\mathcal{P}'$ *that is contained in the constant energy surface* $H' = 0$. *Let* $(x(t), p(t), X(t), P(t))$ *be a path in* $\mathcal{L}'$. *Consider the following statements:*

(i) $(x(t), p(t), X(t), P(t))$ *is a Hamiltonian trajectory for* H'

(ii) $(x(t), p(t))$ *is a Hamiltonian trajectory for* H

(iii) $(X(t), P(t))$ *is a Hamiltonian trajectory for* H

Then (i) $\Leftrightarrow$ *(ii) and (iii). Let* $\pi : \mathcal{L}' \to \mathbb{R}^k \times \mathbb{R}^k$ *be the projection* $(x, p, X, P) \mapsto (X, P)$ *and* $d\pi$ *the induced map on tangent spaces. If* $d\pi$ *is onto (this requires* $k \le n$*) then (ii)* $\Rightarrow$ *(iii) and (i).*

PROOF. Condition (i) can be written

$$\begin{bmatrix}\dot{x}\\ \dot{p}\\ \dot{X}\\ \dot{P}\end{bmatrix} = \begin{bmatrix}J & 0\\ 0 & -J\end{bmatrix}\nabla H' = \begin{bmatrix}J & 0\\ 0 & -J\end{bmatrix}\begin{bmatrix}\dfrac{\partial H}{\partial x}\\ \dfrac{\partial H}{\partial p}\\ -\dfrac{\partial H_1}{\partial X}\\ -\dfrac{\partial H_1}{\partial P}\end{bmatrix}$$

which is equivalent to $((x(t), p(t))$ and $(X(t), P(t))$ satisfying Hamiltions equations for H and H_1 respectively. This shows (i) $\Leftrightarrow$ (ii) and (iii).

To show that (ii) $\Rightarrow$ (iii) and (i) assume that (ii) holds. Let $X' = \begin{bmatrix}a\\ b\\ A\\ B\end{bmatrix}$ be in the tangent space to $\mathcal{L}'$. The constant energy condition yields $\langle \nabla H', X'\rangle = 0$ which implies $\left\langle \nabla H, \begin{bmatrix}a\\ b\end{bmatrix}\right\rangle = \left\langle \nabla H_1, \begin{bmatrix}A\\ B\end{bmatrix}\right\rangle$. Since $\mathcal{L}'$ is Lagrangian and $\begin{bmatrix}\dot{x}\\ \dot{p}\\ \dot{X}\\ \dot{P}\end{bmatrix}$ is also a tangent vector we also know that $\left\langle \begin{bmatrix}\dot{x}\\ \dot{p}\\ \dot{X}\\ \dot{P}\end{bmatrix}, \begin{bmatrix}J & 0\\ 0 & -J\end{bmatrix}\begin{bmatrix}a\\ b\\ A\\ B\end{bmatrix}\right\rangle = 0$ so that $\left\langle \begin{bmatrix}\dot{x}\\ \dot{p}\end{bmatrix}, J\begin{bmatrix}a\\ b\end{bmatrix}\right\rangle = \left\langle \begin{bmatrix}\dot{X}\\ \dot{P}\end{bmatrix}, J\begin{bmatrix}A\\ B\end{bmatrix}\right\rangle$. Condition (ii) says that $(x(t), y(t))$ is Hamiltonian flow for H. Thus

$$\begin{aligned}\left\langle \begin{bmatrix}\dot{X}\\ \dot{P}\end{bmatrix}, J\begin{bmatrix}A\\ B\end{bmatrix}\right\rangle &= \left\langle \begin{bmatrix}\dot{x}\\ \dot{p}\end{bmatrix}, J\begin{bmatrix}a\\ b\end{bmatrix}\right\rangle\\ &= \left\langle J\nabla H, J\begin{bmatrix}a\\ b\end{bmatrix}\right\rangle\\ &= \left\langle \nabla H, \begin{bmatrix}a\\ b\end{bmatrix}\right\rangle\\ &= \left\langle \nabla H_1, \begin{bmatrix}A\\ B\end{bmatrix}\right\rangle\\ &= \left\langle J\nabla H_1, J\begin{bmatrix}A\\ B\end{bmatrix}\right\rangle\end{aligned}$$

The condition on $d\pi$ means that we hit every vector $\begin{bmatrix}A\\ B\end{bmatrix}$ in $\mathbb{R}^k \times \mathbb{R}^k$ as X' ranges over the tangent space of $\mathcal{L}'$. Thus $J\begin{bmatrix}A\\ B\end{bmatrix}$ also ranges over all vectors in $\mathbb{R}^k \times \mathbb{R}^k$. So we find that $\left\langle \begin{bmatrix}\dot{X}\\ \dot{P}\end{bmatrix}\right\rangle = J\nabla H_1$. Thus (iii) holds. We already saw that (ii) and (iii) $\Rightarrow$ (i). □

1.13. Remark on solving for the flow using the Hamilton Jacobi equation. As an aside, we can now see how the Hamilton Jacobi equation can be used to solve for the flow. Suppose we can solve a family of Hamilton Jacobi equations ,

$$H(x, \nabla S(x,P)) - H_1(P) = 0.$$

indexed by $P \in \mathbb{R}^k$ with $k = n$. Then $\mathcal{L}' = \{(x, \nabla_x S(x,P), \nabla_P S(x,P), P), x \in \mathbb{R}^n, P \in \mathbb{R}^k\}$ is a Lagrangian submanifold in the surface $H(x,p) - H_1(P) = 0$. In favourable situations one can solve for (x,p) as as function of (X,P) when $(x,p,X,P) \in \mathcal{L}'$. Call this function $(x(X,P), p(X,P))$. The proposition above tells us that if $(X(t), P(t))$ is a Hamiltonian trajectory for H_1 then the image $(x(X(t),P(t)), p(X(t),P(t))$ is a Hamiltonian trajectory for H. But the equations for $(X(t),P(t))$ are easy to solve. We have $\dot{X} = \frac{\partial H_1}{\partial P}$ and $\dot{P} = 0$. Thus $P = P_0$ is constant in time and $X(t) = \frac{\partial H_1}{\partial P}(P_0)t$.

1.14. An example. We now construct the solution to (10) corresponding to the set of free trajectories starting at a given point x_0 in configuration space. So the Hamiltonian is $H(x,p) = \frac{p^2}{2m}$.

Start with the Lagrangian submanifold of $\mathbb{R}^n \times \mathbb{R}^n$ given by

$$\mathcal{L} = \{(x_0, p) : p \in \mathbb{R}^n)\}$$

Then we imbed this as an isotropic zero energy (for $H' = H(p) - E$) surface in extended phase space $\mathbb{R}^n \times \mathbb{R}^n \times \mathbb{R} \times \mathbb{R}$ via

$$(x_0, p) \mapsto (x_0, p, 0, \frac{p^2}{2m}).$$

Flowing out using the flow for H' from the initial point $(x_0, p_0, 0, \frac{p_0^2}{2m})$ yields the trajectory $(x_0 + \frac{1}{m}p_0 t, p_0, t, \frac{p_0^2}{2m})$. Thus a typical point in $\mathcal{L}'$ can be written

$$(x, p, \tau, E) = (x_0 + \frac{1}{m}p_0\tau, p_0, \tau, \frac{1}{2m}p_0^2)$$

for some choice of p_0 and τ. Here we have written it as a graph over (p_0, τ). Solving $x = x_0 + \frac{1}{m}p_0\tau$ for $p_0 = \frac{m(x-x_0)}{\tau}$ We can also write points in $\mathcal{L}'$ as

$$(x, p, \tau, E) = (x, \frac{m(x-x_0)}{\tau}, \tau, \frac{m(x-x_0)^2}{2\tau^2})$$

A path γ from $(x_0, p_0, 0, \frac{1}{2m}p_0^2)$ to (x, p, τ, E) is provided by

$$(x(t), p(t), \tau(t), E(t)) = (x_0 + \frac{1}{m}p_0 t, p_0, t, \frac{1}{2m}p_0^2), \quad t \in [0, \tau]$$

Since $dS = pdx - Ed\tau$ we find

$$\begin{aligned} S(x,\tau) &= \int_\gamma pdx - Ed\tau \\ &= \int_0^\tau (\langle p_0, \dot{x}(t)\rangle - \frac{1}{2m}p_0^2)dt \\ &= \frac{1}{2m}p_0^2\tau \end{aligned}$$

Since $p_0 = \dfrac{x - x_0}{m\tau}$,

$$S(x,\tau) = m\frac{|x - x_0|^2}{2\tau}.$$

The singularity at $\tau = 0$ is connected with the fact that $\mathcal{L}'$ fails to be a graph over (x,τ) at $\tau = 0$.

There is a connection with the classical action. If we carry out this construction with a more general Hamiltonian we will find that if $x(t)$ is the path in configuration space corresponding to a Hamiltonian trajectory we have

$$S(x(\tau),\tau) = \int_{t=0}^{\tau} \langle p(t), \dot{x}(t)\rangle - H(x(t), p(t))dt$$

where $\dot{x}(t) = \dfrac{\partial H}{\partial p}(x(t), p(t))$. This implies that $\langle p(t), \dot{x}(t)\rangle - H(x(t), p(t))$ is the Legendre transform of $H(x,p)$ with respect to p, namely the Lagrangian $L(x(t), \dot{x}(t))$. Thus $S(x(\tau),\tau)$ is the action integral for this path.

2. Review of probability and operator theory

2.1. Probability. We now want to consider the situation in classical mechanics where we have incomplete knowledge of our system. This incomplete knowledge is described by a probability measure on phase space. Probability measures also are basic to quantum mechanics, where they appear even when we have a complete knowledge of the system. So in this section we will review some basic definitions.

A *probability space* is a triple

$$(\Omega, \mathcal{F}, \mathbb{P})$$

where Ω is a set (e.g., phase space), $\mathcal{F}$ is a σ–algebra of subsets of Ω and $\mathbb{P}$ is a probability measure. A *σ-algebra on Ω* is a collection $\mathcal{F}$ of subsets of Ω satisfying

(i) $\Omega \in \mathcal{F}$, $\emptyset \in \mathcal{F}$,
(ii) $A \in \mathcal{F} \Rightarrow \Omega\backslash A \in \mathcal{F}$ (closed under complements),
(iii) $\{A_i\}_{i\in\mathbb{N}} \subset \mathcal{F} \Rightarrow \bigcup_{i\in\mathbb{N}} A_i \in \mathcal{F}$ (closed under countable unions),
(iv) $\{A_i\}_{i\in\mathbb{N}} \subset \mathcal{F} \Rightarrow \bigcap_{i\in\mathbb{N}} A_i \in \mathcal{F}$ (closed under countable intersections).

In fact, for a non-empty collection of subsets, (ii) and (iii) $\Rightarrow$ (i) and (iv). Subsets in $\mathcal{F}$ are called measurable sets and the pair $(\Omega, \mathcal{F})$ is called a *measurable space*. Sets in $\mathcal{F}$ are also called *events*.

For any collection $\mathcal{F}_0$ of subsets of Ω there is a smallest σ–algebra on Ω containing $\mathcal{F}_0$, namely the intersection of all such σ–algebras. This is the σ–algebra generated by $\mathcal{F}_0$. If Ω has a topology then there is a smallest σ–algebra $\mathcal{B}$ containing the open sets. This is called the *Borel* σ–algebra and its elements are called Borel sets.

A *positive measure* on a measurable space $(\Omega, \mathcal{F})$ is a countably additive function $\mu : \mathcal{F} \to [0, \infty]$. This means

(i) $\mu(\emptyset) = 0$
(ii) $\mu\left(\bigcup_{i\in\mathbb{N}} A_i\right) = \sum_{i\in\mathbb{N}} \mu(A_i)$ for any countable collection sets in $\{A_i\}_{i\in\mathbb{N}} \subset \mathcal{F}$ that is pairwise disjoint, i.e., $A_i \cap A_j = \emptyset$ for $i \neq j$.

If $\mathcal{F} = \mathcal{B}$, the Borel sets, then μ is called a Borel measure.

Example: a point mass (or delta function)

$$\delta_{x_0}(A) = \begin{cases} 1 & \text{if } x_0 \in A \\ 0 & \text{if } x_0 \notin A \end{cases}$$

If $\Omega = \mathbb{R}^n$ this is also denoted $\delta(x - x_0)$. A *pure point* (or discrete) measure is one of the form

$$\mu = \sum_i \alpha_i \delta_{x_i}$$

for some countable collection of points $\{x_i\}$ and positive numbers α_i with $\sum_i \alpha_i < \infty$.

The standard measure on $\mathbb{R}^n$ defined on Borel sets that reproduces the volume of rectangular boxes and is invariant under translations and rotations is called *Lebesgue measure.*

A *probability measure* on a measurable space $(\Omega, \mathcal{F})$ is a positive measure $\mathbb{P}$ with $\mathbb{P}(\Omega) = 1$. The number $\mathbb{P}(A)$ indicates the likelihood that an outcome in A occurs. In this case $(\Omega, \mathcal{F}, \mathbb{P})$ is called a *probability space.*

A function $f : \Omega_1 \to \Omega_2$ where $(\Omega_1, \mathcal{F}_1)$ and $(\Omega_2, \mathcal{F}_2)$ are measurable spaces is called *measurable* if the pre-image of any set in $\mathcal{F}_2$ is contained in $\mathcal{F}_1$, that is

(i) $f^{-1}(F_2) \in \mathcal{F}_1$ for any $F_2 \in \mathcal{F}_2$.

Measurable functions on probability spaces are also called *random variables.*

If $f : \Omega_1 \to \Omega_2$ is a measurable function with respect to $(\Omega_1, \mathcal{F}_1)$ and $(\Omega_2, \mathcal{F}_2)$ and μ is a measure on $\mathcal{F}_1$ then we can define the image measure $f_*[\mu]$ as

$$f_*[\mu](F_2) = \mu(f^{-1}(F_2))$$

If $f : \Omega_1 \to \Omega_2$ is a measurable function with respect to $(\Omega_1, \mathcal{F}_1)$ and $(\Omega_2, \mathcal{F}_2)$ then the σ–algebra generated by f is $\{f^{-1}(A) : A \in \mathcal{F}_2\}$. It is the smallest σ-algebra for which f is measurable.

An important case is when $\Omega_2 = \mathbb{R}$ and $\mathbb{P}$ is a probability measure. Then the image measure $f_*[\mathbb{P}]$ is a measure on $\mathbb{R}$ and called the distribution measure of f. The number $f_*[\mathbb{P}](I)$ gives the probability that the value of f lies in I.

We now discuss *integration.* Let $(\Omega, \mathcal{F}, \mu)$ be a measure space. Define the indicator function

$$\chi_A(x) = \begin{cases} 1 & \text{if } x \in A \\ 0 & \text{if } x \notin A \end{cases}.$$

Then function of the form $s = \sum_{i=1}^n \alpha_i \chi_{A_i}$, for $\alpha_i \in \mathbb{R}$ and $A_i \in \mathcal{F}$ a finite collection of disjoint sets, is called a simple function. The integral of s is defined as

$$\int_\Omega s d\mu = \sum_{i=1}^n \alpha_i \mu(A_i)$$

The integral of a positive measurable function f is defined using monotone limits of simple functions. For a measurable function f that is not necessary positive we can write $f = f_+ - f_-$ and define the integral of f is the difference of the integrals of f_- and f_+.

If f is a random variable on a probability space the *expected value* of f, denoted $\mathcal{E}[f]$ is the average or mean

$$\mathcal{E}[f] = \int_\Omega f d\mathbb{P}.$$

We have

$$\mathcal{E}[\chi_A] = \mathbb{P}(A).$$

The *variance* of f is the expected value of $(f - \mathcal{E}[f])^2$

$$\mathrm{Var}(f) = \mathcal{E}[(f - \mathcal{E}[f])^2] = \mathcal{E}[f^2] - (\mathcal{E}[f])^2.$$

Two events A and B are *independent* if

$$\mathbb{P}[A \cap B] = \mathbb{P}[A]\,\mathbb{P}[B].$$

If A is an event with $\mathbb{P}[A] \neq 0$ then the *conditional probability* of an event B given A is

$$\mathbb{P}[B|A] = \frac{\mathbb{P}[A \cap B]}{P[A]}$$

In this case A and B are independent if and only if $\mathbb{P}[B|A] = \mathbb{P}[B]$. Conditional probabilities appear again in the last lecture where the idea of conditioning with respect to a σ–algebra is also discussed.

2.2. Two theorems in measure theory. We will refer to the following results:

THEOREM 1 (The Riesz-Markov theorem). *Let X be a locally compact Hausdorff space. Every positive linear functional λ on $C_c(X)$, is represented by a unique Borel measure μ on X. This means*

$$\lambda(f) = \int_X f(x)d\mu(x)$$

for allf $\in C_c(X)$.

A Borel measure μ on $\mathbb{R}^n$ is *absolutely continuous* with respect to Lebesgue measure m if

$$m(A) = 0 \Rightarrow \mu(A) = 0.$$

In this case we write $\mu << m$. If $\mu << m$ then there exists an integrable function $f(x)$ such that for every Borel set A

$$\mu(A) = \int_A f(x)dm(x).$$

Two measures μ_1 and μ_2 on $(\Omega, \mathcal{F})$ are *singular*, denoted $\mu_1 \perp \mu_2$ if there are disjoint sets $A, B \in \mathcal{F}$ with $A \cup B = \Omega$ such that $\mu_1(A) = \mu_2(B) = 0$. A measure μ is *continuous* if $\mu(\{x\}) = 0$ for every $x \in \Omega$.

THEOREM 2 (Lebesgue decomposition theorem). . *Let μ be a regular Borel measure on $\mathbb{R}^n$ and let m denote Lebesgue measure. Then*

$$\mu = \mu_{ac} + \mu_{sc} + \mu_{pp}$$

where $\mu_{ac} << m$, μ_{sc} is continuous with $\mu_{sc} \perp m$ and μ_{pp} is pure point.

2.3. Mixed states in classical mechanics. We now define a *mixed state* in classical mechanics to be a probability measure μ on phase space. This represents a situation when we have incomplete knowledge of the system. The pairing of an observable f and a state μ now does not yield a real number. Instead the value is a measure on $\mathbb{R}$, namely the distribution measure of f:

$$\langle f|\mu\rangle = f_*[\mu]$$

A pure state (x, p) now identified with the mixed state $\delta_{(x,p)}$.

The time evolution of a mixed state is defined in the obvious way. If Φ_t is the Hamiltonian flow on phase space, then the flow on mixed states is Φ_{t*}. In other words

$$\mu_t(A) = \mu(\Phi_t^{-1}(A))$$

2.4. Background and self-adjoint operators. I'll assume you know the definitions of *Hilbert space* (with complex scalars), *bounded operators*, *adjoints*, *projections*, *spectrum*, *linear functionals* and *unitary operators*.

A bounded operator A is self-adjoint if $A^* = A$. I won't give the definition of unbounded self-adjoint operators, and will try to get by with examples. An unbounded self-adjoint operator A acting in a Hilbert space $\mathcal{H}$ is only defined on a dense set $\mathcal{D}(A) \subset \mathcal{H}$ called the domain. The spectrum $\sigma(A)$ of a self-adjoint operator A is a closed subset of $\mathbb{R}$. The set $\sigma(A)$ is bounded iff the operator A is bounded. All eigenvalues of A lie in $\sigma(A)$, but there may be more points. Here are some examples.

EXAMPLES. (i) $\mathcal{H} = \mathbb{C}^n$, A is Hermitian $n \times n$, $\sigma(A) = \{\text{eigenvalues}\}$.

(ii) $\mathcal{H} = L^2(\mathbb{R}^n, d^n x)$, $A\psi = V(x)\psi(x)$ is multiplication by a bounded measureable function $V(x)$, $\sigma(A) = \text{ess range}(V)$ (or $\text{range}(V)$ if V is continuous).

(iii) $\mathcal{H} = L^2(\mathbb{R}^n, d^n x)$, $A\psi = x_i\psi(x)$ is multiplication by x_i, $\mathcal{D}(A) = \{\psi \in \mathcal{H} : x_i\psi \in \mathcal{H}\}$, $\sigma(A) = \mathbb{R}$.

(iv) $\mathcal{H} = L^2(\mathbb{R}^n, d^n k)$, $A\phi = -i\partial\phi/\partial k_i$, $\mathcal{D}(A) = \{\phi : x_i(\mathcal{F}^{-1}\phi)(x) \in L^2(\mathbb{R}^n, d^n x\}$, $\sigma(A) = \mathbb{R}$.

In example (iv), $\mathcal{F}$ is the Fourier transform (see below). If ϕ is not differentialble $\partial\phi/\partial k_i$ can be taken to mean $\mathcal{F}x_i\mathcal{F}^{-1}$, where x_i is the multiplication operator.

2.5. Spectral theorem for self-adjoint operators. Let A be self-adjoint and (for the moment) bounded. If $p(x)$ is a polynomial then we may define $p(A)$ in the obvious way by substituting A for x. This is possible for any bounded A, but if A is self-adjoint then one can show that

$$\|p(A)\| \leq \sup_{x \in \sigma(A)} |p(x)|.$$

This alllows us to define $f(A)$ for any $f \in C(\sigma(A))$ by taking limits. Now pick $\psi \in \mathcal{H}$. Then the map

$$f \mapsto \langle \psi, f(A)\psi \rangle$$

defines a bounded positive linear functional on $C(\sigma(A))$. Thus, by the Riesz-Markov theorem, there exists a regular positive Borel measure μ_ψ^A such that

$$\langle \psi, f(A)\psi \rangle = \int_{\sigma(A)} f(\lambda)\, d\mu_\psi^A(\lambda).$$

By definition, μ_ψ^A is the spectral measure for ψ and A. Now notice that the right side of this equation still makes sense if f is bounded Borel function. This allows us to use the equation (in reverse) to define and operator $f(A)$ for any bounded Borel functions. The map $f \mapsto f(A)$ is a $*$–algebra homomorphism. This means (among other things) that $\bar{f}(A) = f(A)^*$ and $(fg)(A) = f(A)g(A)$. Now let $f(x) = \chi_I(x)$ be the indicator function for a Borel set I. Then the operator $\chi_I(A)$ is a projection, called the spectral projection for I and A.

The proof of these facts for unbounded self-adjoint A is different, but the results are the same.

EXAMPLES.
(i) $\mathcal{H} = \mathbb{C}^n$, A is Hermitian $n \times n$: Let $\{\lambda_k\}$ be the eigenvalues and $\{e_k\}$ an orthonormal basis of eigenvectors. Then $f(A) = \sum_k f(\lambda_k)\, e_k \otimes e_k^*$ so

$$\begin{aligned} \langle \psi, f(A)\psi \rangle &= \sum_k f(\lambda_k) \mid \langle e_k, \psi \rangle \mid^2 \\ \mu_\psi^A &= \sum_k \delta(\lambda - \lambda_k) \mid \langle e_k, \psi \rangle \mid^2 \\ \chi_I(A) &= \sum_{\{k : \lambda_k \in I\}} e_k \otimes e_k^* \end{aligned}$$

(ii) $\mathcal{H} = L^2(\mathbb{R}, dx)$, A is multiplication by x then

$$\begin{aligned} \langle \psi, f(A)\psi \rangle &= \int_{\mathbb{R}} f(x)\, |\psi(x)|^2 dx \\ \mu_\psi^A &= |\psi(x)|^2 dx \\ \chi_I(A) &= \text{multiplication by } \chi_I(x) \end{aligned}$$

2.6. The Fourier transform. Let $\mathcal{S}$ be the Schwartz space of rapidly decreasing functions. The Fourier transform $\mathcal{F}\psi$ of $\psi \in \mathcal{S}$ is the function

$$(\mathcal{F}\psi)(k) = (2\pi)^{-n/2} \int_{\mathbb{R}^n} e^{-i\langle k,x\rangle} \psi(x) d^n x.$$

The Fourier transform maps $\mathcal{S} \to \mathcal{S}$ and has an inverse given by

$$(\mathcal{F}^{-1}\phi)(x) = (2\pi)^{-n/2} \int_{\mathbb{R}^n} e^{i\langle x,k\rangle} \phi(k) d^n k.$$

We will use the notation $\widehat{\psi}$ for $\mathcal{F}\psi$ amd $\widetilde{\phi}$ for $\mathcal{F}^{-1}\phi$. Here are some properties of $\mathcal{F}$.

(i) $i\dfrac{d}{dk}\widehat{\psi}(k) = \widehat{x\psi(x)}$,
(ii) $k\widehat{\psi}(k) = \widehat{-i\dfrac{d}{dx}\psi(x)}$,
(iii) $(\widetilde{\widehat{\psi}})(x) = \psi(x)$, $(\widehat{\widetilde{\psi}})(k) = \psi(k)$,
(iv) $\int_{\mathbb{R}^n} |\psi(x)|^2 d^n x = \int_{\mathbb{R}^n} |\hat{\psi}(k)|^2 d^n k$
(v) Let $\varphi_0(x)$ be the Gaussian $\varphi_0(x) = \pi^{-1/4} e^{-x^2/2}$. Then $\mathcal{F}\varphi_0 = \varphi_0$.

Properties (i) and (ii) say that the Fourier transform exchanges multiplication and differentiation by the co-ordinate functions. Property (iii) says that $\mathcal{F}$ and $\mathcal{F}^{-1}$ really are inverses. Property (iv) lets us extend the definition of $\mathcal{F}$ from $\mathcal{S}$ to $L^2(\mathbb{R}^n, d^n x)$ by taking limits. The extension is a unitary operator $\mathcal{F} : L^2(\mathbb{R}^n, d^n x) \to L^2(\mathbb{R}^n, d^n k)$.

The unitary operator $\mathcal{F}$ allows us to define differentiation operators. Let p_i denote the operator acting in $\mathcal{H} = L^2(\mathbb{R}^n, d^n x)$ as

$$(p_i\psi)(x) = \mathcal{F}^{-1}k_i\mathcal{F}\psi$$

with domain $\mathcal{D}(p_i) = \{\psi \in L^2(\mathbb{R}^n, d^n x) : k_i\widehat{\psi} \in L^2(\mathbb{R}^n, d^n k)$. Then p_i is unitarily equivalent to a multiplication operator by a real valued function, and hence self-adjoint. If $\psi \in \mathcal{S}$, then $p_i\psi = -i\partial\psi/\partial x_i$.

We now define three families of unitary operators T_x, M_k and D_a as

$$\begin{aligned}(T_{x_0}\psi)(x) &= \psi(x - x_0) \quad \text{(translation by } x_0 \in \mathbb{R}^n\text{)}\\ (M_{k_0}\psi)(x) &= e^{i\langle k_0,x\rangle}\psi(x) \quad \text{(multiplication by } e^{i\langle k_0,x\rangle},\ p_0 \in \mathbb{R}^n\text{)}\\ (D_a\psi)(x) &= a^{-n/2}\psi(x/a) \quad \text{(dilation by } a > 0\text{)}\end{aligned}$$

and note that

(i) $T_{x_0}M_{k_0} = e^{-i\langle k_0,x_0\rangle}M_{k_0}T_{x_0}$
(ii) $\mathcal{F}T_{x_0} = M_{-x_0}\mathcal{F}$
(iii) $\mathcal{F}M_{k_0} = T_{k_0}\mathcal{F}$
(iv) $\mathcal{F}D_a = D_{a^{-1}}\mathcal{F}$.
We also have (at least when acting on suitable ψ),
(v) $T_{x_0}xT_{-x_0} = x - x_0, \quad T_{x_0}pT_{-x_0} = p$
(vi) $M_{k_0}pM_{-k_0} = p - k_0, \quad M_{k_0}xM_{-k_0} = x$
(vii) $D_axD_{a^{-1}} = a^{-n/2}x, \quad D_apD_{a^{-1}} = a^{n/2}p$

2.7. Uncertainty principle. For simplicity let $n=1$ and $\psi \in \mathcal{S}$ with $\|\psi\|=1$. Define

$$\begin{aligned}\mathrm{Var}_\psi(x) &= \langle\psi, x^2\psi\rangle - \left(\langle\psi, x\psi\rangle\right)^2\\ \mathrm{Var}_\psi(p) &= \langle\psi, p^2\psi\rangle - \left(\langle\psi, p\psi\rangle\right)^2 = \left\langle\widehat{\psi}, k^2\widehat{\psi}\right\rangle - \left(\left\langle\widehat{\psi}, k\widehat{\psi}\right\rangle\right)^2\end{aligned}$$

Then

$$\mathrm{Var}_\psi(x)\mathrm{Var}_\psi(p) \geq 1/4$$

and equality holds iff $\psi = M_{p_0}T_{x_0}D_a\varphi_0$ (up to a possible phase factor $e^{i\eta}$) where φ_0 is the normalized Gaussian.

Proof. Using (iv), (v) and (vi) above we can find x_0, k_0 and a such that $\phi = D_{a^{-1}}T_{-x_0}M_{-k_0}\psi$ verifies $\mathrm{Var}_\psi(x) = \langle\phi, x^2\phi\rangle = 1/2$ and $\mathrm{Var}_\psi(p) = \langle\phi, p^2\phi\rangle$. By the unitarity of the operators, $\|\phi\| = 1$.

The key ingredient is the commutator formula $i[p,x] = 1$. This implies

$$1 = \langle\phi,\phi\rangle = \langle\phi, i[p,x]\phi\rangle = -\Re\langle\phi', x\phi\rangle \leq \|\phi'\|\|x\phi\| = \sqrt{\mathrm{Var}_\psi(x)\mathrm{Var}_\psi(p)}$$

To get equality in the Cauchy-Schwarz inequality we must have $\langle\phi', x\phi\rangle$ real and the two factors proportional. This implies $\phi' = -\lambda x\phi$ for $\lambda > 0$. Integrating, we get $\phi(x) = Ce^{-\lambda x^2/2}$ and then $\|\phi\| = 1$ and $\|x\phi\|^2 = 1/2$ imply $\lambda = 1$ and $C = \pi^{-1/4}$. So $\phi = \varphi_0$ and $\psi = M_{p_0}T_{x_0}D_a\varphi_0$. □

2.8. One parameter strongly continuous unitary groups and Stone's theorem. A map $\mathbb{R} \to \{\text{unitary operators on } \mathcal{H}\}$ is called a one parameter strongly continuous unitary group if $U(0) = I$, $U(t_1 + t_2) = U(t_1)U(t_2)$ and $t \mapsto U(t)\psi$ is continuous in $\mathcal{H}$ for any fixed $\psi \in \mathcal{H}$.

THEOREM 3 (Stone's theorem). *Let $U(t)$ be a one parameter strongly continuous unitary group. Then there is a unique self-adjoint operator A such that $U(t) = e^{itA}$. Conversely, every self-adjoint operator A generates a one parameter strongly continuous unitary group $U(t) = e^{itA}$. The limit $\lim_{t\to 0} \dfrac{(U(t) - U(0))\psi}{t}$ exists and equals $iA\psi$ iff $\psi \in \mathcal{D}(A)$. So we see that in general $U(t)\psi$ is continuous, but if $\psi \in \mathcal{D}(A)$ then $U(t)\psi$ is differentiable with $\dfrac{d}{dt}U(t)\psi = iAU(t)\psi$.*

PROBLEM 7. The family of operators T_{ta} for $a \in \mathbb{R}^n$ and M_{tb} for $b \in \mathbb{R}^n$ are both one parameter strongly continuous unitary groups. Their generators are $\langle a, p\rangle$ and $\langle b, x\rangle$. What can you say about $D(a)$ for $a > 0$?

3. Quantum mechanics

3.1. Abstract quantum description of a physical system. Start with a Hilbert space $\mathcal{H}$. *Observables* are self-adjoint operators acting in $\mathcal{H}$. For an observable A, the spectrum $\sigma(A)$ represents possible outcomes of a measurement. *Pure states* are one dimensional subspaces of $\mathcal{H}$. We can identify these with normalized vectors $\psi \in \mathcal{H}$ with $\|\psi\| = 1$, but then ψ and $e^{i\eta}\psi$ (for real η) represent the same state. Given an observable A and a state ψ, the pairing is given by the spectral measure.

$$\langle A|\psi\rangle = \mu_\psi^A$$

This is interpreted as the distribution measure for the value of A. So the expected value of A in the state ψ is

$$\int \lambda d\mu_\psi^A(\lambda) = \langle \psi, A\psi\rangle$$

If we repeatedly measure A when the system is in the state ψ this will be the average value. Similarly the probability of finding the value of A in a Borel set I is

$$\mathbf{P}_\psi[A \in I] = \langle \psi, \chi_I(A)\psi\rangle\,.$$

We are using the symbol $\mathbf{P}_\psi$ rather than $\mathbb{P}$ is to emphasize that there is no underlying probability measure on the set of states. There is only a single state ψ involved. This number represents the proportion of times a repeated experiment to measure A lands in I when the system is in the state ψ.

It is important to realize that a pure state $\psi \in \mathcal{H}$, represents complete knowledge of the system, despite the fact that we cannot predict with certainty the outcome of all measurements of observables. In situations where we have incomplete knowledge, one can define quantum mixed states, as we did in classical mechanics. Briefly, a mixed state is given by a trace class positive definite operator M with trace equal to 1, called a density matrix. Such an operator has a spectral representation $M = \sum_{i=1}^\infty \mu_i\psi_i \otimes \psi_i^*$, where $\psi_i \otimes \psi_i^*$ denotes the projection given by $\psi_i \otimes \psi_i^*\phi = \langle \psi_i, \phi\rangle\psi_i$, and the μ_i are positive numbers summing to 1. The number μ_i is interpreted as the probabilty that the system is in state ψ_i. In this setup, a pure state is represented by a rank one density matrix.

We will not consider mixed quantum states in this course, except to point out that the superposition of two states ψ_1 and ψ_2, given by $\dfrac{\psi_1+\psi_2}{\|\psi_1+\psi_2\|}$ is a pure state, represented by a single unit vector and not a mixed state.

Every observable A generates a flow on states given by e^{-itA}. There is a distinguished observable, the Hamiltonian H that generates the time evolution e^{-itH}. The trajectory $\psi_t = e^{-itH}\psi$ is defined for all states, but if $\psi \in \mathcal{D}(H)$ then ψ_t is differentiable and satisfies the Schrödinger equation

$$i\frac{\partial}{\partial t}\psi_t = H\psi_t.$$

We can transfer the time evolution to the observables in such a way that

$$\langle \psi_t, A\psi_t\rangle = \langle \psi, A_t\psi\rangle$$

by choosing

$$A_t = e^{itH}Ae^{-itA}$$

Notice that

$$\dot{A}_t = i[H, A_t]$$

where $[A, B] = AB - BA$.

Now we consider the case of commuting observables. Two bounded operators are said to commute if $[A, B] = 0$. Two unbounded self-adjoint operators A and B commute if $f(A)$ and $g(B)$ commute for bounded Borel functions f and g. For commuting self-adjoint operators A and B, $\chi_I(A)\chi_J(B)$ is a projection and

$$\mathbf{P}[A \in I \text{ and } B \in J] = \langle \psi, \chi_I(A)\chi_J(B)\psi\rangle$$

gives the joint distribution for A and B if we measure them simultaneously. If A and B don't commute, then $\chi_I(A)\chi_J(B)$ need not be a projection. We interpret this as saying that we cannot measure A and B simultaneously.

3.2. Quantum particle in an external potential. To describe the motion of a particle we need to choose a Hilbert space, and operators to represent the important observables. For a particle moving in configuration space $\mathbb{R}^n$ we may choose

$$\mathcal{H} = L^2(\mathbb{R}^n, d^n x).$$

Then it is natural to take the position operator x_i to be multiplication by x_i. Classical observables that are functions of x are then represented by multiplication by the same function. The probability of finding the particle in the set I when the system is in the state ψ is

$$\langle \psi, \chi_I(x)\psi\rangle = \int_I |\psi(x)|^2 d^n x.$$

In other words, the distribution measure for position is $|\psi(x)|^2 d^n x$. To determine what operator should represent momentum p_i we recall that in classical mechanics the flow generated by p is a translation in x. Therfore it is natural to choose p_i to be the generator of the unitary group of translations. Thus we set

$$p_i = -i\frac{\partial}{\partial x_i} = \mathcal{F}^{-1}k_i\mathcal{F}.$$

The distribution measure for the momentum will then be $|\widehat{\psi}(k)|^2 d^n k$.

We can already see peculiar feature about quantum mechanics: the distributions for position and momentum are not independent. The uncertainty principle for the Fourier transform implies that if the distribution for the position is sharply peaked in some state ψ, so that $\mathrm{Var}_\psi(x)$ is small, then $\mathrm{Var}_\psi(p)$ will have to be large.

Finally we define the Hamiltonian operator as

$$H = \frac{p^2}{2m} + V = -\frac{1}{2m}\Delta + V$$

where $p^2 = -\Delta$ is (minus) the Laplacian operator and V is multiplication by $V(x)$.

The first task is to find a domain that makes H self-adjoint. This can be complicated if V is unbounded. But if V is bounded H is self-adjoint on the domain of Δ given by

$$\mathcal{D}(H) = \{\psi \in L^2(\mathbb{R}^n, d^n x) : k^2\widehat{\psi}(k) \in L^2(\mathbb{R}^n, d^n k)\}.$$

Once we have defined H as a self-adjoint operator, we can ask about the properties of the time evolution e^{-itH}. These are closely connected to the spectral properties of H.

For example, if H has an eigenvalue E so that $H\psi = E\psi$, then the eigenfunction ψ evolves in time as

$$e^{itH}\psi = e^{itE}\psi.$$

This expression represents the same state for all t, since e^{itE} is a number with modulus 1. So eigenvalues correspond to bound states.

What can we say about the time evolution of ψ if μ_ψ^H is absolutely continuous (with respect to Lebesgue measure)?

To answer this question, consider the observable given by the projection $P\psi = \langle\phi, \psi\rangle\,\phi$ onto the state ϕ. This observable has spectrum $\{0, 1\}$ corresponding to whether or not a state in our system is in the state ϕ. So the expected value of P, which is the probability that ψ is in the state ϕ is

$$\langle\psi, P\psi\rangle = |\,\langle\psi, \phi\rangle\,|^2.$$

If ψ is some initial state evolving as $\psi_t = e^{itH}\psi$, the probability that ψ_t remains in the initial state after time t is $|\,\langle\psi, \psi_t\rangle\,|^2 = |\,\langle\psi, e^{itH}\psi\rangle\,|^2$. For example, if ψ is an eigenfunction then this quantity is 1 for all time. On the other hand, if the spectral measure μ_ψ^H is absolutely continuous, then by the spectral theorem

$$\langle\psi, e^{itH}\psi\rangle = \int_{\sigma(H)} e^{it\lambda}\, d\mu_\psi^H(\lambda).$$

The Riemann Lebesgue lemma, a result in measure theory, says that this tends to zero as $t \to \infty$.

But is such a state really a scattering state? Does it leave bounded regions of space in the sense that for a bounded region I in configuration space $\langle\psi_t, \chi_I(x)\psi_t\rangle \to 0$? And if so, supposing that the expected value of x^2 given by $\langle\psi, x^2\psi\rangle$ is initially finite, how does it change in time?

There are many interesting questions like this and answering them for systems of interest can be challenging. There isn't time in this course to make a systematic study. Instead we will pick out a few examples for the simple systems we considered in the classical case. To simplify the notation, we will stick to one dimensional systems.

3.3. Example 1: Free motion. The Hilbert space is $\mathcal{H} = L^2(\mathbb{R}, dx)$ and

$$H = \frac{p^2}{2m} = -\frac{1}{2m}\frac{d^2}{dx^2}.$$

This operator is easy to study since it is diagonalized by the Fourier transform. This means that $\mathcal{F}H\mathcal{F}^{-1}$ is multiplication by $\frac{k^2}{2m}$, which implies that the spectrum is $[0, \infty)$ and is purely absolutely continuous. Moreover the time evolution ψ_t is given by

$$\psi_t = e^{-itH}\psi = \mathcal{F}^{-1}e^{-itk^2/(2m)}\mathcal{F}\psi.$$

Notice that the momemtum distribution $|\widehat{\psi}_t(k)|^2 dk$ doesn't change in time.

To begin lets take our initial state to be a Gaussian $\varphi_0 = \pi^{-1/4}e^{-x^2/2}$. Then $\widehat{\varphi}_0(k) = \varphi_0(k)$ and the constant momentum distribution is $\pi^{-1/2}e^{-k^2}dk$. The expected values for both x and p in this state are zero. So this corresponds to a classical particle sitting at the origin.

How does this state evolve? We can calculate

$$\begin{aligned}\mathcal{F}(e^{-itH}\varphi_0)(k) &= \pi^{-1/4}e^{-itk^2/(2m)}e^{-k^2/2}\\ &= \pi^{-1/4}e^{-(1+it/m)k^2/2}\\ &= \pi^{-1/4}e^{-\alpha k^2/2}\end{aligned}$$

where $\alpha = 1 + it/m$ is complex. Thus the time evolution is

$$e^{-itH}\varphi_0 = \pi^{-1/4}\mathcal{F}^{-1}e^{-\alpha k^2/2}$$

PROBLEM 8. Even though the complex dilation $D(\alpha^{-1/2})\psi$ has no meaning for a general $\psi \in L^2(\mathbb{R}, dk)$, show that $D(\alpha^{-1/2})\varphi_0$ makes sense and verifies $\mathcal{F}^{-1}D(\alpha^{-1/2})\varphi_0 = D(\alpha^{1/2})\mathcal{F}^{-1}\varphi_0$.

Using this result we continue the calculation.

$$\begin{aligned}(e^{-itH}\varphi_0)(x) &= \alpha^{-1/4}\mathcal{F}^{-1}D(\alpha^{-1/2})\varphi_0(k)\\ &= \alpha^{-1/4}D(\alpha^{1/2})\mathcal{F}^{-1}\varphi_0(k)\\ &= \alpha^{-1/2}\varphi_0(x/\sqrt{\alpha})\end{aligned}$$

This yields the distribution

$$\frac{1}{\sqrt{\pi}\sqrt{1+(t/m)^2}}e^{\frac{-x^2}{(1+(t/m)^2)}}dx$$

for the position x in the state $e^{-itH}\varphi_0$. What is happening here? The state is not stationary as in the classical case, but it is also not moving in any direction. It is just slowly spreading out as time goes on.

To see some motion, lets start with an initial state where the momentum distribution has been shifted up. Let $\psi = M[k_0]\varphi_0$ so that $\widehat{\psi} = \mathcal{F}M[k_0]\varphi_0 = T[k_0]\mathcal{F}\varphi_0$. Then the initial momentum distribution $|\widehat{\psi}(k)|^2 dk = |\varphi_0(k - k_0)|^2 dk$ is centered at k_0. Now

$$e^{-itH}\psi = \pi^{-1/4}\mathcal{F}^{-1}e^{-itk^2/(2m)}e^{-(k-k_0)^2/2}$$

PROBLEM 9. Using the expansion

$$\frac{tk^2}{2m} = \frac{t(k_0 + k - k_0)^2}{2m} = \frac{t(k_0)^2}{2m} + \frac{tk_0(k - k_0)}{m} + \frac{t(k - k_0)^2}{2m}$$

and the commutation properties of $T[k_0]$, $M[tk_0/m]$ with $\mathcal{F}^{-1}$ show that

$$|\psi_t(x)|^2 = \frac{1}{\sqrt{\pi}\sqrt{1+(t/m)^2}} e^{\frac{-(x-tk_0/m)^2}{(1+(t/m)^2)}}$$

Now we see the particle moving to the right with velocity k_0/m while still spreading at the same rate.

Even when the initial state is not a Gaussian we can still use the Fourier transform to compute the time evolution. To begin we assume that $\psi \in \mathcal{S}$. Then one can compute

$$(e^{-itH}\psi)(x) = (2\pi it/m)^{-1/2} \int_{\mathbb{R}} e^{im\frac{(x-y)^2}{2t}} \psi(y)dy$$

Then, as with the Fourier transform, we can extend the definition to $L^2(\mathbb{R}, dx)$ since we know the operator is unitary.

Now think of expanding $(x-y)^2 = x^2 - 2xy + y^2$ in the exponent and throwing out the y^2 term. The resulting operator is

$$\tilde{U}(t)\psi = \frac{1}{(t/m)^{1/2}} e^{i\frac{mx^2}{2t}} \widehat{\psi}(xm/t)$$

If the momentum distribution $\widehat{\psi}$ is sharply peaked at k_0 then $\widehat{\psi}(xm/t)$ is concentrated near where $xm/t = k_0$, or $x = k_0t/m$ which is the classical trajectory starting at $x_0 = 0$. Moreover the exponent looks like the classical action we computed before.

PROBLEM 10. Show that $\|(\tilde{U}(t) - e^{-itH})\psi\| \to 0$ as $t \to \infty$.

3.4. Using the Hamilton Jacobi equation. In this section we will construct approximate scattering solutions to the Schrödinger equation using the Hamilton Jacobi equation. For the free motion this may seem pointless since we can already compute the exact solution using the Fourier transform! But there are two reasons to proceed. So far we have just recognized the classical behaviour in solutions constructed another way. But now we will use the classical flow to actually build the solution. Secondly, this procedure works in situations where the Fourier transform is not available. In this section we will go back to general dimension n.

To get an approximate solution to the time dependent Schrödinger equation $\left(H - i\frac{\partial}{\partial t}\right)\psi = 0$ of the form $\psi(x,t) = e^{iS(x,t)}u(x,t)$, we substitute this expression into the left side of the equation. This is the key calculation, which we do for $V \neq 0$.

$$\left(\frac{-1}{2m}\Delta + V(x) - i\frac{\partial}{\partial t}\right) e^{iS}u = e^{iS}\left(\frac{|\nabla S|^2}{2m} + V(x) + \frac{\partial S}{\partial t}\right)$$
$$- ie^{iS}\left(\frac{1}{m}\nabla S \cdot \nabla u + \frac{\partial u}{\partial t} + \frac{1}{2m}(\Delta S)u\right) - e^{iS}\frac{1}{2m}\Delta u$$

Then the first term on the right side vanishes if $S(x,t)$ solves the Hamilton Jacobi equation

$$\frac{|\nabla S|^2}{2m} + V(x) + \frac{\partial S}{\partial t} = 0.$$

From our previous work, we know that given a Lagrangian submanifold $\mathcal{L}$ in extended phase space that (i) is contained in the constant energy surface $\{(x,p,\tau,E): H(x,p)-E=0\}$ and (ii) is a graph over (x,t), then the generating function (action) $S(x,t)$ is a solution to the Hamilton Jacobi equation. Moreover, we can find S by integrating the 1–form $\sum_i p_i dx_i - E d\tau$ over paths in $\mathcal{L}$. We constructed such a Lagrangian using a flow out. In this case $(x,p,\tau,E)\in\mathcal{L}$ if there is an point (x_0,p_0) in the initial constant energy Lagrangian manifold $\mathcal{L}_0$ in the original phase space (this implies $E=H(x_0,p_0)$), and a τ such that the Hamiltonian trajectory $(x(t),p(t))$ with initial condition (x_0,p_0) arrives at (x,p) at time τ.

Now suppose we have solved the Hamilton Jacobi equation so that we know $S(x,t)$. Then our next task is to solve the equation

$$\frac{1}{m}\nabla S\cdot\nabla u+\frac{\partial u}{\partial t}+\frac{1}{2m}(\Delta S)u=0$$

for $u(x,t)$. This is a first order equation, called a transport equation, that we can solve using flows. The standard trick for such equations is to assume we have a solution and determine its values along the flow associated to the vector field in the equation. Of course, here these flows will be Hamiltonian trajectories. So assume that we have a solution $u(x,t)$. Fix a starting point (x_0,p_0) the original Lagrangian of initial conditions. For notational simplicity lets suppose this Lagrangian is a graph over p so a unique starting point is determined by p_0. Now let us determine the values of u along this trajectory. Setting $v(t)=u(x(t),t)$ we find that

$$\dot v(t)=\dot x(t)\cdot\nabla u(x(t),t)+\frac{\partial u}{\partial t}(x(t),t)$$

But

$$\dot x(t)=p(t)/m=\nabla S(x(t),t)/m.$$

The first equality is Hamilton's equation and the second follows from the fact that S is a generating function. So

$$\begin{aligned}\dot v(t)&=\nabla S\cdot\nabla u+\frac{\partial u}{\partial t}\\&=\nabla S\cdot\nabla u+\frac{\partial u}{\partial t}+\frac{1}{2m}(\Delta S)u-\frac{1}{2m}(\Delta S)u\\&=-\frac{1}{2m}(\Delta S)(x(t),t)v(t).\end{aligned}$$

Since we know $S(x,t)$, the function $\frac{1}{2m}(\Delta S)(x(t),t)$ is a known function of t, once we specify p_0, which tells us the orbit we are on. Thus we know $\alpha(p_0,t)=\frac{1}{2m}(\Delta S)(x(t),t)$ and

$$\dot v(t)=-\alpha(p_0,t)v(t)$$

Integrating, and keeping in mind that the constant of integration may depend on the orbit we are on, we find

$$v(\tau)=f(p_0)e^{-\int_0^\tau\alpha(p_0,t)dt}.$$

Now we can verify that any smooth function whose values along the trajectory $(x(t),t)$ labelled by p_0 are given by

$$u(x(t),t)=f(p_0)e^{-\int_0^\tau\alpha(p_0,t)dt}$$

where $\alpha(p_0,t)$ is given above, solves the equation.

If we construct $S(x,t)$ and $u(x,t)$ according to this scheme we find

$$\left(\frac{-1}{2m}\Delta + V(x) - i\frac{\partial}{\partial t}\right) e^{iS} u = -e^{iS}\frac{1}{2m}\Delta u$$

The term remaining on the left is an error term which may become negligible for large t.

To illustrate this procedure, let us sketch the constructing of scattering solutions for the free motion. In this case we found that

$$S(x,t) = \frac{m|x-x_0|^2}{2t}$$

Thus

$$\frac{1}{2m}\Delta S(x,t) = \frac{n}{2t}$$

which implies that in this case $\alpha(p_0, t) = n/(2t)$ is independent of the trajectory and so

$$u(x,t) = f(p_0)e^{-n\int_1^t \log(\tau)d\tau/2} = f(p_0)t^{-n/2}.$$

To complete the construction we must determine p_0, the initial momentum, if we know (x,t). The Hamiltonian trajectory is $x = x_0 + p_0 t/m$, so $p_0 = m(x - x_0)/t$. Thus our approximate solutions have the form

$$\psi(x,t) = e^{im(x-x_0)^2/(2t)} f\left(\frac{m(x-x_0)}{t}\right) t^{\ n/2}$$

Now suppose we want to show that

$$\lim_{t\to\infty} e^{itH}\psi(x,t)$$

exists. This is like a wave operator in scattering theory. We can show existence by writing

$$e^{itH}\psi(x,t) = \psi(x,1) + \int_1^t \frac{d}{dt}e^{itH}\psi(x,t)$$

Using what is known as Cook's method, it suffices to prove

$$\int_1^\infty \left\|\frac{d}{dt}e^{itH}\psi(x,t)\right\| < \infty.$$

By the calculation above

$$\begin{aligned}\left\|\frac{d}{dt}e^{itH}\psi(x,t)\right\| &= \left\|e^{itH}\left(iH + \frac{\partial}{\partial t}\right)\psi(x,t)\right\| \\ &= \left\|\left(H - i\frac{\partial}{\partial t}\right)\psi(x,t)\right\| \\ &= \left\|\frac{1}{2m}\Delta u\right\|\end{aligned}$$

where $u(x,t) = f\left(\frac{m(x-x_0)}{t}\right) t^{-n/2}$. So

$$\begin{aligned} \|\Delta u(x,t)\|^2 &\le C \int_{\mathbb{R}^n} \left| t^{-n/2-2} f\left(\frac{m(x-x_0)}{t}\right)\right|^2 d^n x \\ &= t^{-4} \left| f\left(m(y)\right)\right|^2 d^n y \end{aligned}$$

This shows $\|\Delta u(x,t)\| \le Ct^{-2}$ and so the integral is finite.

Now given a choice of f set

$$\phi = \lim_{t\to\infty} e^{itH}\psi(x,t).$$

Then

$$\lim_{t\to\infty} \left\|e^{-itH}\phi - \psi(x,t)\right\| = \lim_{t\to\infty} \left\|\phi - e^{itH}\psi(x,t)\right\| = 0$$

and we have constructed an approximate scattering solution. A harder question is whether every scattering solution can be obtained by such a construction. This the question of asymptotic completeness.

I learned about this construction from Ira Herbst who used related more complicated constructions in his work with Erik Skibsted. For the case of Schrödinger operators see [DG].

3.5. Example 2: The harmonic oscillator. We will again restrict ourselves to one dimension. Then the Hilbert space is $L^2(\mathbb{R}, dx)$ and the Hamiltonian operator is

$$H = \frac{1}{2}(p^2 + x^2)$$

(We have set $m = 1$). The classical motion consist of only bound states. My original intention was to present a proof of a theorem here that states that if $\lim_{|x|\to\infty} V(x) = \infty$ then the spectrum of H only contains isolated eigenvalues with finite multiplicity. This can be found, for example, in Reed and Simon IV [RSIV] Theorem XIII.67.

The harmonic oscillator is special in many ways, and we can proceed differently using creation and anihilation operators. This is explained in almost every book on quantum mechanics! Define the (differential) operators

$$\begin{aligned} a &= \frac{1}{\sqrt{2}}\left(x + \frac{d}{dx}\right) && \text{anihilation operator} \\ a^* &= \frac{1}{\sqrt{2}}\left(x - \frac{d}{dx}\right) && \text{creation operator} \\ N &= a^* a && \text{number operator} \end{aligned}$$

and let $\varphi_0 = \pi^{-1/4}e^{-x^2/2}$ be the normalized Gaussian. Then

$$\begin{aligned} H &= N + \frac{1}{2} \\ [a, a^*] &= 1 \\ a\varphi_0 &= 0 \end{aligned}$$

and one defines $\psi_0 = \varphi_0$ and for $n = 1, 2, 3, \ldots$,

$$\psi_n = \frac{1}{\sqrt{n!}}(a^*)^n \psi_0.$$

These can be shown to form an orthonormal set of eigenfunctions for H with eigenvalues $E_n = n + \frac{1}{2}$. The have the explicit form

$$\psi_n(x) = \frac{1}{\sqrt{2^n n!}} \pi^{-1/4} H_n(x)$$

where

$$H_n(z) = (-1)^n e^{z^2} \frac{d^n}{dz^n} e^{-z^2}$$

are the classical Hermite polynomials. There is a theorem that states that the set $\{p(x)\varphi_0(x) : p(x) \text{ is a polynomial}\}$ is dense in $L^2(\mathbb{R}, dx)$. This implies that $\{\psi_n : n \in \mathbb{N}\}$ is an orthonormal basis of eigenvectors for H.

The first thing I want to do is to fix an energy $E_n = n + \frac{1}{2}$ and compare the classical and quantum states with this energy. The classical state is a a circular trajectory in phase space with radius $\sqrt{2E_n}$. The quantum bound state is stationary. It doesn't move in time and the probability distribution for the position of the particle is the constant measure $|\psi_n(x)|^2 dx$ which looks like this (for $n = 2$ and $n = 50$):

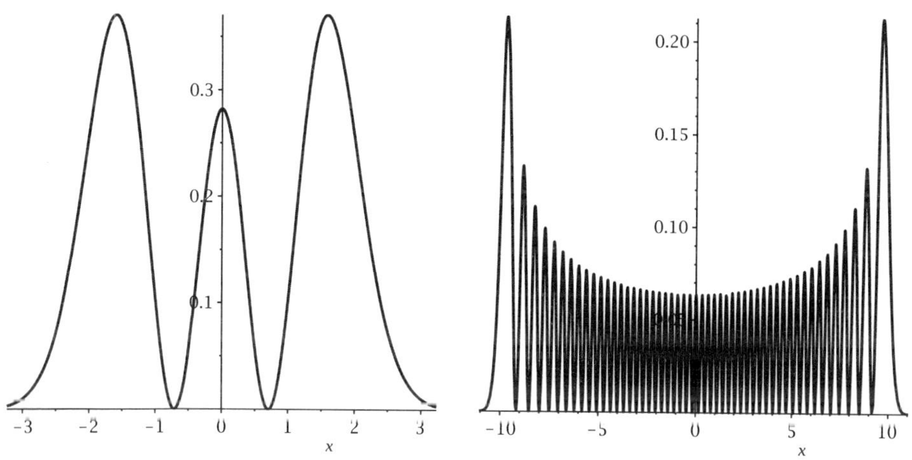

What classical quantity could one compare to these distributions? Maybe the quantum distribution represents a sort of average of the classical motion. To test this we must compute measure the on $[-\sqrt{2E_n}, \sqrt{2E_n}]$ defined by

$$m_n(I) = \lim_{T\to\infty} T^{-1} \int_{-T}^{T} \chi_I(t)\,dt.$$

This gives the asymptotic proportion of the time that the classical particle spends in the set I. A calculation yields

$$dm_n = \frac{1}{\pi\sqrt{2E_n - x^2}} dx$$

Here is what we get if we superimpose the classical and quantum densities.

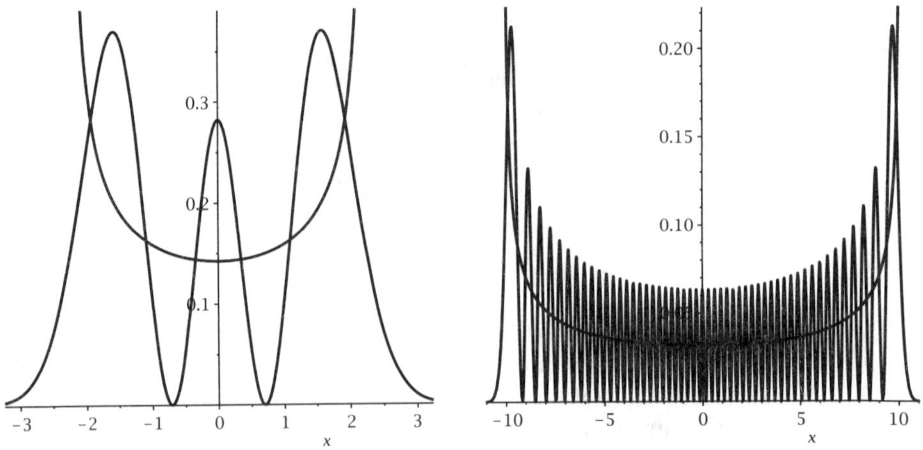

When $n = 2$ they are pretty different, but when $n = 50$ the classical density looks like some sort of average.

PROBLEM 11. Is it true that for $x \in [-\sqrt{2E_n}, \sqrt{2E_n}]$

$$\lim_{n\to\infty} \left| \int_{-\infty}^{x} |\psi_n(y)|^2 dy - \int_{-\sqrt{2E_n}}^{x} m_n(y)dy \right| = 0?$$

The above discussion might leave you with the impression that there is no motion in the quantum harmonic oscillator. But when you take a linear combination, or superposition, of eigenstates with different eigenvalues the result is no longer a stationary state. Amazingly, one can construct states which exhibit exactly the same motion as the classical particle! These states are called coherent states. They were introduced by Klauder and have been greatly generalized. For an exposition see, for example, the second volume of Simon's Course in Analysis [S2]

A simple way to define them is as eigenstates of the annihilation operator. We already know that $a\varphi_0 = 0$ so 0 is an eigenvalue of a with eigenvector φ_0.

PROPOSITION 5. *Every complex number α is an eigenvalue of the annihilation operator a. If $\alpha = \frac{1}{\sqrt{2}}(x_0 + ip_0)$, then the normalized eigenvector φ_α is given by*

$$\varphi_\alpha(x) = M[p_0]T[x_0]\varphi_0 = e^{ip_0x}\varphi_0(x - x_0).$$

We can think of α as a point in classical phase space. Then $|\varphi_\alpha(x)|^2 = \pi^{-1/2}e^{-(x-x_0)^2}$ while $|\widehat{\varphi}_\alpha(k)|^2 = \pi^{-1/2}e^{-(k-p_0)^2}$. So the position and momentum distributions for this state are Gaussians centered at the classical values x_0 and p_0.

PROOF. Writing out the equation $a\varphi_\alpha = \alpha\varphi_\alpha$ yields $x\varphi_\alpha + \varphi_\alpha' = \sqrt{2}\alpha\varphi_\alpha$. whose solution is

$$\begin{aligned}\varphi_\alpha(x) &= Ce^{-x^2/2+\sqrt{2}\alpha x}\\ &= Ce^{-x^2/2+(x_0+ip_0)x}\\ &= \tilde{C}e^{-(x-x_0)^2/2}e^{ip_0x}.\end{aligned}$$

The normalization condition then implies $\tilde{C} = 1$ (up to an inessential phase). $\square$

Recall that the harmonic oscillator time evolution of a point α in classical phase space, thought of as a point in $\mathbb{C}$, is $\alpha(t) = e^{-it}\alpha$. The quantum time evolution of φ_α completely classical in the following sense.

PROPOSITION 6.

$$e^{-itH}\varphi_\alpha = e^{-it/2}\varphi_{\alpha(t)}$$

Of course the phase factor $e^{-it/2}$ doesn't change the state.

PROOF. Expand φ_α in the orthonormal basis of eigenvectors. We have

$$\varphi_\alpha = \sum_n \langle \psi_n, \varphi_\alpha \rangle \psi_n$$

where

$$\langle \psi_n, \varphi_\alpha \rangle = \frac{1}{\sqrt{n!}} \langle (a^*)^n \varphi_0, \varphi_\alpha \rangle = \frac{1}{\sqrt{n!}} \langle \varphi_0, a^n \varphi_\alpha \rangle = \frac{\alpha^n}{\sqrt{n!}} \langle \varphi_0, \varphi_\alpha \rangle$$

To determine $\langle \varphi_0, \varphi_\alpha \rangle$ use

$$1 = \|\varphi_\alpha\|^2 = \sum_n |\langle \psi_n, \varphi_\alpha \rangle|^2 = \sum_n \frac{|\alpha|^{2n}}{n!} |\langle \varphi_0, \varphi_\alpha \rangle|^2 = e^{|\alpha|^2} |\langle \varphi_0, \varphi_\alpha \rangle|^2.$$

So $|\langle \varphi_0, \varphi_\alpha \rangle| = e^{-|\alpha|^2/2}$ and (up to a phase factor)

$$\varphi_\alpha = e^{-|\alpha|^2/2} \sum_n \frac{\alpha^n}{\sqrt{n!}} \psi_n$$

Since each ψ_n is an eigenvector of H with eigenvalue $E_n = n + \frac{1}{2}$ we know that $e^{-itH}\psi_n = e^{-it(n+\frac{1}{2})}\psi_n$. Therefore

$$\begin{aligned}
e^{-itH}\varphi_\alpha &= e^{-|\alpha|^2/2} \sum_n \frac{\alpha^n}{\sqrt{n!}} e^{-itH} \psi_n \\
&= e^{-|\alpha|^2/2} \sum_n \frac{\alpha^n}{\sqrt{n!}} e^{-it(n+\frac{1}{2})} \psi_n \\
&= e^{-it/2} e^{-|\alpha|^2/2} \sum_n \frac{(e^{-it}\alpha)^n}{\sqrt{n!}} \psi_n \\
&= e^{-it/2} e^{-|\alpha(t)|^2/2} \sum_n \frac{\alpha(t)^n}{\sqrt{n!}} \psi_n \\
&= e^{-it/2} \varphi_{\alpha(t)}
\end{aligned}$$

$\square$

The final topic I intended to cover for the harmonic oscillator is the Maslov-WKB construction of approximate eigenfunctions using the Hamilton-Jacobi equation (11). Again, this construction is only important as an illustration of the method, as we know the eigenfunctions exactly. In this case, one needs to take large E to get a good approximation.

The difficulty is that the Lagrangian manifold is now a circle, which is not a graph over (x, E). We can define $S = \int p dx$ as a smooth function on the circle, but the approximate eigenfunction we get as a function of x blows up at the turning points $x = \pm\sqrt{2E}$ where the manifold fails to be a graph. The solution is to work locally. Near the turning points project onto p instead. This is particularly easy for the harmonic oscillator since p and x appear symmetrically in Hamiltonian. The resulting function of p is then changed to a function of x by taking the Fourier transform. Combining these local function into a smooth function requires the introduction of the Maslov index at each turning point. Moreover the final definition of the approximate eigenfunction gives zero unless the Bohr-Sommerfeld quantization condition holds.

There was no time for this, but fortunately there is a beautiful pedagogical paper by Eckmann and Seneor [ES] that explains the procedure.

3.6. Example 3: The two bump potential. The two bump potential we considered before illustrates some interesting differences between classical and quantum mechanics. Recall that the classical particle could be trapped by the potential well. What can we say about bound and scattering states for the quantum Hamiltonian $H = p^2 + V(x)$ acting in $L^2(\mathbb{R})$?

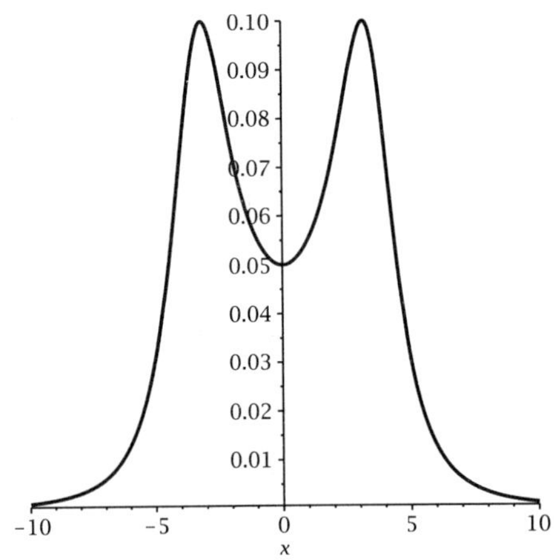

Two bump potential

There was no time for this either in the course, but here are some results about this operator that we could have considered. For simplicity let us assume that $V(x)$ is smooth and compactly supported.

THEOREM 4. *The spectrum of H is $\sigma(H) = [0, \infty)$ and is absolutely continuous.*

So in contrast to the classical case, where there are bound states with positive energy corresponding to particles trapped by the well, there are no bound states in the quantum case. The physical explanation for this is that the particle can tunnel through the barrier. The mathematical theory of the absence of eigenvalues imbedded in the continuous spectrum is harder. There are examples of potentials (e.g., the Wigner–von Neumann potential) which tend to zero (slowly) for large $|x|$ but do have a positive eigenvalue imbedded in the continuous spectrum. See, for example, Chapter 4 of [CFKS].

What would happen if we moved the bottom of the central well so that it dipped below zero? Since we are working in one dimension, H would immediately have a negative eigenvalue (bound state) in addition to the absolutely continuous spectrum on the postive half line. The proof of this is a nice application of min-max methods (see [RSIV]).

What would happen if we moved the bottom of the central well so that it's minimum was exactly zero, and then multiplied $V(x)$ by a large positive coupling constant μ? If the minimum was non-degenerate, then the potential would look more and more like a harmonic oscillator as μ goees to infinity. Yet for each finite μ the potential would satisfy the hypothesis of the theorem. In this case the spectrum would be absolutely continuous in $[0, \infty)$ for every μ. However there would be *resonances* in the lower half plane converging to the harmonic oscillator eigenvalues. For a definition and introduction to resonances, see e.g., [CKFS] [HS]. Resonances could be the subject of another course like this!

4. Hidden variables and non-locality

4.1. Introduction. In general, quantum mechanics only provides probability distributions for the measured values of a quantum observables., and cannot predict the outcomes with certainty. In the early days of quantum mechanics, Einstein strongly resisted the idea that a physical theory could be essentially probabilistic in this way. Experiments like the one described by Einstein, Podolsky and Rosen in 1935 were proposed to show that measurement of spatially separated but quantum mechanically entangled particles leads to action at a distance. In the intervening eighty years there has been much philosophical debate on the foundations of quantum mechanics and the nature of physical reality. A readable account is *The Infamous Boundary* by David Wick [W]. Remarkably, recent experiments support the idea that action at a distance is real [F"TZWF]. In this lecture, we will consider an example of Lucien Hardy [H] that is an outgrowth of the fundamental work of J.S. Bell [B]. This account is based on Faris [F1] [F2] and conversations with David Brydges.

It is possible (and usual) to work on the mathematics of quantum mechanics without worrying about the philosophical underpinnings. Still, I think it is worthwhile to spend some time to appreciate the essential weirdness of this theory.

Recall that for a system in the pure state ψ, given a Borel set $I \subseteq \mathbb{R}$, the probability that the measured value of a quantum observable A will lie in I is

$$\mathbf{P}_\psi[A \in I] = \langle \psi, \chi_I(A)\psi \rangle , \tag{13}$$

where $\chi_I(A)$ is the spectral projection for the operator A corresponding to the set I.

Two commuting A and B can in principle be measured simultaneously. In this case $\chi_I(A)$ and $\chi_J(B)$ commute and $\chi_I(A)\chi_J(B)$ is again a projection. The probability that the simultaneously measured values of A and B lie in I and J respectively is

$$\mathbf{P}_\psi[A \in I \text{ and } B \in J] = \langle \psi, \chi_I(A)\chi_J(B)\psi \rangle \tag{14}$$

Equations (13) and (14) can be taken as definitions of the quantum probabilities $\mathbf{P}_\psi[A \in I]$ and $\mathbf{P}_\psi[A \in I \textbf{ and } B \in J]$ on the left sides. Alternatively, we can think of $\mathbf{P}_\psi[A \in I]$ and $\mathbf{P}_\psi[A \in I \textbf{ and } B \in J]$ as quantities that can be determined experimentally by running the same experiment many times. Then (13) and (14) express the agreement of theory and experiments.

It is natural to ask whether a more complete theory could predict outcomes with certainty. Perhaps quantum observables have exact values, and saying that a system is in a quantum state ψ only gives partial information about them. In other words, maybe the situation is like a mixed state in classical mechanics, where there is a probability measure on some underlying state space.

Could there be a quantum state space Ω with quantum observables $A, B, \ldots$ corresponding to functions $a, b, \ldots : \Omega \to \mathbb{R}$ with $a(\omega), b(\omega), \ldots$ giving the exact value of the observables if the system is in the state ω. Then each quantum state ψ would correspond to a probability measure $\mathbb{P}_\psi$ on Ω such that the spectral measures for $A, B, \ldots$ are the image measures of $\mathbb{P}_\psi$ under $a(\omega), b(\omega), \ldots$. Or perhaps this is true for some subset $\mathcal{A}$ of observables. Thus we are led to the following question.

QUESTION 1. Given a quantum system and a collection $\mathcal{A}$ of observables does there exist

- a measureable space $(\Omega, \mathcal{F})$ and
- for each $A \in \mathcal{A}$ a random variable $a : \Omega \to \mathbb{R}$

such that every pure state ψ is associated with a probability measure $\mathbb{P}_\psi$ that reproduces the quantum distributions for observables, that is,

$$\langle \psi, P_I(A)\psi \rangle = \mathbb{P}_\psi[\{a \in I\}] \tag{15}$$

and if A and B commute,

$$\langle \psi, P_I(A)P_j(B)\psi \rangle = \mathbb{P}_\psi[\{a \in I\} \cap \{b \in J\}]? \tag{16}$$

EXAMPLE 1. If $\mathcal{A} = \{A\}$ contains only one observable, then the answer to Question 1 is obviously yes. We can take $\Omega = \mathbb{R}$ and $\mathbb{P}_\psi = \mu_\psi^A$.

EXAMPLE 2. If $\mathcal{A}$ a collection of non-commuting observables, then the answer to Question 1 is also yes. We don't have to verify (16) and can satisfy (15) by assuming that measurements of different observables are independent.

In general the answer to Question 1 is no. Our first goal is to construct an example to illustrate this.

4.2. Spin observables A and B. Our examples are constructed from observables A, B given by 2×2 Hermitian matrices acting on the Hilbert space $\mathbb{C}^2$. We will assume that they have the same two distinct eigenvalues λ_1 and λ_2 but different eigenvectors.

Such matrices occur in physics when describing the spin of a spin $\frac{1}{2}$ particle. In units where $\frac{\hbar}{2} = 1$ these have the form

$$S_e = \sum_{i=1}^{3} e_i \sigma_i$$

where $e = [e_1, e_2, e_3]$ is a unit vector (the spin direction) in $\mathbb{R}^3$, and σ_i are Pauli spin matrices

$$\sigma_1 = \begin{bmatrix} 0 & 1 \\ 1 & 0 \end{bmatrix} \quad \sigma_2 = \begin{bmatrix} 0 & -i \\ i & 0 \end{bmatrix} \quad \sigma_3 = \begin{bmatrix} 1 & 0 \\ 0 & -1 \end{bmatrix}.$$

For any choice or direction e the matrix S_e has eigenvalues -1 and 1. These are the possible outcomes of measuring the spin in this direction. To make the examples easier to follow, we will think of A and B as spin matrices.

Let $\{\phi_1, \phi_2\}$ be an orthonormal basis of eigenvectors for A and denote by P_i, $i = 1, 2$ the corresponding spectral projections. Similarly, let $\{\psi_1, \psi_2\}$ be an orthonormal basis of eigenvectors for B and denote by Q_i, $i = 1, 2$ the corresponding spectral projections. We will assume that the bases are related by the unitary change of basis matrix $U = [\langle \psi_i, \phi_j \rangle]$ given by the rotation matrix

$$U = \begin{bmatrix} \alpha & -\beta \\ \beta & \alpha \end{bmatrix},$$

with $\alpha, \beta \in [-1, 1]$, $\alpha^2 + \beta^2 = 1$. We will assume that neither α nor β is zero.

To begin, let us illustrate explicitly the construction of independent random variables in Example 2 above. This requires a product measure. We can use $\mathcal{A} = \{A, B\}$. Consider a state $\psi = \frac{1}{\sqrt{2}}\phi_1 + \frac{1}{\sqrt{2}}\phi_2$. The quantum probabilities are

$$\begin{aligned}
\mathbf{P}[A = \lambda_1] &= |\langle \psi, P_1 \psi \rangle|^2 = \frac{1}{2} \\
\mathbf{P}[A = \lambda_2] &= |\langle \psi, P_2 \psi \rangle|^2 = \frac{1}{2} \\
\mathbf{P}[B = \lambda_1] &= |\langle \psi, Q_1 \psi \rangle|^2 = \frac{(\alpha - \beta)^2}{2} \\
\mathbf{P}[B = \lambda_2] &= |\langle \psi, Q_2 \psi \rangle|^2 = \frac{(\alpha + \beta)^2}{2}
\end{aligned}$$

Let Ω be the finite probability space

$$\Omega = \{\lambda_1, \lambda_2\} \times \{\lambda_1, \lambda_2\} = \{(\lambda_1, \lambda_1), (\lambda_1, \lambda_2), (\lambda_2, \lambda_1), (\lambda_2, \lambda_2)\},$$

$\mathcal{F}$ be all subsets of Ω and define the random variables a and b by

$$a((\lambda_i, \lambda_j)) = \lambda_i, \quad b((\lambda_i, \lambda_j)) = \lambda_j.$$

Then the probability measure generated by

$$\begin{aligned}
\mathbb{P}_\psi[\{(\lambda_1, \lambda_1)\}] &= \frac{1}{2} \cdot \frac{(\alpha - \beta)^2}{2} \\
\mathbb{P}_\psi[\{(\lambda_1, \lambda_2)\}] &= \frac{1}{2} \cdot \frac{(\alpha + \beta)^2}{2} \\
\mathbb{P}_\psi[\{(\lambda_2, \lambda_1)\}] &= \frac{1}{2} \cdot \frac{(\alpha - \beta)^2}{2} \\
\mathbb{P}_\psi[\{(\lambda_2, \lambda_2)\}] &= \frac{1}{2} \cdot \frac{(\alpha + \beta)^2}{2}
\end{aligned}$$

will work.

4.3. Tensor product. We will need to use the tensor product $\mathbb{C}^2 \otimes \mathbb{C}^2$ in the following examples. Tensor products are important in describing multi-particle systems in physics and situations where two systems interact. In our examples they describe two spin measurements, done on the left and the right.

One way to describe the tensor product is with bases. If $\{e_1, e_2\}$ is a basis for the copy of $\mathbb{C}^2$ on the left and $\{f_1, f_2\}$ is a basis for the copy of $\mathbb{C}^2$ on the right, then $\mathbb{C}^2 \otimes \mathbb{C}^2$ is a 4 dimensional space with basis $\{e_1 \otimes f_1, e_2 \otimes f_1, e_2 \otimes f_1, e_2 \otimes f_2\}$. The inner product is defined on basis vectors as $\langle e_i \otimes f_j, e_k \otimes f_l \rangle = \langle e_i, e_k \rangle \langle f_j, f_l \rangle$ and is then extended by linearity to the whole space. Similarly, if A, B are operators on $\mathbb{C}^2$ (i.e., 2×2 matrices) then $A\otimes B$ acts on basis vectors via $(A\otimes B)(e_i\otimes f_j) = Ae_i\otimes Bf_j$ and can be extended by linearity to the whole space.

One way to think about the tensor product is to think of the elements of $\mathbb{C}^2$ on the left as column vectors $\begin{bmatrix} a_1 \\ a_2 \end{bmatrix}$ and elements of $\mathbb{C}^2$ on the right as row vectors $[b_1, b_2]$. Then $\begin{bmatrix} a_1 \\ a_2 \end{bmatrix} \otimes [b_1, b_2]$ corresponds to the 2×2 matrix $\begin{bmatrix} a_1 \\ a_2 \end{bmatrix} [b_1, b_2] = \begin{bmatrix} a_1 b_1 & a_1 b_2 \\ a_2 b_1 & a_2 b_2 \end{bmatrix}$ and $\mathbb{C}^2\otimes\mathbb{C}^2$ can be identified with the space of 2×2 matrices. Product states, that is, states of the form $\begin{bmatrix} a_1 \\ a_2 \end{bmatrix} \otimes [b_1, b_2]$ correspond to matrices of rank 1. However a generic 2×2 has rank 2 and cannot be factored in this way. Such a state is called entangled.

Recall that if A, B are operators on $\mathbb{C}^2$ then $A \otimes B$ acts on $\mathbb{C}^2 \otimes \mathbb{C}^2$. If we represent a vector in $\mathbb{C}^2 \otimes \mathbb{C}^2$ as a 2×2 matrix M, then the action of $A \otimes B$ on M is AMB^T.

4.4. Hardy's example. Now we present Hardy's example of a system where a random variable description is not possible. This system models spin measurement for two particles in an entangled state. The Hilbert space for this system is $\mathbb{C}^2\otimes\mathbb{C}^2$. We can imagine the two particles as being separated in space, one on the left and one on the right, having been sent there after the state is prepared in some central location. Let A and B be the spin matrices defined above. We consider the four observables given by

$$\begin{aligned} A^L &= A \otimes I \\ A^R &= I \otimes A \\ B^L &= B \otimes I \\ B^R &= I \otimes B \end{aligned}$$

and notice that A^L commutes with A^R and B^R. This agrees with the idea that measurements on the left do not interfere with measurements on the right. Thus we may simultaneously measure A^L and A^R, and (in a different experiment) simultaneously measure A^L and B^R. Similarly A^R commutes with A^L and B^L, etc.

We have four bases for $\mathbb{C}^2 \otimes \mathbb{C}^2$ at our disposal, namely $\{\phi_i \otimes \phi_j\}$, $\{\psi_i \otimes \phi_j\}$, $\{\phi_i \otimes \psi_j\}$, and $\{\psi_i \otimes \psi_j\}$, for $i, j = 1, 2$. If we expand a state $\Psi \in \mathbb{C}^2 \otimes \mathbb{C}^2$ in the first of these as

$$\Psi = \sum_{i,j} c_{i,j} \phi_i \otimes \phi_j$$

and write the coefficients as a matrix

$$C = \begin{bmatrix} c_{1,1} & c_{1,2} \\ c_{2,1} & c_{2,2} \end{bmatrix},$$

then the corresponding coefficient matrices for the other basis are UC for $\{\psi_i\otimes\phi_j\}$, CU^T for $\{\phi_i \otimes \psi_j\}$ and UCU^T for $\{\psi_i \otimes \psi_j\}$.

To calculate the quantum probability that a simultaneous measurement of A^L and A^R for two particles in the state Ψ yields $A^L = \lambda_1$ and $A^R = \lambda_1$ we must take the product of the corresponding spectral projections for A^L and A^R (which is again a projection since A^L and A^R commute)

$$(P_1 \otimes I)(I \otimes P_1) = P_1 \otimes P_1.$$

Then the quantum probability is given by

$$\mathbf{P}[A^L = \lambda_1 \textbf{ and } A^R = \lambda_1] = |\langle \Psi, P_1 \otimes P_1 \Psi \rangle|^2 = |c_{1,1}|^2.$$

For our example we take the coefficient matrix for the first basis to be

$$C = \frac{1}{N}\begin{bmatrix} 0 & -\alpha \\ -\alpha & \beta \end{bmatrix},$$

where N is the normalization constant $N = \sqrt{2\alpha^2 + \beta^2} = \sqrt{1 + \alpha^2}$. Then the corresponding coefficient matrices for the other bases are

$$\begin{aligned} UC &= \frac{1}{N}\begin{bmatrix} \alpha\beta & -1 \\ -\alpha^2 & 0 \end{bmatrix} \\ CU^T &= \frac{1}{N}\begin{bmatrix} \alpha\beta & -\alpha^2 \\ -1 & 0 \end{bmatrix} \\ UCU^T &= \frac{1}{N}\begin{bmatrix} \alpha^2\beta + \beta & \alpha\beta^2 - \alpha \\ -\alpha^2 & -\beta\alpha^2 \end{bmatrix} \end{aligned}$$

The key point is the position of the zero entries, and the fact that the $2, 2$ entry of the final matrix is non-zero. These lead to the quantum probabilities

$$\begin{aligned} \mathbf{P}[A^L = \lambda_1 \textbf{ and } A^R = \lambda_1] &= 0 \\ \mathbf{P}[B^L = \lambda_2 \textbf{ and } A^R = \lambda_2] &= 0 \\ \mathbf{P}[A^L = \lambda_2 \textbf{ and } B^R = \lambda_2] &= 0 \\ \mathbf{P}[B^L = \lambda_2 \textbf{ and } B^R = \lambda_2] &= \frac{\alpha^4\beta^2}{N^2} > 0 \end{aligned}$$

We now suppose that there is an underlying probability space such that

$$\begin{aligned} \mathbb{P}[\{a^L = \lambda_1\} \cap \{a^R = \lambda_1\}] &= 0 \\ \mathbb{P}[\{b^L = \lambda_2\} \cap \{a^R = \lambda_2\}] &= 0 \\ \mathbb{P}\{[a^L = \lambda_2\} \cap \{b^R = \lambda_2\}] &= 0 \\ \mathbb{P}[\{b^L = \lambda_2\} \cap \{b^R = \lambda_2\}] &= \frac{\alpha^4\beta^2}{N^2} > 0 \end{aligned}$$

and derive a contradiction. Here is the key point. When we assume that a probability space exists, we can write the set $\{a^L = \lambda_1\} \cap \{a^R = \lambda_1\}$ as a disjoint union

$$\begin{aligned} &\{a^L = \lambda_1\} \cap \{a^R = \lambda_1\} = \\ &\quad \{a^L = \lambda_1\} \cap \{b^L = \lambda_1\} \cap \{a^R = \lambda_1\} \cup \{a^L = \lambda_1\} \cap \{b^L = \lambda_2\} \cap \{a^R = \lambda_1\} \end{aligned}$$

This is saying is that even though we don't know the value of b^L, it has some value, and we can decompose an event into a disjoint union of events depending on that value. We can do this one more time with the value of b^R to produce the disjoint union

$$\begin{aligned}\{a^L=\lambda_1\}\cap\{a^R=\lambda_1\} = &\{a^L=\lambda_1\}\cap\{b^L=\lambda_1\}\cap\{a^R=\lambda_1\}\cap\{b^R=\lambda_1\}\cup\\ &\{a^L=\lambda_1\}\cap\{b^L=\lambda_1\}\cap\{a^R=\lambda_1\}\cap\{b^R=\lambda_2\}\cup\\ &\{a^L=\lambda_1\}\cap\{b^L=\lambda_2\}\cap\{a^R=\lambda_1\}\cap\{b^R=\lambda_1\}\cup\\ &\{a^L=\lambda_1\}\cap\{b^L=\lambda_2\}\cap\{a^R=\lambda_1\}\cap\{b^R=\lambda_2\}\cup\end{aligned}$$

Now let

$$p(i,j,k,l)=\mathbb{P}\left[\{a^L=\lambda_i\}\cap\{b^L=\lambda_j\}\cap\{a^R=\lambda_k\}\cap\{b^R=\lambda_l\}\right]$$

Then we can rewrite our equations as

$$\begin{array}{lllllllll} p(1,1,1,1) & + & p(1,1,1,2) & + & p(1,2,1,1) & + & p(1,2,1,2) & & =0\\ p(1,2,2,1) & + & p(1,2,2,2) & + & p(2,2,2,1) & + & p(2,2,2,2) & & =0\\ p(2,1,1,2) & + & p(2,1,2,2) & + & p(2,2,1,2) & + & p(2,2,2,2) & & =0\\ p(1,2,1,2) & + & p(1,2,2,2) & + & p(2,2,1,2) & + & p(2,2,2,2) & & =\alpha^4\beta^2/N^2>0 \end{array}$$

But the $p(i,j,k,l)$ are non-negative numbers. So all the ones that appear in the first three equations are zero. But this includes all all the terms in the last equation. So they cannot add up to something positive. This contradiction shows that the probability space cannot exist.

4.5. Bell's second theorem: non-locality. This section follows Faris' appendix in the book by Wicks. He includes a more careful explanation of the needed concepts from probability aimed at non-mathematicians.

Since we have to set up different experiments for each pair of measurements in the example above, maybe the probability measure which reproduces the quantum results depends on which experiment we are performing. In other words we might have a probability space Ω and σ-algebra $\mathcal{F}$ with four different probability measures $\mathbb{P}_{AA}$, $\mathbb{P}_{AB}$, $\mathbb{P}_{BA}$ and $\mathbb{P}_{BB}$ depending on which spin direction we are measuring on the left and on the right.

In this case there are some natural restrictions if we want to disallow action at a distance. Measurements which only involve the equipment on the left should not be influenced by what happens on the right, and vice versa. Furthermore, simultaneous measurements on the left and right should be conditionally independent, if we take into account events in the preparation stage. The idea is that events in the preparation stage in the experiment, for example preparing a singlet state where the spins on the left and right always have opposite values, can result in correlations for the measurements on the left and the right. However, if we factor out these effects, the residual randomness leads to independent outcomes on the left and right.

To put this in a mathematical form we need some more ideas from probability theory. We assume that there are three sub σ-algebras of $\mathcal{F}$, denoted $\mathcal{F}^L$, $\mathcal{F}^R$ and $\tilde{\mathcal{F}}$. The events in $\mathcal{F}^L$ describe what happens on the left. Knowing whether a experimental outcome is contained in an event from $\mathcal{F}^L$ only tells us about measurements performed on the left. An example would be $\{a^L=1\}$. Similarly, the events in $\mathcal{F}^R$ are associated with what happens on the right. The events in $\tilde{\mathcal{F}}$ are associated what happened in the preparation stage of the experiment. Assume we are given four probability measures $\mathbb{P}_{AA}$, $\mathbb{P}_{AB}$, $\mathbb{P}_{BA}$ and $\mathbb{P}_{BB}$ such that

1. If $E \in \mathcal{F}^L \cup \tilde{\mathcal{F}}$ then $\mathbb{P}_{AA}(E) = \mathbb{P}_{AB}(E)$ and $\mathbb{P}_{BA}(E) = \mathbb{P}_{BB}(E)$.

In other words, events on the left or from the preparation stage have probabilities which don't change when we move around the equipment on the right.

2. Similarly, if $E \in \mathcal{F}^R \cup \tilde{\mathcal{F}}$ then $\mathbb{P}_{AA}(E) = \mathbb{P}_{BA}(E)$ and $\mathbb{P}_{AB}(E) = \mathbb{P}_{BB}(E)$.

The third assumption expresses the conditional independence of events on the right and left. Recall that if E is an event and G is another event with $\mathbb{P}(G) \neq 0$ then the conditional probability of F given G is the number $\mathbb{P}[E|G] = \dfrac{\mathbb{P}(E \cap G)}{\mathbb{P}(G)}$. If $\tilde{\mathcal{F}}$ is a sub σ-algebra of $\mathcal{F}$ then the conditional probability $\mathbb{P}[E|\tilde{\mathcal{F}}]$ is not a number, but a random variable. The idea (which is actually correct if Ω is finite and there are no non-empty events with zero probability) is that

$$\mathbb{P}[E|\tilde{\mathcal{F}}](\omega) = \mathbb{P}[E|G_\omega] \quad \text{where} \quad G_\omega = \bigcap_{G \in \tilde{\mathcal{F}} : \omega \in G} G.$$

In other words, $\mathbb{P}[E|\tilde{\mathcal{F}}](\omega)$ is the probability of E given all the information in $\tilde{\mathcal{F}}$ about ω. The situation is more complicated when there are set of measure zero and we can't talk about the value of a random variable at a point. We won't worry about this. What we use the fact that $\mathbb{E}[\mathbb{P}[E|\tilde{\mathcal{F}}]] = \mathbb{P}[E]$. The third assumption is

3. If $E \in \mathcal{F}^L$ and $F \in \mathcal{F}^R$ then $P_{xy}[E \cap F|\tilde{\mathcal{F}}] = P_{xy}[E|\tilde{\mathcal{F}}]\ P_{xy}[F|\tilde{\mathcal{F}}]$. Here x and y are any choice of A, B.

These assumptions are reasonable if we don't believe in action at a distance. Yet we can show that there can be no hidden variable description of the experiment unless they are violated.

PROPOSITION 7. *Let the matrices A and B and the coefficients α, β and N be as defined in the previous section. Suppose that there are probability measures $\mathbb{P}_{AA}$, $\mathbb{P}_{AB}$, $\mathbb{P}_{BA}$ and $\mathbb{P}_{BB}$ and sub σ-algebras $\mathcal{F}^L$, $\mathcal{F}^R$ and $\tilde{\mathcal{F}}$ satisfying the assumptions above. Then there are no random variables a^L, a^R, b^L, b^R such that*

$$\mathbb{P}_{AA}[\{a^L = \lambda_1\} \cap \{a^R = \lambda_1\}] = 0$$
$$\mathbb{P}_{BA}[\{b^L = \lambda_2\} \cap \{a^R = \lambda_2\}] = 0$$
$$\mathbb{P}_{AB}[\{a^L = \lambda_2\} \cap \{b^R = \lambda_2\}] = 0$$
$$\mathbb{P}_{BB}[\{b^L = \lambda_2\} \cap \{b^R = \lambda_2\}] = \frac{\alpha^4\beta^2}{N^2} > 0$$

PROOF. If the probability of an event is zero, then so is the conditional probability. Thus the first three equations, together with the conditional independence, imply

$$\left(\mathbb{P}_{AA}[\{a^L = \lambda_1\}|\tilde{\mathcal{F}}]\right)\left(\mathbb{P}_{AA}[\{a^R = \lambda_1\}|\tilde{\mathcal{F}}]\right) = 0$$
$$\left(\mathbb{P}_{BA}[\{b^L = \lambda_2\}|\tilde{\mathcal{F}}]\right)\left(\mathbb{P}_{BA}[\{a^R = \lambda_2\}|\tilde{\mathcal{F}}]\right) = 0$$
$$\left(\mathbb{P}_{AB}[\{a^L = \lambda_2\}|\tilde{\mathcal{F}}]\right)\left(\mathbb{P}_{AB}[\{b^R = \lambda_2\}|\tilde{\mathcal{F}}]\right) = 0$$

almost everywhere. Then

$$\begin{aligned}
&\mathbb{P}_{BB}[\{b^L = \lambda_2\} \cap \{b^R = \lambda_2\}|\tilde{\mathcal{F}}] \\
&\qquad = \left(\mathbb{P}_{BB}[\{b^L = \lambda_2\}|\tilde{\mathcal{F}}]\right)\left(\mathbb{P}_{BB}[\{b^R = \lambda_2\}|\tilde{\mathcal{F}}]\right) \\
&\qquad = \left(\mathbb{P}_{BA}[\{b^L = \lambda_2\}|\tilde{\mathcal{F}}]\right)\left(\mathbb{P}_{AB}[\{b^R = \lambda_2\}|\tilde{\mathcal{F}}]\right) \\
&\qquad = \left(\mathbb{P}_{BA}[\{b^L = \lambda_2\} \cap \{a^R = \lambda_1\}|\tilde{\mathcal{F}}] + \mathbb{P}_{BA}[\{b^L = \lambda_2\} \cap \{a^R = \lambda_2\}|\tilde{\mathcal{F}}]\right) \\
&\qquad\quad \left(\mathbb{P}_{AB}[\{a^L = \lambda_1\} \cap \{b^R = \lambda_2\}|\tilde{\mathcal{F}}] + \mathbb{P}_{AB}[\{a^L = \lambda_2\} \cap \{b^R = \lambda_2\}|\tilde{\mathcal{F}}]\right) \\
&\qquad = \left(\mathbb{P}_{BA}[\{b^L = \lambda_2\} \cap \{a^R = \lambda_1\}|\tilde{\mathcal{F}}]\right)\left(\mathbb{P}_{AB}[\{a^L = \lambda_1\} \cap \{b^R = \lambda_2\}|\tilde{\mathcal{F}}]\right) \\
&\qquad \le \mathbb{P}_{BA}[\{a^R = \lambda_1\}|\tilde{\mathcal{F}}]\mathbb{P}_{AB}[\{a^L = \lambda_1\}|\tilde{\mathcal{F}}] \\
&\qquad = \mathbb{P}_{AA}[\{a^R = \lambda_1\}|\tilde{\mathcal{F}}]\mathbb{P}_{AA}[\{a^L = \lambda_1\}|\tilde{\mathcal{F}}] \\
&\qquad = 0
\end{aligned}$$

almost everywhere. This implies

$$\mathbb{P}_{BB}[\{b^L = \lambda_2\} \cap \{b^R = \lambda_2\}] = \mathbb{E}[\mathbb{P}_{BB}[\{b^L = \lambda_2\} \cap \{b^R = \lambda_2\}|\tilde{\mathcal{F}}]] = 0,$$

which contradicts the last line in our original set of equations. □

References

[A] V. I. Arnold, *Mathematical Methods of Classical Mechanics, 2nd Ed.*, Springer 1997.

[B] J. S. Bell, *Speakable and Unspeakable in Quantum Mechanics,* Cambridge University Press, 2004.

[CFKS] H. L. Cycon, R. G. Froese, W. Kirsch, and B. Simon, *Schrödinger Operators, with Applications to Quantum Mechanics and Global Geometry,* 2nd ed., Springer, 2007

[DG] Jan Dereziński and Christian Gérard, *Scattering theory of classical and quantum N-particle systems,* Texts and Monographs in Physics, Springer-Verlag, Berlin, 1997. MR1459161

[ES] J.-P. Eckmann and R. Sénéor, *The Maslov-WKB method for the (an-)harmonic oscillator,* Arch. Rational Mech. Anal. **61** (1976), no. 2, 153–173, DOI 10.1007/BF00249703. MR0406147

[F1] William G. Faris, *Outline of quantum mechanics,* Entropy and the quantum, Contemp. Math., vol. 529, Amer. Math. Soc., Providence, RI, 2010, pp. 1–52, DOI 10.1090/conm/529/10426. MR2681767

[F2] William Faris, Appendix to Wick's book [W].

[FTZWF] Maria Fuwa, Shuntaro Takeda, Marcin Zwierz, Howard M. Wiseman & Akira Furusawa, *Experimental proof of nonlocal wavefunction collapse for a single particle using homodyne measurements,* Nature Communications, 2015 DOI: 10.1038/ncomms7665

[H] Lucien Hardy, *Nonlocality for two particles without inequalities for almost all entangled states,* Phys. Rev. Lett. **71** (1993), no. 11, 1665–1668, DOI 10.1103/PhysRevLett.71.1665. MR1234454

[HS] P. D. Hislop and I. M. Sigal, *Introduction to spectral theory,* Applied Mathematical Sciences, vol. 113, Springer-Verlag, New York, 1996. With applications to Schrödinger operators. MR1361167

[MR] Jerrold E. Marsden and Tudor S. Ratiu, *Introduction to mechanics and symmetry,* 2nd ed., Texts in Applied Mathematics, vol. 17, Springer-Verlag, New York, 1999. A basic exposition of classical mechanical systems. MR1723696

[RSIV] Michael Reed and Barry Simon, *Methods of modern mathematical physics. IV. Analysis of operators,* Academic Press [Harcourt Brace Jovanovich, Publishers], New York-London, 1978. MR0493421

[S2] Barry Simon, *Basic complex analysis*, A Comprehensive Course in Analysis, Part 2A, American Mathematical Society, Providence, RI, 2015. MR3443339

[W] David Wick, *The infamous boundary*, Birkhäuser Boston, Inc., Boston, MA, 1995. Seven decades of controversy in quantum physics; With a mathematical appendix by William Farris. MR1352686

Department of Mathematics, University of British Columbia, Vancouver, British Columbia V6T 1Z2, Canada

Email address: rfroese@math.ubc.ca

Contemporary Mathematics
Volume **720**, 2018
https://doi.org/10.1090/conm/720/14523

Numerical methods for spectral theory

Felix Kwok

ABSTRACT. The spectrum of the Laplacian operator can only be computed analytically for a few simple geometries. When direct computation is not possible, numerical methods can be useful for finding approximations of the spectrum. In this paper, we present two major techniques for approximating the spectrum of the Laplacian operator via discretization, namely the finite difference and finite element methods. We show how to construct the matrices whose eigenvalues approximate the desired spectrum, and we discuss the convergence of these methods as the grid is refined. We also introduce a few numerical methods that are commonly used for solving the associated matrix eigenvalue problems. Finally, we illustrate how the above techniques can be used to handle higher order operators, such as the biharmonic operator.

1. Introduction

The purpose of these notes is to introduce a few numerical methods for approximating the eigenvalues and eigenfunctions of partial differential operators. Such numerical approximations are often important in physics and engineering, as they can be used to describe the vibrations of a physical structure.

1.1. Vibrations and eigenvalues. Consider a thin membrane modelled as a two-dimensional bounded, open subset Ω with boundary $\partial\Omega$. We describe its vibrating motion over a time interval $(0, t_f)$ in terms of its vertical displacement $z : \Omega \times (0, t_f) \to \mathbb{R}$, starting from an initial displacement $z_0(x, y)$ and initial velocity $v_0(x, y)$. Assume that the boundary of the membrane is fixed, so that its displacement is zero at all times. If there are no external forces acting on the membrane, i.e., the only forces driving the vibrating motion are the restoring force due to the deformation of the membrane, then Newton's second law states that

$$\frac{\partial^2 z}{\partial t^2} = -\mathscr{L} z, \qquad \text{in } \Omega \times (0, t_f],$$

where $\mathscr{L}$ is a spatial differential operator acting on $z(x, y, t)$. The precise form of $\mathscr{L}$ depends on the model used for the material and can often be derived from energy considerations; see Section 5 for the vibrating plate. For the vibrating membrane, a simple model uses the scaled *Laplacian operator*

$$\mathscr{L} z = -c\Delta z := -c\Big(\frac{\partial^2 z}{\partial x^2} + \frac{\partial^2 z}{\partial y^2}\Big), \qquad c > 0,$$

2010 *Mathematics Subject Classification.* Primary 65N06, 65N25, 65N30, 65F15.

which is a second-order linear elliptic differential operator. As a result, we get the linear initial-boundary value problem

$$
\begin{aligned}
&\frac{\partial^2 z}{\partial t^2} = c\Delta z \qquad \text{in } \Omega \times (0, t_f) \\
(1) \qquad &z|_{t=0} = z_0(x,y), \qquad \left.\frac{\partial z}{\partial t}\right|_{t=0} = v_0(x,y) \\
&z|_{\partial\Omega} = 0, \qquad t \in (0, t_f).
\end{aligned}
$$

The requirement that z vanish along the boundary is known as a *homogeneous Dirichlet boundary condition* on z.

1.2. Separation of variables. A *standing wave* is any solution of (1) of the form

$$z(x,y,t) = u(x,y) \cdot T(t).$$

Because of linearity, any linear combination of standing waves is also a solution of (1). A classical technique for solving linear problems of this type is to first seek a family of standing wave solutions, and then find a linear combination of these solutions that satisfies the initial conditions. This procedure is known as *separation of variables.*

More concretely, if $z(x,y,t) = u(x,y)T(t)$ is a standing wave solution, then substituting into the PDE yields

$$T''(t)u(x,y) = -\mathscr{L}u(x,y)T(t)$$

or

$$(2) \qquad -\frac{T''(t)}{T(t)} = \frac{\mathscr{L}u(x,y)}{u(x,y)},$$

where we assumed that $u(x,y)$ and $T(t)$ do not vanish identically. Since the left-hand side is independent of x and y, and the right-hand side independent of t, we see that both sides are in fact constant; in particular, we have

$$(3) \qquad \frac{\mathscr{L}u(x,y)}{u(x,y)} = \lambda \qquad \text{or} \qquad \mathscr{L}u = \lambda u,$$

i.e., u is an eigenfunction of $\mathscr{L}$ with eigenvalue λ, which are assumed to exist.[1] Moreover, we see that λ is non-negative, since we can multiply (3) by u and integrate to obtain

$$\int_\Omega \lambda u^2\,dx = -c\int_\Omega u\Delta u\,dx = c\int_\Omega |\nabla u|^2\,dx,$$

where the boundary terms in Green's identity vanish because $u|_{\partial\Omega} = 0$. Thus, we see that

$$\lambda = \frac{c\int_\Omega |\nabla u|^2\,dx}{\int_\Omega u^2 dx} \geq 0.$$

Let $u_n(x,y)$ be an eigenfunction with eigenvalue λ_n. Then (2) implies that the corresponding time-dependent function $T = T_n(t)$ satisfies

$$T_n''(t) = -\lambda T_n(t) \implies T_n(t) = a_n\cos(\sqrt{\lambda_n}t) + b_n\sin(\sqrt{\lambda_n}t).$$

[1] For the existence and uniqueness of eigenvalues and eigenvectors for the Laplacian operator with Dirichlet boundary conditions, see Chapter 5 of Richard Laugesen's notes on the spectral theory of PDEs [**31**].

Thus, the eigenvalues λ_n can be interpreted as the square of the frequencies of vibration. Taking linear combinations over all such frequencies, we can write the solution $z(x, y, t)$ as

$$z(x,y,t) = \sum_{n=1}^{\infty} a_n \cos(\sqrt{\lambda_n} t) u_n(x,y) + b_n \sin(\sqrt{\lambda_n} t) u_n(x,y),$$

where the coefficients a_n and b_n are determined by the initial conditions

$$z(x,y,0) = \sum_{n=1}^{\infty} a_n u_n(x,y) = z_0(x,y),$$
$$\frac{\partial z}{\partial t}(x,y,0) = \sum_{n=1}^{\infty} \sqrt{\lambda_n} b_n u_n(x,y) = v_0(x,y).$$

Since the eigenfunctions $\{u_n\}_{n=1}^{\infty}$ form an orthonormal basis of $L^2(\Omega)$ by the spectral theorem [**31**, Theorem 4.1], it is always possible to find a_n and b_n such that the above equality holds in the sense of L^2-convergence for every $z_0, v_0 \in L^2(\Omega)$.

For special geometries and boundary conditions, it is possible to solve the eigenvalue problem analytically. For example, if $\Omega = (0,1) \times (0,1)$ is the unit square and $u = 0$ on the boundary $\partial\Omega$, then one can again use separation of variables and find that the eigenfunctions of the negative Laplacian $\mathscr{L} = -\Delta$ (with $c = 1$) are of the form

$$u_{mn}(x,y) = \sin(m\pi x)\sin(n\pi y),$$

with corresponding eigenvalues $\lambda_{mn} = \pi^2(m^2 + n^2)$. Analytical solutions are also available for rectangles, discs and sectors [**40**, Chapter 11]. For more complicated geometries, however, one can only obtain approximations to the eigenvalues by replacing the PDE eigenvalue problem by a finite-dimensional, approximate matrix eigenvalue problem. As with any approximation, we are interested in the following questions:

(1) How to approximate the continuous problem by a discrete one?
(2) How good is the approximation?
(3) How to solve the associated finite dimensional problem?

To answer the first question, we will consider two methods in these notes: the finite difference method will be discussed in Section 2, and the finite element method will be discussed in Section 3. For each method, we will study its approximation properties and establish some convergence results for the Laplace problem. We will briefly consider the third question in Section 4, where we explain the fundamentals of numerical methods for matrix eigenvalue problems. Finally, in Section 5, we consider a more complicated example with a fourth order biharmonic operator and different types of boundary conditions. Section 6 provides some suggestions for further reading, and Section 7 contains practice problems given during the CRM Summer School on Spectral Theory and Applications, held in Quebec City, Canada in August 2016.

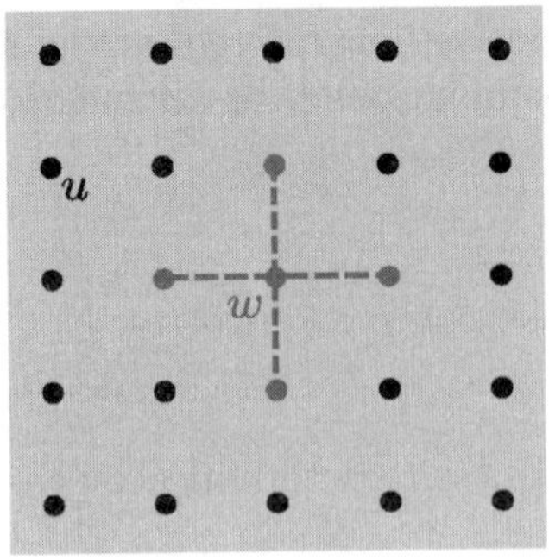

FIGURE 1. A five-point finite difference stencil on a regular grid, connected by dashed lines.

2. Finite difference methods

The finite difference method is perhaps the simplest discretization that can be used to approximate a PDE eigenvalue problem. In essence, one uses a Taylor-based approximation to replace partial derivatives by partial differences, and then solves the transformed eigenvalue problem, which is now finite dimensional. As we will see, the method works best for domains whose boundaries align with a rectangular grid.

2.1. Finite difference for the 2D Laplacian operator. Consider the eigenvalue problem on the unit square $\Omega = (0,1) \times (0,1)$

$$-\Delta u = \lambda u, \qquad u = 0 \text{ on } \partial\Omega$$

Instead of seeking $u(x,y)$ everywhere, we seek an approximation of it on a *uniform grid*

$$u_{ij} \approx u(x_i, y_j), \qquad 1 \leq i, j, \leq n-1$$

with $x_i = ih$, $y_j = jh$, $h = 1/n$. This $\{u_{ij}\}_{i,j=0}^N$ is called a *grid function*, since it is a function defined on the grid $\{(x_i, y_j)\}_{i,j=0}^N$. If u_{ij} represents exact values of u at the grid points, then the second derivatives u_{xx} and u_{yy} can be written as

$$u_{xx} = \frac{u_{i-1,j} - 2u_{ij} + u_{i+1,j}}{h^2} + O(h^2), \qquad u_{yy} = \frac{u_{i,j-1} - 2u_{ij} + u_{i,j+1}}{h^2} + O(h^2),$$

which can be verified readily using Taylor expansions. Substituting these approximations into the PDE eigenvalue problem and dropping the $O(h^2)$ terms, we obtain the *discrete* eigenvalue problem

$$\frac{4u_{ij} - u_{i-1,j} - u_{i+1,j} - u_{i,j-1} - u_{i,j+1}}{h^2} = \tilde{\lambda} u_{ij} \tag{4}$$

for $1 \leq i, j \leq n-1$, with boundary conditions

$$u_{i,0} = u_{i,n} = u_{0,j} = u_{n,j} = 0.$$

In Figure 1, we show the location of the five unknowns involved in the equation (4) associated with the point w. If we gather all $(n-1)^2$ linear equations and write them in matrix form, we obtain the *matrix eigenvalue problem*

$$A\mathbf{u} = \tilde{\lambda}\mathbf{u},$$

where $\mathbf{u}$ is a vector obtained from the grid function u_{ij} by removing the boundary nodes (whose values are known to be zero), and by imposing a contiguous ordering on the remaining nodes, from 1 to $(n-1)^2$. For example, if we use the lexicographical ordering

$$\mathbf{u}^T = (u_{11}, \ldots, u_{n-1,1} \mid u_{12}, \ldots, u_{n-1,2} \mid \cdots \mid u_{1,n-1}, \ldots, u_{n-1,n-1})^T,$$

then the matrix A has the form

$$A = \begin{bmatrix} T & -I & & & \\ -I & T & -I & & \\ & \ddots & \ddots & \ddots & \\ & & -I & T & -I \\ & & & -I & T \end{bmatrix}, \quad \text{where} \quad T = \begin{bmatrix} 4 & -1 & & & \\ -1 & 4 & -1 & & \\ & \ddots & \ddots & \ddots & \\ & & -1 & 4 & -1 \\ & & & -1 & 4 \end{bmatrix}$$

and I is the $(n-1) \times (n-1)$ identity matrix. Note that A is a *sparse matrix* with at most five entries per row, since each equation only contains five unknowns. This motivates the use of eigenvalue solvers for sparse matrices, which we will discuss in Section 4.

2.2. Discrete eigenvalues. For the simple case of a unit square, it is in fact possible to calculate the eigenvalues and eigenvectors analytically. Let $\mathbf{u}$ be the vector corresponding to the grid function (u_{ij}) of the form

$$u_{ij} = \sin(i\theta)\sin(j\phi),$$

where θ and ϕ are fixed constants. Note that this vector satisfies the boundary conditions

$$u_{i,0} = u_{0,j} = 0.$$

Using some elementary trigonometric identities, we can show that

$$4u_{ij} - u_{i-1,j} - u_{i+1,j} - u_{i,j-1} - u_{i,j+1} = (4 - 2\cos(\theta) - 2\cos(\phi))u_{ij},$$

so that (4) implies, for $u_{ij} \neq 0$, that

$$\tilde{\lambda} = \frac{4 - 2\cos(\theta) - 2\cos(\phi)}{h^2} = \frac{4}{h^2}\left[\sin^2(\theta/2) + \sin^2(\phi/2)\right].$$

To determine θ and ϕ, we use the remaining boundary conditions

$$\begin{aligned} u_{n,j} = 0 \quad &\Longrightarrow \quad \sin(n\theta) = 0 \\ u_{i,n} = 0 \quad &\Longrightarrow \quad \sin(n\phi) = 0 \end{aligned}$$

Thus, the possible θ and ϕ values are

$$\theta_k = \frac{k\pi}{n}, \quad \phi_\ell = \frac{\ell\pi}{n} \quad \text{for } 1 \le k, l \le n-1.$$

It is possible to show that for different (k, ℓ) pairs, the resulting eigenfunctions are mutually orthogonal, so they are linearly independent. The associated eigenvalues are

$$\tilde{\lambda}_{k\ell} = 4n^2\left[\sin^2\left(\frac{k\pi}{2n}\right) + \sin^2\left(\frac{\ell\pi}{2n}\right)\right]$$

for $k = 1, \ldots, n-1$, $\ell = 1, \ldots, n-1$. Because discrete sine functions of different frequencies are mutually orthogonal, we have found $(n-1)^2$ linearly independent eigenfunctions, so we have found all the eigenvalues. Note that some of them have multiplicities higher than 1, because of the symmetry between k and ℓ.

We now compare the discrete eigenvalues with those of the **continuous** problem, which we found in Section 1 to be

$$\lambda_{k\ell} = \pi^2(k^2 + \ell^2).$$

THEOREM 2.1 (Convergence of discrete eigenvalues). *For the Laplacian problem on the unit square, the discrete eigenvalues $\tilde{\lambda}_{k\ell}$ obtained from a finite difference discretization with mesh size $h = 1/n$, $n > k, l$, satisfy $\tilde{\lambda}_{k\ell} \le \lambda_{k\ell}$ with*

$$\frac{|\tilde{\lambda}_{k\ell} - \lambda_{k\ell}|}{|\lambda_{k\ell}|} \le C(k,\ell)h^2,$$

where $C(k,\ell) = \frac{1}{12}\max(k^2,\ell^2)$ is independent of h but dependent on k and ℓ.

PROOF. Using the definition $h = 1/n$, we calculate

$$(5)\quad \lambda_{k\ell} - \tilde{\lambda}_{k\ell} = k^2\pi^2\left[1 - \left(\frac{2}{k\pi h}\sin(k\pi h/2)\right)^2\right] + \ell^2\pi^2\left[1 - \left(\frac{2}{\ell\pi h}\sin(\ell\pi h/2)\right)^2\right].$$

Noting that for $0 < \theta < \pi/2$, we have

$$1 - \frac{\theta^2}{6} \le \frac{\sin\theta}{\theta} \le 1,$$

we immediately deduce that $\tilde{\lambda}_{k\ell} \le \lambda_{k\ell}$ for any choice of k and ℓ. Moreover, we have

$$1 - \frac{\sin^2\theta}{\theta^2} = \left(1 - \frac{\sin\theta}{\theta}\right)\left(1 + \frac{\sin\theta}{\theta}\right) \le \frac{\theta^2}{6}\cdot 2 = \frac{\theta^2}{3},$$

so applying this estimate to (5) with $\theta = k\pi h/2$ and $\theta = \ell\pi h/2$ leads to

$$0 \le \lambda_{k\ell} - \tilde{\lambda}_{k\ell} \le \frac{(k\pi)^4 + (\ell\pi)^4}{12}h^2 \le \frac{\max(k^2,\ell^2)\pi^2}{12}\lambda_{k\ell}h^2,$$

from which the conclusion follows. □

To illustrate Theorem 2.1, we plot in Figure 2 the relative error $|\lambda_{k\ell} - \tilde{\lambda}_{k\ell}|/|\lambda_{k\ell}|$ of discrete eigenvalues $\tilde{\lambda}_{k\ell}$ for different values of n. We see that as we increase n, the relative error becomes smaller, meaning they become more and more accurate. Nonetheless, the smallest eigenvalues are much better approximated than the larger ones, especially for small n. Indeed, the relative error is much worse for the larger eigenvalues than for the small ones. This is consistent with the statement of the theorem, where the constant in front of the relative error grows with k and ℓ. Thus, for the larger eigenvalues, a much finer grid is needed to obtain a good approximation.

2.3. L-shaped domain. The real power of numerical methods lies in their ability to produce approximate eigenvalues when they cannot be computed analytically. Consider the L-shaped domain shown in the left panel of Figure 3. We wish to study the eigenvalues of the Laplacian on this domain with Dirichlet boundary conditions; in other words, we consider

$$-\Delta u = \lambda u,$$
$$u|_{\partial\Omega} = 0.$$

The finite difference method with a regular grid (4) can be used here, with the only differences being that there are no unknowns corresponding to points in the second quadrant, and that all nodes along the edges $\{0\} \times [0,1]$ and $[-1,0] \times \{0\}$ need

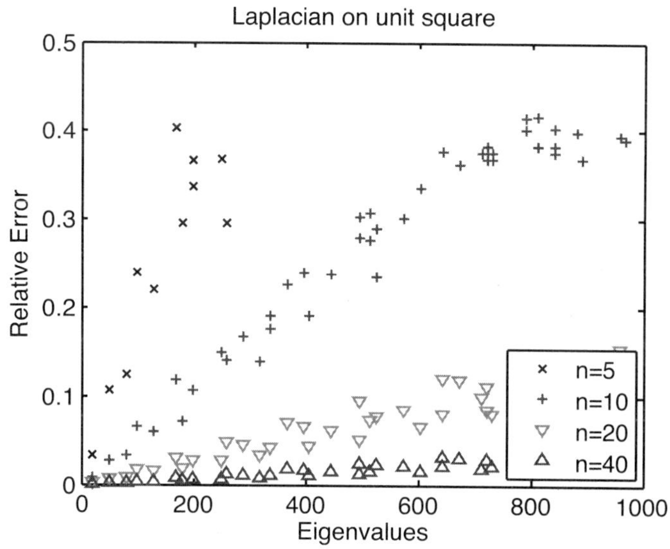

FIGURE 2. Relative error of the eigenvalues on the unit square obtained from a finite difference discretization for different mesh sizes.

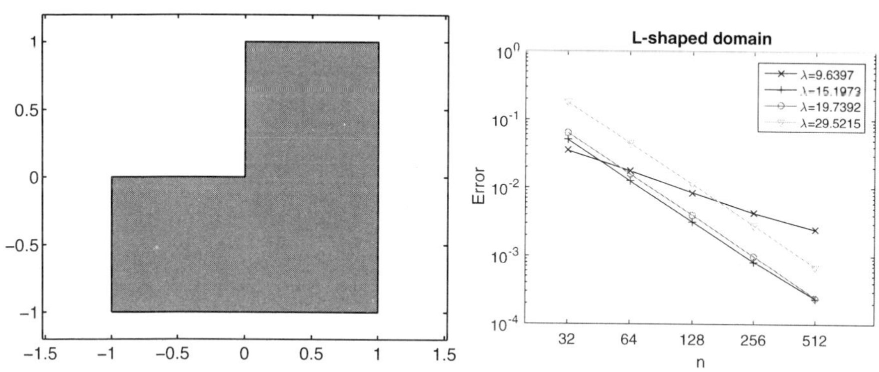

FIGURE 3. Left: an L-shaped domain. Right: Error of the first four eigenvalues as a function of the grid size n.

to take on the value zero. In Matlab, the matrix A can be generated conveniently using the commands

```
G = numgrid('L',n+1);
A = n^2/4*delsq(G);
```

We then calculate the four smallest eigenvalues of the resulting matrix for mesh sizes $h = 1/n$. Since we do not know the exact eigenvalues in the continuous case, we approximate them by *extrapolation*: assuming that each discrete eigenvalue behaves like $\tilde{\lambda}_{k\ell} = \lambda_{k\ell} + Ch^{\alpha}$, we use the ε algorithm by P. Wynn (cf. [**21**, §5.2.4]) to obtain its limit as $h \to 0$. We then subtract this value from the discrete eigenvalues to compute the error, which is plotted on the right panel of Figure 3. Moreover, we plot

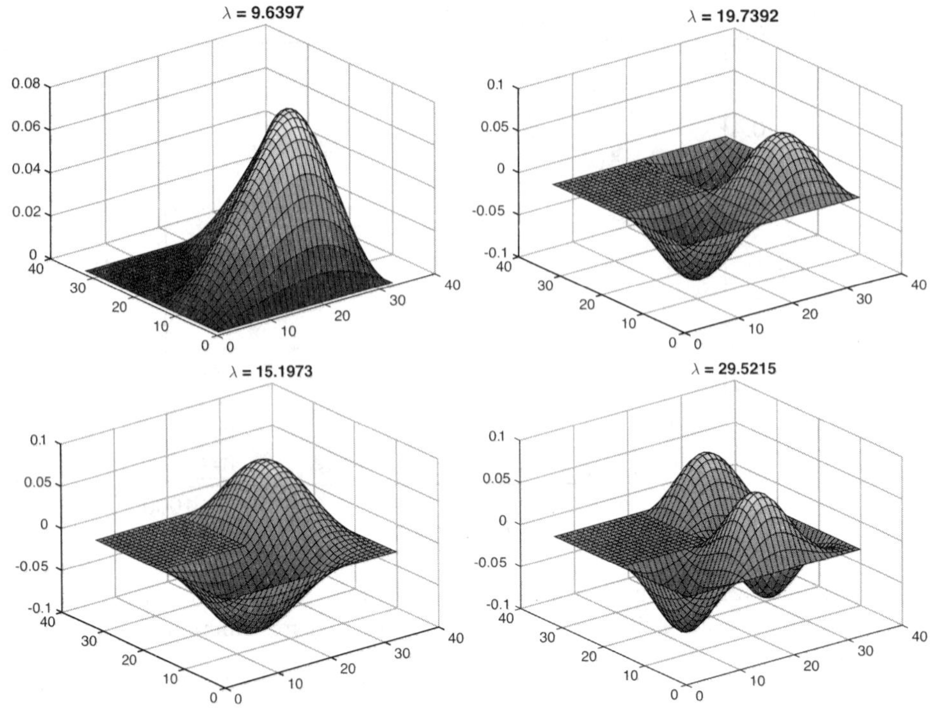

FIGURE 4. Eigenfunctions corresponding to the first four eigenvalues of the L-shaped domain.

the corresponding eigenfunctions in Figure 4. Note the resemblance between the first eigenfunction and the MATLAB logo. The latter is in fact an approximation of the eigenfunction, obtained using the *Method of Particular Solutions*; see Section 6 for more details and references.

Regarding the approximation errors, we observe that the first eigenvalue behaves differently from the other three: whereas the error curves for the other three have the same slope, and behaves like $O(h^2)$, the first eigenvalue converges more slowly, behaving more like $O(h^\alpha)$ with $1 < \alpha < 2$. In fact, the reason has to do with the regularity of the eigenfunction itself: whereas the other three eigenfunctions appear to be smooth, the first eigenfunction has a fold near the re-entrant corner and is not sufficiently differentiable near it. As a result, the Taylor expansion that was used to derive the finite difference method is no longer valid near the re-entrant corner, and the method suffers from a loss of accuracy as a result.

2.4. More general boundary conditions. Let us now consider the Laplace eigenvalue problem on the unit square $\Omega = (0,1) \times (0,1)$ with *Neumann boundary conditions*

$$-\Delta u = \lambda u, \qquad \nabla u \cdot \mathbf{n} = 0 \text{ on } \partial\Omega.$$

The main difference between this problem and the one with Dirichlet boundary conditions is that the nodes on the boundary no longer assume the value zero, but are instead unknown quantities that need to be solved, just like interior points. One possibility is to use ghost point techniques, such as the one introduced in

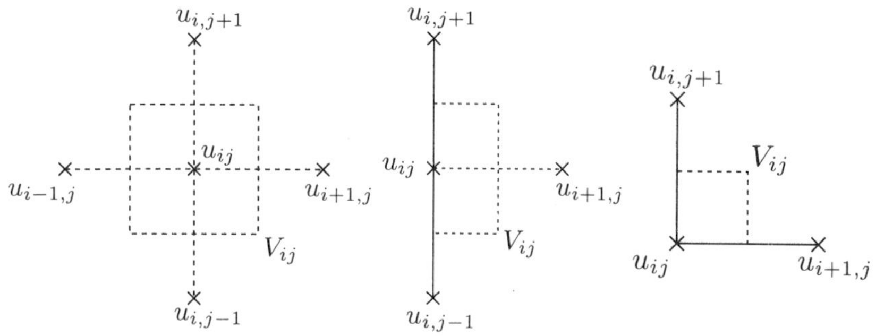

FIGURE 5. Control volume for an interior point (left panel), an edge point (middle panel), and a corner point (right panel).

Section 5.2; however, unless a special scaling is used, this approach leads to a non-symmetric matrix A, which is undesirable for a normal operator, especially from a numerical point of view, see Section 4. Here, we consider a different possibility, known as a *finite volume discretization*, to handle boundary conditions involving normal derivatives.

The idea of a finite volume discretization to simply integrate the equation $-\Delta u = \lambda u$ over a small *control volume* around a grid point u_{ij}. On a regular grid in two dimensions, the control volume V_{ij} for an interior point is simply an $h \times h$ square cell centered at (x_i, y_j), see the left panel of Figure 5. Integrating the left hand side over V_i and applying the divergence theorem gives

$$-\int_{V_{ij}} \Delta u \, dx = \int_{\partial V_{ij}} \nabla u \cdot \mathbf{n} \, dS(x)$$
$$\approx h \left[\frac{u_{ij} - u_{i+1,j}}{h} + \frac{u_{ij} - u_{i-1,j}}{h} + \frac{u_{ij} - u_{i,j+1}}{h} + \frac{u_{ij} - u_{i,j-1}}{h} \right],$$

where we approximate the integrals along each edge by the appropriate finite difference, multiplied by the length of the edge. The right hand side is simply approximated by

$$\int_{V_{ij}} \lambda u \, dx \approx \lambda h^2 u_{ij}.$$

Thus, we get the same equation as in the finite difference method, up to a multiplicative factor of h^2.

On the left boundary, we know that the normal derivative is zero, so there is one fewer term to approximate in the divergence theorem. Using the same notation as in the middle panel of Figure 5, we get for the left hand side

$$\int_{\partial V_{ij}} \nabla u \quad \mathbf{n} \, dS(x) \approx \frac{h}{2} \left[\frac{u_{ij} - u_{i,j+1}}{h} + \frac{u_{ij} - u_{i,j-1}}{h} \right] + h \cdot \frac{u_{ij} - u_{i+1,j}}{h},$$

whereas the right hand side reads

$$\int_{\partial V_{ij}} \lambda u \, dx \approx \frac{h^2}{2} u_{ij},$$

where the factor $1/2$ comes from the fact that the area of V_{ij} is only half of that of the other control volumes. The other edge points can be handled similarly. Finally, for the corner grid point shown on the right panel of Figure 5 , the stencil reads

$$\int_{\partial V_{ij}} \nabla u \cdot \mathbf{n}\, dS(x) \approx \frac{h}{2}\left[\frac{u_{ij}-u_{i+1,j}}{h} + \frac{u_{ij}-u_{i,j+1}}{h}\right],$$

and the right-hand side carries a factor of 1/4. Collecting all these equations and using the matrix notation, we obtain the *generalized eigenvalue problem*

$$Au = \tilde{\lambda} B\mathbf{u}, \tag{6}$$

where $B = \text{diag}(b_i)$ is diagonal with entries

$$b_i = \begin{cases} h^2, & \text{at interior points,} \\ h^2/2, & \text{at edge points,} \\ h^2/4, & \text{at corners points.} \end{cases}$$

Note that both A and B are symmetric matrices. Moreover, we can rewrite B as $B = D^2$, where $D = \text{diag}(\sqrt{b_i})$ is a non-singular diagonal matrix. Using the transformation $\mathbf{v} = D\mathbf{u}$, we obtain the equivalent eigenvalue problem

$$D^{-1}AD^{-1}\mathbf{v} = \tilde{\lambda}\mathbf{v}, \tag{7}$$

where the matrix $D^{-1}AD^{-1}$ remains symmetric. Note that although (7) has the same eigenvalues as (6), the eigenvectors are not identical; in fact, the entries of the eigenvectors $\mathbf{v}$ do not approximate the corresponding continuous eigenfunction at the grid points, because of the extra scaling by D. Fortunately, the Matlab built-in commands `eig` and `eigs` know how to solve generalized as well as standard eigenvalue problems, so users do not need to transform (6) into (7) manually.

3. Finite element methods

The finite element method is perhaps one of the most versatile and commonly used discretizations in the numerical solution of partial differential equations. It is based on a different formulation of the PDE, the so-called *weak* or *variational formulation*, which is obtained from integration by parts.

3.1. Variational formulation. Consider the following Laplace eigenvalue problem: find a scalar λ and a non-zero function u such that

$$-\Delta u = \lambda u \quad \text{on } \Omega, \qquad u|_{\partial\Omega} = 0. \tag{8}$$

To obtain the variational formulation, let v be another sufficiently smooth function on Ω such that $v|_{\partial\Omega} = 0$. Then multiplying (8) by v and integrating over Ω, we get

$$-\int_\Omega v\Delta u = \lambda \int_\Omega uv$$

$$\underbrace{-\int_{\partial\Omega} v\nabla u \cdot n}_{=0} + \int_\Omega \nabla u \cdot \nabla v = \lambda \int_\Omega uv, \tag{9}$$

where the first term vanishes because v is zero on the boundary. Note that (9) holds for many functions v; in fact, we can define a linear space of functions V such that (9) holds as long as $v \in V$. This function space V would include all v such that

(1) $\int_\Omega v^2 < \infty$ (RHS integral defined),
(2) $\int_\Omega |\nabla v|^2 < \infty$ (LHS integral defined),
(3) $v|_{\partial\Omega} = 0$ (boundary conditions satisfied).

The space of functions satisfying (1) and (2), i.e., functions whose values and partial derivatives are square integrable, is known as the Sobolev space $H^1(\Omega)$, see [**18**, Chapter 5]; if we require all partial derivatives up to order k to be square integrable, then the resulting space is known as $H^k(\Omega)$. The subspace of $H^1(\Omega)$ of functions that also satisfy (3) is called $H_0^1(\Omega)$, where the subscript zero indicates that functions in this space vanish on the boundary.[2] Thus, the *weak* or *variational form* of the eigenvalue problem is: find $u \in V = H_0^1(\Omega)$ such that

$$\text{(W)} \qquad a(u, v) = \lambda(u, v) \quad \text{for all } v \in V,$$

where $a(u, v) = \int_\Omega \nabla u \cdot \nabla v$ on the left is a bilinear form with domain $V \times V$, λ is the eigenvalue, and $(\cdot, \cdot)$ on the right denotes the standard L^2 inner product on Ω, i.e., $(u, v) = \int_\Omega uv$. The function space in which we look for solution candidates u is called the *trial space*, and the function space of all v is called the *test space*. In the above weak formulation, the test and trial spaces are the same, although this is not always required.

Variational formulations can also be derived for problems related to (8). The following are a few examples. We leave their derivations as an exercise.

1. Neumann boundary conditions: for the problem

$$-\Delta u = \lambda u \quad \text{on } \Omega, \qquad \nabla u \cdot n|_{\partial\Omega} = 0,$$

the weak form is: find $u \in H^1(\Omega)$ such that

$$a(u, v) = \lambda(u, v) \qquad \forall v \in H^1(\Omega).$$

Note that the only difference between this and the Dirichlet problem is in the trial and test spaces: for the Neumann problem, it is all of $H^1(\Omega)$, rather than $H_0^1(\Omega)$. Also note that the solution u of this problem automatically satisfies the Neumann boundary condition, and it is not necessary to include a restriction in the trial space. Such conditions are called *natural* boundary conditions. On the other hand, Dirichlet conditions need to be imposed explicitly in the trial and test spaces; these are known as *essential* boundary conditions.

2. Robin boundary conditions: for the problem

$$-\Delta u = \lambda u \quad \text{on } \Omega, \qquad \nabla u \cdot n + pu|_{\partial\Omega} = 0,$$

the weak form is: find $u \in H^1(\Omega)$ such that

$$\tilde{a}(u, v) = \lambda(u, v) \qquad \forall v \in H^1(\Omega)$$

where

$$\tilde{a}(u, v) = \int_\Omega \nabla u \cdot \nabla v + \int_{\partial\Omega} puv.$$

---

[2] In reality, the definitions of $H^1(\Omega)$ and $H_0^1(\Omega)$ involve weak derivatives and are somewhat more technical, but for the purpose of these notes, the informal definitions given above will suffice.

3. Variable diffusivity: for the problem

$$-\nabla \cdot (\kappa(\mathbf{x})\nabla u) = \lambda u \quad \text{on } \Omega, \qquad u|_{\partial\Omega} = 0,$$

the weak form is: find $u \in H_0^1(\Omega)$ such that

$$a_\kappa(u, v) = \lambda(u, v) \qquad \forall v \in H_0^1(\Omega)$$

where

$$a_\kappa(u, v) = \int_\Omega \kappa(\mathbf{x})\nabla u \cdot \nabla v.$$

3.2. Finite elements. To obtain a numerical method for the problem (W), we replace the infinite dimensional space $H_0^1(\Omega)$ by a *finite dimensional* subspace V_h; this is known as a *Ritz–Galerkin Approximation*. Thus, the problem becomes: Find $u_h \in V_h$ such that

$$(\mathrm{W}_h) \qquad a(u_h, v_h) = \lambda \int_\Omega u_h v_h \quad \text{for all } v_h \in V_h,$$

where $a(u, v) = \int_\Omega \nabla u \cdot \nabla v$. The above problem is finite dimensional and can thus be rewritten (and solved) as a matrix eigenvalue problem, which we will show in detail in the next section. Different *finite element methods* correspond to different choices of the subspace V_h; for a general reference on finite element methods, see [**27**].

EXAMPLE 3.1. Let $\Omega = (0, 1) \subset \mathbb{R}$ be equipped with the partition

$$0 = x_0 < x_1 < \cdots < x_n < x_{n+1} = 1.$$

If we choose V_h to be the set of continuous functions that is linear within each interval $I_j = (x_{j-1}, x_j)$, $j = 1, \dots, n+1$, then V_h is a finite dimensional space spanned by the "hat functions" $(\varphi_i)_{i=1}^n$, where

$$\varphi_i(x) = \begin{cases} \frac{x - x_{i-1}}{x_i - x_{i-1}}, & x \in [x_{i-1}, x_i], \\ \frac{x_{i+1} - x}{x_{i+1} - x_i}, & x \in [x_i, x_{i+1}], \\ 0 & \text{otherwise.} \end{cases}$$

Figure 6 shows a typical hat function. These hat functions are also known as P^1 finite element shape functions, because they are piecewise linear. Note that the degrees of freedom are located at the grid points $x_1, \dots, x_n$, and the derivative of any $\varphi_h \in V_h$ is a piecewise constant function. Higher order elements, e.g., piecewise quadratic, cubic, etc., are also possible.

EXAMPLE 3.2. Suppose $\Omega \subset \mathbb{R}^2$ is a 2D polygonal domain. Then it can be *triangulated*, i.e., decomposed into a union of disjoint triangles, which we denote by $\mathcal{T}_h$, see Figure 7 for an example. Then we can define a set of hat functions similarly to the one-dimensional case: at each node i of the triangulation, we associate the hat function ϕ_i that is linear on each triangle, whose value is 1 at node i and zero at all the other nodes. Not surprisingly, they are also called P^1 finite element shape functions, see Figure 8 for an illustration. These hat functions span V_h, the set of all continuous functions that is piecewise linear on each triangle in $\mathcal{T}_h$. V_h is again a finite dimensional space, and the Ritz–Galerkin problem of finding $u_h \in V_h$ such that

$$a(u_h, v_h) = \lambda \int_\Omega u_h v_h \qquad \text{for all } v_h \in V_h$$

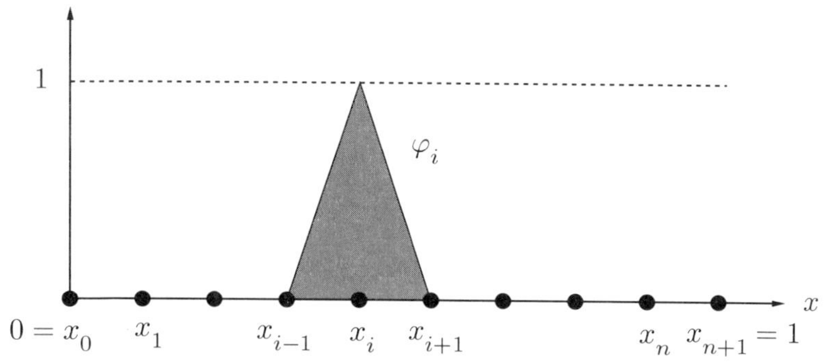

FIGURE 6. A hat function in 1D.

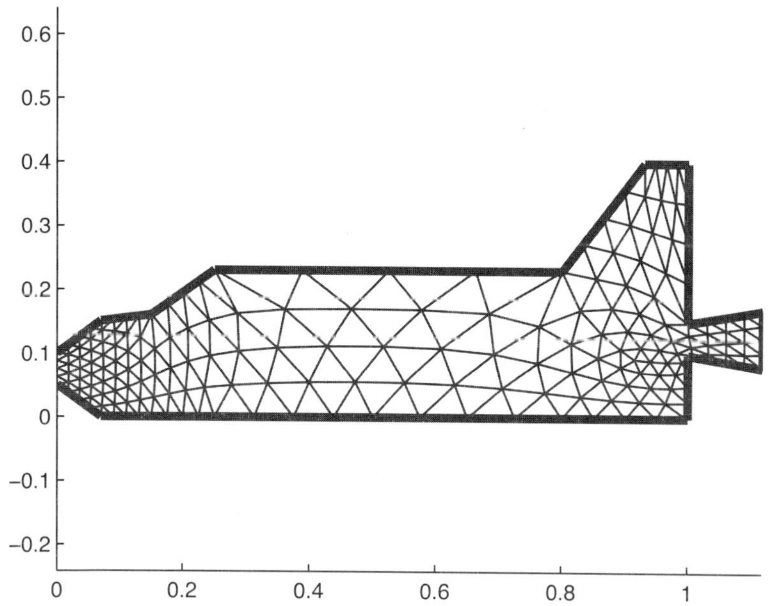

FIGURE 7. Triangulation of a polygonal domain.

can again be rewritten as a matrix eigenvalue problem, as we will show below.

3.3. Matrix problem. Let $V_h = \text{Span}\{\varphi_1, \ldots, \varphi_N\}$. Letting $u_h = \sum_j u_j \varphi_j$ and $v_h = \varphi_i$ in the weak form, we see that

$$\sum_j u_j a(\varphi_j, \varphi_i) = \lambda \sum_j u_j \int_\Omega \varphi_i \varphi_j, \qquad i = 1, \ldots, N.$$

Rewriting the above equation in matrix form yields the *generalized eigenvalue problem*

$$K\mathbf{u} = \lambda M \mathbf{u},$$

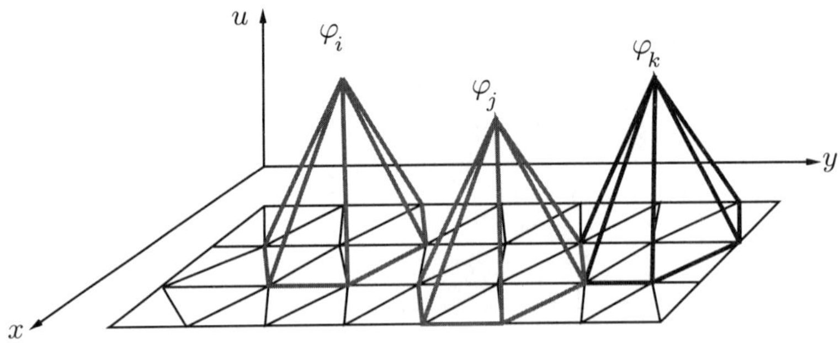

FIGURE 8. Hat functions in 2D.

where $K = (K_{ij})_{i,j=1}^N$ and $M = (M_{ij})_{i,j=1}^N$ with

$$K_{ij} = a(\varphi_i, \varphi_j) = \int_\Omega \nabla\varphi_i \cdot \nabla\varphi_j, \quad M_{ij} = \int_\Omega \phi_i\phi_j.$$

K is often called the *stiffness* matrix, and M the *mass* matrix. These matrices enjoy the following properties:

- M and K are symmetric;
- M is positive definite;
- K is positive definite for the Dirichlet and Robin problems, and is positive semi-definite for the Neumann problem.

Moreover, M and K are *sparse*. Recall the definition of the (i, j)-th entries of K and M:

$$K_{ij} = \int_\Omega \nabla\varphi_i \cdot \nabla\varphi_j, \quad M_{ij} = \int_\Omega \varphi_i\varphi_j.$$

If nodes i and j do not share a common triangle (or interval in 1D), then the support of φ_i and φ_j are disjoint. In this case, the corresponding integrals must vanish, so $K_{ij} = M_{ij} = 0$ there. In the 1D case, this means a degree of freedom can only be coupled to itself and its neighbours, so the matrices K and M are *tridiagonal* if the degrees of freedom are numbered contiguously from left to right.

EXAMPLE 3.3. For the Dirichlet problem in 1D, the hat functions with uniform mesh size h generate a mass matrix M with

$$M_{ij} = \begin{cases} 0, & |i-j| > 1, \\ \int_0^h h^{-2}x(h-x)\,dx = h/6, & |i-j| = 1, \\ 2\int_0^h h^{-2}x^2\,dx = 2h/3, & i = j. \end{cases}$$

In 2D, the matrices K and M can be *assembled* from individual contributions from each triangle. To do so, note that the integrals in the definition of K_{ij} and M_{ij} can be written as

$$K_{ij} = \sum_{T\in\mathcal{T}_h} \int_T \nabla\varphi_i \cdot \nabla\varphi_j, \quad M_{ij} = \sum_{T\in\mathcal{T}_h} \int_T \varphi_i\varphi_j.$$

Thus, if integrals of the form $\int_T \varphi_i\varphi_j$ and $\int_T \nabla\varphi_i \cdot \nabla\varphi_j$ are known for all triangles T, then it suffices to put these quantities in the right positions in the matrices K and M and add up all the contributions. On the other hand, for a fixed triangle T, the only non-zero integrals are those for which i and j are vertices of T; all the other integrals vanish, because T would be outside the support of φ_i or φ_j in that case. Thus, at most $12|\mathcal{T}_h|$ integrals are needed in order to compute K and M, where $|\mathcal{T}_h|$ is the number of triangles in the triangulation. This leads to the following *assembly process* for calculating K and M:

1. Define the mesh and number the nodes.
2. Calculate the *element* stiffness and mass matrices
$$[K_i]_{jk} = \int_{\mathcal{T}_i} \nabla\varphi_j \cdot \nabla\varphi_k.$$
3. Add the element matrices into the right places in the assembled matrices, i.e., if
$$K_i := \begin{bmatrix} (\nabla\varphi_j, \nabla\varphi_j)_{T_i} & (\nabla\varphi_j, \nabla\varphi_k)_{T_i} & (\nabla\varphi_j, \nabla\varphi_l)_{T_i} \\ & (\nabla\varphi_k, \nabla\varphi_k)_{T_i} & (\nabla\varphi_k, \nabla\varphi_l)_{T_i} \\ \text{Sym.} & & (\nabla\varphi_l, \nabla\varphi_l)_{T_i} \end{bmatrix} =: \begin{bmatrix} p_{11} & p_{12} & p_{13} \\ & p_{22} & p_{23} \\ \text{Sym.} & & p_{33} \end{bmatrix},$$
update the global stiffness matrix K by
$$K \leftarrow K + \begin{bmatrix} 0 & & & & & & 0 \\ & p_{11} & & p_{12} & & p_{13} & \\ & & 0 & & 0 & & \\ & p_{12} & & p_{22} & & p_{23} & \\ & & 0 & & 0 & & \\ & p_{13} & & p_{23} & & p_{33} & \\ 0 & & & & & & 0 \end{bmatrix} \begin{matrix} \\ j \\ \\ k \\ \\ l \\ \\ \end{matrix},$$
$$\qquad\qquad j \qquad\quad k \qquad\quad l$$
and similarly for M.
4. Repeat for all the other triangles.
5. Remove rows and columns associated with Dirichlet boundary nodes.

A sample finite element code is provided at `http://www.math.hkbu.edu.hk/~felix_kwok/crm/`. More information can be found in the Section 7.

3.4. Minimax principle. A powerful characterization of eigenvalues, in terms of extrema of certain expressions over subspaces, is known as the *minimax principle*. Let $\mathscr{L}$ be a self-adjoint, positive definite operator in a Hilbert space H, equipped with the inner product $(\cdot,\cdot)$, and let $\lambda_1 \le \lambda_2 \le \cdots \le \lambda_k \le \cdots$ be the eigenvalues of $\mathscr{L}$.

THEOREM 3.1 (Minimax principle). *The kth eigenvalue λ_k of $\mathscr{L}$ satisfies*

$$(*) \qquad \lambda_k = \min_{\dim(U)=k} \max_{\substack{u\in U \\ u\neq 0}} R(u),$$

where $R(u) := \dfrac{(u, \mathscr{L}u)}{\|u\|^2}$ *is called the* Rayleigh quotient.

PROOF. Let $u_1, u_2, \ldots$ denote the eigenvectors of $\mathscr{L}$ (for existence, see [**31**]). If $U = \mathrm{Span}(u_1, \ldots, u_k)$, then

$$\max_{\substack{u\subset U \\ u\neq 0}} R(u) = \frac{(u_k, \mathscr{L}u_k)}{\|u_k\|^2} = \frac{(u_k, \lambda_k u_k)}{\|u_k\|^2} = \lambda_k.$$

TABLE 1. Convergence of the first two eigenvalues of the Laplacian on the unit square, discretized using a P^1 finite element method. The ratios between errors in successive rows indicate $O(h^2)$ convergence.

h	$\tilde{\lambda}_1$	$\tilde{\lambda}_1 - \lambda_1$	Ratio	$\tilde{\lambda}_2$	$\tilde{\lambda}_2 - \lambda_2$	Ratio
1	22.8658	3.1266		62.5602	13.2122	
1/2	20.5055	0.7663	0.2451	52.6298	3.2818	0.2484
1/4	19.9298	0.1906	0.2487	50.1664	0.8184	0.2494
1/8	19.7868	0.0476	0.2497	49.5525	0.2045	0.2499
1/16	19.7511	0.0119	0.2499	49.3991	0.0511	0.2500

To show that this is the minimum over all choices of k-dimensional subspace, let us consider another $U = \text{Span}(w_1, \ldots, w_k)$, this time an arbitrary k-dimensional subspace. We choose $w = \alpha_1 w_1 + \cdots + \alpha_k w_k$ such that $w \neq 0$ and

$$(w, u_1) = \cdots = (w, u_{k-1}) = 0.$$

This is possible because this is in fact a system of $k-1$ equations in k unknowns, so a non-trivial solution always exists. Now since $u_1, u_2, \ldots$ is a basis of H, we can write

$$w = \sum_{i=k}^{\infty} \beta_i u_i,$$

where we used the fact that $\beta_1 = \cdots = \beta_{k-1} = 0$ because of the constraints. Then

$$(w, \mathscr{L}w) = \sum_{i=k}^{\infty} \beta_i^2 (u_i, \mathscr{L}u_i) \geq \lambda_k \sum_{i=k}^{\infty} \beta_i^2 = \lambda_k \|w\|^2.$$

Thus, $R(w) \geq \lambda_k$, so

$$\min_{\dim(U)=k} \max_{\substack{u \in U \\ u \neq 0}} R(u) = \lambda_k.$$

□

COROLLARY 3.2. *If $V_h \subset H_0^1(\Omega)$ is a finite element subspace, λ_k and $\tilde{\lambda}_k$ are the k-th smallest eigenvalues of the problems (W) and (W_h) respectively, then $\lambda_k \leq \tilde{\lambda}_k$.*

PROOF. Let $\tilde{u}_1, \ldots, \tilde{u}_k \in V_h \subset H_0^1(\Omega)$ be the eigenfunctions corresponding to the k smallest eigenvalues of the problem (W_h). Then $U = \text{Span}\{\tilde{u}_1, \ldots, \tilde{u}_k\}$ is a k-dimensional subspace of $H_0^1(\Omega)$ with $\max_{u \in U \setminus \{0\}} R(u) = \tilde{\lambda}_k$. Thus, by the minimax principle, we have $\lambda_k \leq \tilde{\lambda}_k$, as required. □

EXAMPLE 3.4. Consider the first two eigenvalues of unit square, $\lambda_1 = 2\pi^2$, $\lambda_2 = 5\pi^2$. In Table 1, we discretize the square using a finite element method with a regular triangulation. We successively refine the mesh by a factor of 2, so that the space V_h used in each row is a subspace of the V_h in the subsequent rows, and they are all subspaces of $H_0^1(\Omega)$. We see that all the $\tilde{\lambda}_i$ are larger than λ_i, and the $\tilde{\lambda}_i$ decrease as the mesh is refined. We also observe that the error behaves like $O(h^2)$, just like in the finite difference case. This will be proved in the next section.

REMARK 3.1. Corollary 3.2 shows that the finite element method always produces *over-estimations* of the exact eigenvalues. This is in contrast with finite difference methods in Chapter 1, which can produce approximations that are smaller than the exact eigenvalues, e.g. for the unit square, cf. Theorem 2.1. This is because finite difference approximations cannot be interpreted as the exact Laplacian applied to a subspace of the continuous functions, so the minimax principle does not apply.

3.5. Convergence of the finite element method. Let Ω be a bounded open subset of $\mathbb{R}^2$ and consider the Dirichlet eigenvalue problem

$$-\Delta u = \lambda u \quad \text{on } \Omega, \qquad u|_{\partial\Omega} = 0.$$

Let $0 < \lambda_1 \leq \lambda_2 \leq \cdots$ be the exact eigenvalues, and $0 < \tilde{\lambda}_1 \leq \tilde{\lambda}_2 \leq \cdots$ be the eigenvalues of the finite element approximation over the conforming, quasi-uniform mesh $\mathcal{T}_h$.[3] We define the mesh parameter h by

$$h = \max_{T \in \mathcal{T}_h} \operatorname{diam}(T),$$

i.e., h is the maximum diameter of all the triangles in $\mathcal{T}_h$. Our goal in this section is to prove the following theorem:

THEOREM 3.3. *Let $\Omega \subset \mathbb{R}^2$ be a convex polygon, $\mathcal{T}_h$ be a conforming, quasi-uniform triangulation of Ω, and V_h be the P^1 finite element space associated with $\mathcal{T}_h$. If $\tilde{u}_k$ is an eigenfunction corresponding to the k-th smallest eigenvalue $\tilde{\lambda}_k$ of the discrete problem*

$$a(\tilde{u}_k, v_h) = \tilde{\lambda}_k(\tilde{u}_k, v_h) \quad \forall v_h \in V_h,$$

then for $h > 0$ small enough, we have $|\tilde{\lambda}_k - \lambda_k| \leq C(k)h^2$, where $C(k)$ is a constant depending on k and Ω but independent of h. In particular, $\tilde{\lambda}_k$ converges to the exact eigenvalue λ_k as $h \to 0$.

To do so, we need to understand how well the finite element space V_h approximates the invariant subspaces of $-\Delta$, which are spanned by the eigenfunctions $u_1, u_2, \ldots$. This leads us to consider the following projection operator.

DEFINITION 3.1. The operator $P_h : H_0^1(\Omega) \to V_h$ is defined by

$$a(P_h u, v_h) = a(u, v_h) \quad \forall v_h \in V_h.$$

This operator is well defined because the stiffness matrix K is symmetric positive definite. The following properties of P_h can be readily verified:

(i) P_h is a *projection* onto V_h,
(ii) $a(u - P_h u, v_h) = 0$ for all $v_h \in V_h$,
(iii) $a(P_h u, P_h u) \leq a(u, u)$ for all $u \in H_0^1(\Omega)$.

The next step is to estimate the norm of $\|u - P_h u\|_{L^2(\Omega)}$. This will require the following tools:

1. A best approximation estimate,
2. An interpolation estimate,
3. A duality argument.

[3] For a precise definition, see [**13**].

1. Best approximation estimate. The best approximation result is due to the following lemma.

THEOREM 3.4 (Céa's lemma). *For all $w_h \in V_h$, we have*

$$a(u - P_h u, u - P_h u) \le a(u - w_h, u - w_h).$$

PROOF. For any $w_h \in V_h$, we have

$$\begin{aligned} a(u - w_h, u - w_h) &= a(u - P_h u + \underbrace{(P_h u - w_h)}_{\in V_h}, u - P_h u + (P_h u - w_h)) \\ &= a(u - P_h u, u - P_h u) + \underbrace{a(P_h u - w_h, P_h u - w_h)}_{\ge 0} \\ &\ge a(u - P_h u, u - P_h u). \end{aligned}$$

□

Since $a(u - P_h u, u - P_h u) = \|\nabla(u - P_h u)\|^2_{L^2(\Omega)}$, Céa's lemma says that the elliptic projector chooses from the subspace V_h the function u_h that minimizes the gradient of the error in the L^2 sense. On the other hand, the error can still be large if u is very far away from V_h. Thus, we need to quantify how well the function u is approximated by functions in V_h. This is known as an *interpolation estimate.*

2. Interpolation estimate. If u is a continuous function, then we define u_I to be its piecewise polynomial interpolant of degree r on each element:

$$u_I = \sum_j u(\mathbf{x}_j)\varphi_j,$$

where $\mathbf{x}_j$ denotes the j-th node in the grid. Note carefully that u_I is in general different from the $P_h u$ obtained from the elliptic projector: although both functions are in V_h, $P_h u$ is defined by a minimization problem and need not interpolate the continuous function exactly, whereas u_I is an interpolant that need not minimize anything. For functions in $H_0^1(\Omega)$ that are discontinuous, there are ways of defining a quasi-interpolant based on averaged Taylor polynomials. For details, see [**13**].

With the above definitions, we have the following general approximation result.

THEOREM 3.5 (Interpolation error, cf. [**41**, Lemma B.6]). *Let $\mathcal{T}_h$ be a conforming, shape-regular triangulation of $\Omega \subset \mathbb{R}^d$ with maximal element diameter h. Let $0 \le m \le s$ and $d/2 < s \le r+1$, where r is the degree of of the piecewise polynomial on each element. Then there exists a constant C depending only on m, s and the aspect ratio (i.e., the ratio of radii between the circumscribed and inscribed circles) of the elements in $\mathcal{T}_h$, such that*

$$\|D^m(u - u_I)\|_{L^2(\Omega)} \le C h^{s-m} \|D^s u\|_{L^2(\Omega)},$$

where D^k denotes the vector of all k-th order (weak) partial derivatives.

The proof of the above theorem in its full generality is rather technical and can be found in [**13**, Theorem 4.4.4]. Nonetheless, we are able to show the result for the simple one-dimensional case $\Omega = (0, 1)$ and piecewise linear interpolation as follows. In an interval $K = [x_{i-1}, x_i]$ of length h, we have

$$\begin{aligned}\|u' - u_I'\|_{L^2(K)}^2 &= \int_{x_{i-1}}^{x_i} (u' - u_I')^2\, d\xi \\ &= -\int_{x_{i-1}}^{x_i} (u - u_I)(u'' - u_I'')\, d\xi \le \|u - u_I\|_{L^2(K)}\, \|u''\|_{L^2(K)},\end{aligned}$$

where the boundary terms vanish because of the interpolation property $u(x_i) = u_I(x_i)$. On the other hand, since for all $x \in K$ we have

$$|u(x) - u_I(x)|^2 = \left| \int_{x_{i-1}}^{x} (u' - u_I')\, d\xi \right|^2 \le (x - x_{i-1}) \|u' - u_I'\|_{L^2(K)}^2,$$

we can estimate $\|u - u_I\|_{L^2(K)}$ by

$$\|u - u_I\|_{L^2(K)}^2 \le \int_{x_{i-1}}^{x_i} (\xi - x_{i-1}) \|u' - u_I'\|_{L^2(K)}^2\, d\xi = \frac{h^2}{2} \|u' - u_I'\|_{L^2(K)}^2.$$

Thus, summing over all intervals gives

$$\|u' - u_I'\|_{L^2(\Omega)} \le \frac{h}{\sqrt{2}} \|u''\|_{L^2(\Omega)}, \qquad \|u - u_I\|_{L^2(\Omega)} \le \frac{h^2}{2} \|u''\|_{L^2(\Omega)},$$

which is just a special case of Theorem 3.5 for $d = 1$, $s = 2$, $r = 1$ and $m = 0, 1$.

3. Duality argument. We are now ready to estimate $\|u - P_h u\|_{L^2(\Omega)}$ when Ω is a convex polygon. We know from elliptic regularity theory [**18**, Chapter 6] that if $f \in L^2(\Omega)$ and $a(u, v) = (f, v)$ for all $v \in H_0^1(\Omega)$, then $u \in H^2(\Omega)$, and

$$\|D^2 u\|_{L^2(\Omega)} \le C \|f\|_{L^2(\Omega)}.$$

Thus, if $u \in H_0^1(\Omega)$ is an eigenfunction with eigenvalue λ, then letting $f = \lambda u$ shows that in fact $u \in H^2(\Omega)$. Using the best approximation property and then the interpolation estimate, we get

$$a(u - P_h u, u - P_h u) \le \|\nabla(u - u_I)\|_{L^2(\Omega)}^2 \le C h^2 \|D^2 u\|_{L^2(\Omega)}^2 \le C' h^2 \|f\|_{L^2(\Omega)}^2.$$

To estimate $\|u - P_h u\|_{L^2}$, we use a duality argument known as the *Aubin–Nitsche* trick: consider the auxiliary problem

$$-\Delta \psi = u - P_h u, \qquad \psi|_{\partial\Omega} = 0, \tag{10}$$

or, in weak form,

$$a(\psi, v) = (u - P_h u, v) \quad \forall v \in H_0^1(\Omega).$$

Choosing $v = u - P_h u$ in the weak form yields

$$\begin{aligned}(u - P_h u, u - P_h u) &= a(u - P_h u, \psi) = a(u - P_h u, \psi - P_h \psi) \\ &\le a(u - P_h u, u - P_h u)^{1/2} a(\psi - P_h \psi, \psi - P_h \psi)^{1/2} \\ &\le \sqrt{CC'} h^2 \|u - P_h u\|_{L^2(\Omega)} \|D^2 u\|_{L^2(\Omega)},\end{aligned}$$

where we used the fact that ψ satisfies (10) to deduce that $a(\psi - P_h\psi, \psi - P_h\psi)^{1/2} \le \sqrt{C'} h \|u - P_h u\|_{L^2(\Omega)}$. Thus, we have the estimate

$$\|u - P_h u\|_{L^2(\Omega)} \le \sqrt{CC'} h^2 \|D^2 u\|_{L^2(\Omega)},$$

provided that Ω is convex, which is needed to guarantee that $D^2 u$ is in $L^2(\Omega)$. We are finally ready to prove Theorem 3.3.

PROOF OF THEOREM 3.3. Consider the k-dimensional subspaces

$$E_k = \text{Span}\{u_1, \dots, u_k\}, \quad \tilde{E}_k = P_h E_k,$$

where u_k is the eigenfunction of the continuous operator corresponding to the k-th smallest eigenvalue λ_k. For h small enough, we have for all non-zero $v \in E_k$,

$$\|P_h v\| \geq \|v\| - \|v - P_h v\| \geq (1 - C(k)h^2)\|v\| > 0.$$

Thus, P_h is an isomorphism between E_k and $\tilde{E}_k$, so both subspaces have dimension k. By the minimax principle, we have

$$\begin{aligned} \lambda_k \leq \tilde{\lambda}_k \leq \max_{\substack{v_h \in \tilde{E}_k \\ v_h \neq 0}} \frac{a(v_h, v_h)}{\|v_h\|^2} &= \max_{\substack{v \in E_k \\ v \neq 0}} \frac{a(P_h v, P_h v)}{\|P_h v\|^2} \\ \leq \max_{\substack{v \in E_k \\ v \neq 0}} \frac{a(v,v)}{\|P_h v\|^2} &\leq \underbrace{\max_{\substack{v \in E_k \\ v \neq 0}} \frac{a(v,v)}{\|v\|^2}}_{=\lambda_k} \cdot \max_{\substack{v \in E_k \\ v \neq 0}} \frac{\|v\|^2}{\|P_h v\|^2}. \end{aligned}$$

But since $\|P_h v\| \geq (1 - C(k)h^2)\|v\|$, we conclude that

$$\lambda_k \leq \tilde{\lambda}_k \leq \lambda_k(1 + 2C(k)h^2) + O(h^4),$$

which implies $|\tilde{\lambda}_k - \lambda_k| \leq \tilde{C}(k)h^2 \to 0$ as $h \to 0$. □

Remarks:

1. If the eigenfunction u is not smooth enough, i.e., if it does not belong to $H^2(\Omega)$, like in the case for the L-shaped domain, we still have in general [**35**]

$$\tilde{\lambda}_k \leq \lambda_k \Big(1 + C(k) \sup_{\substack{v \in E_k \\ \|v\|=1}} \|v - P_h v\|^2_{H^1(\Omega)}\Big).$$

However, because of the lack of regularity, the contraction rates will in general be of the form $O(h^\alpha)$ with $\alpha < 2$, i.e., convergence will be worse than in the regular case. In order to recover higher order convergence, various methods can be used, such as adaptive grid refinement, see [**17**] and the references therein.

2. For the convergence of eigenvectors, see [**11**].

4. Solution of matrix eigenvalue problems

We have seen in Sections 2 and 3 that the we can approximate PDE eigenvalue problems by algebraic eigenvalue problems by discretizing the differential operator in various ways. It remains to solve the algebraic eigenvalue problems to obtain the actual eigenvalues and eigenvectors. In theory, the eigenvalues of a matrix A are given by the roots of its characteristic polynomial $p_A(\lambda) = \det(A - \lambda I)$. It would be tempting to calculate this polynomial explicitly and find its roots using a numerical method.[4] Unfortunately, this is a terrible idea because the roots of $p_A(\lambda)$ are extremely sensitive to small perturbations in the coefficients, as one can see in the example below.

[4]From Galois theory, we know there is no explicit formula for the roots when A is a 5×5 matrix or larger! Fortunately, efficient numerical methods exist, such as `roots` in Matlab, which relies on calculating the eigenvalues of the companion matrix. See also [**4**] for a recent improvement.

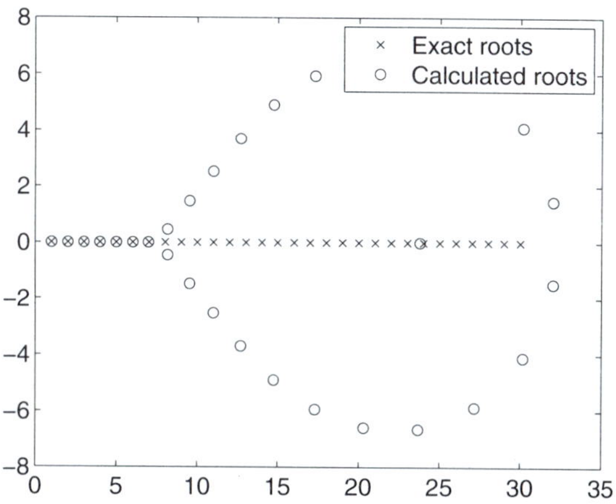

FIGURE 9. The exact and numerically calculated roots of a polynomial of degree 30.

EXAMPLE 4.1. Consider the polynomial[5] $p(\lambda) = (\lambda - 1)(\lambda - 2)\cdots(\lambda - 30)$. We compute the coefficients of this polynomial in Matlab, and then ask for the roots of this polynomial using the following commands:

```
A = diag((1:30),0);          % Matrix with 1,2,...,30 on the diagonal
p = poly(A);                 % Characteristic polynomial of A
lambda = roots(p);           % Roots of p
```

We plot the roots `lambda` thus obtained in Figure 9. We see that the numerically calculated roots are very far from the exact ones, even though Matlab computes with 16 digits of accuracy. The reason is that the largest coefficient in `p` is approximately 2×10^{33}, so even a perturbation in the 16th significant digit can introduce a huge error in the calculated roots, as illustrated by the figure.

Thus, numerical algorithms for finding matrix eigenvalues never form the characteristic polynomial, but instead attempt to transform the matrix directly into diagonal form using a sequence of similarity transformations. There are different methods that are commonly used, depending on the properties of the matrix A and how many eigenvalues we seek:

- The *QR iteration* is generally used when we seek *all* eigenvalues of a *small* and/or *dense* matrix:
- If we seek only a *few* eigenvalues of a dense or sparse matrix, commonly used methods include the power method, shift-and-invert, bisection, etc.
- For large *sparse* matrices from which only a few eigenvalues are sought, methods such as Lanczos and Jacobi-Davidson may be more efficient.

In Matlab, there are two different functions for finding eigenvalues and eigenvectors of a matrix: one uses `eig` to find all eigenvalues of a dense matrix, whereas one uses `eigs` to find a few eigenvectors of a sparse matrix. Before discussing a few of the

[5]This is adapted from Wilkinson's example in [**44**], where we have changed the degree from 20 to 30 to make it work in double-precision arithmetic.

aforementioned algorithms, we give a quick reminder of the properties of matrix eigenvalue problems:

- In an eigenvalue problem, one seeks $\mathbf{v} \neq 0$ such that $A\mathbf{v} = \lambda\mathbf{v}$, where A is an $n \times n$ matrix.
- If A is *diagonalizable*, this is equivalent to finding a *diagonal* matrix Λ and a *non-singular* V such that
$$AV = V\Lambda \iff A = V\Lambda V^{-1}.$$
- For *symmetric* A, V can be taken to be *orthogonal*, i.e., $V = Q$, $Q^TQ = QQ^T = I$.

A related problem is the *generalized* eigenvalue problem: solve $A\mathbf{v} = \lambda B\mathbf{v}$, $A, B =$ $n \times n$ matrices. If B is symmetric and positive definite, then this is equivalent to

$$G^{-T}AG^{-1}\mathbf{z} = \lambda\mathbf{z},$$

where $B = G^TG$ is the Cholesky (or any other related) factorization, and $\mathbf{z} = G\mathbf{v}$. Thus, any method for finding the eigenvalues of symmetric A can be used on $G^{-T}AG^{-1}$, but there are often more efficient and numerically stable methods that do not require computing $G^{-T}AG^{-1}$ explicitly. In these notes, we will concentrate on the case of symmetric matrices. For the non-symmetric case, see [**23**].

4.1. Power iterations. Possibly the simplest method for calculating an eigenvalue of a matrix A is *the power method.* Given an initial vector $\mathbf{x}^{(0)}$, the method produces an estimate $\lambda^{(k)}$ and a new vector $\mathbf{x}^{(k)}$ for $k = 1, 2, 3, \ldots$:

1. $\mathbf{x}^{(k)} = A\mathbf{x}^{(k-1)}$
2. $\lambda^{(k)} = \dfrac{(\mathbf{x}^{(k)})^TA\mathbf{x}^{(k)}}{\|\mathbf{x}^{(k)}\|_2^2}$

In finite precision arithmetic, the first step is usually replaced by

$$\mathbf{x}^{(k)} = \frac{A\mathbf{x}^{(k-1)}}{\|A\mathbf{x}^{(k-1)}\|_2}$$

in order to avoid overflow. If $\mathbf{x}^{(k)}$, after suitable normalization, is a fixed point of the iteration, then it is clearly an eigenvector of A with eigenvalue $\lambda^{(k)}$. Even when $\mathbf{x}^{(k)}$ is not a fixed point, this definition of $\lambda^{(k)}$ minimizes $\|A\mathbf{x}^{(k)} - \lambda\mathbf{x}^{(k)}\|_2$ over all values of λ, so the formula returns the best possible eigenvalue approximation in the L^2 sense.

To understand the behavior of the method, suppose A is diagonalizable with eigenvectors $\mathbf{v}_i$, ordered such that $|\lambda_n| > |\lambda_{n-1}| \geq \cdots \geq |\lambda_1|$, where $A\mathbf{v}_i = \lambda_i\mathbf{v}_i$. Then if $\mathbf{x}^{(0)} = \sum_i \alpha_i\mathbf{v}_i$, the power iteration generates

$$\mathbf{x}^{(k)} = \frac{\sum_i \alpha_i\lambda_i^k\mathbf{v}_i}{\|\sum_i \alpha_i\lambda_i^k\mathbf{v}_i\|_2} = \alpha_n\mathbf{v}_n + O\left(\left|\frac{\lambda_{n-1}}{\lambda_n}\right|\right)^k,$$

so $\lambda^{(k)}$ converges to the largest eigenvalue in magnitude λ_n provided $\alpha_n \neq 0$, with the estimate

$$\lambda^{(k)} = \lambda_n + O\left(\left|\frac{\lambda_{n-1}}{\lambda_n}\right|\right)^k.$$

If the matrix A is symmetric, the exponent can be improved to $2k$ from k using the fact that the eigenvectors form an orthonormal basis.

The power method admits a few variants that can be quite useful for calculating a single eigenvalue. If one would like to calculate the smallest rather than the largest eigenvalue in magnitude, then one would use the *inverse power method*

$$\mathbf{x}^{(k)} = A^{-1}\mathbf{x}^{(k-1)},$$

with the approximated eigenvalue $\mathbf{x}^{(k)}$ calculated the same way as in the power method. Similarly, if we wish to find the eigenvalue closest to a given value μ, then one would use the *shifted inverse iteration*, or the *shift-and-invert method*

$$\mathbf{x}^{(k)} = (A-\mu I)^{-1}\mathbf{x}^{(k-1)}.$$

Of course, one does not calculate the matrix $(A-\mu I)^{-1}$ explicitly in practice; one would instead pre-calculate the LU factorization of $A-\mu I$ and solve at each iteration the linear system

$$(A-\mu I)\mathbf{x}^{(k)} = \mathbf{x}^{(k-1)}$$

using this factorization. This presents significant savings in terms of computational cost, especially when A is a sparse matrix.

EXAMPLE 4.2. Let

$$A = \begin{bmatrix} 4 & 1 & & \\ 1 & 3 & 1 & \\ & 1 & 2 & 1 \\ & & 1 & 1 \end{bmatrix},$$

whose eigenvalues are

$$\frac{1}{2}\left(5+\sqrt{11\pm 2\sqrt{21}}\right) \approx \{0.25471876, 1.8227171, 3.1772829, 4.7452812\}.$$

We apply the power, inverse power and shift-and-invert methods to A, with $\mu = 3$ for the last method. Then we know that the three methods converge to the eigenvalues $\lambda_4 \approx 4.7453$, $\lambda_1 \approx 0.2548$ and $\lambda_3 \approx 3.1773$ respectively. Figure 10 shows the convergence of the different methods. We see that the error of the power method decreases from about 2×10^{-1} at iteration 1 to about 2×10^{-9} at iteration 24, so its contraction factor is approximately

$$\left(\frac{2\times 10^{-9}}{2\times 10^{-1}}\right)^{1/23} = 0.4489 \approx 0.4484 = \left(\frac{3.1773}{4.7453}\right)^2.$$

This nicely matches the stated convergence estimate. Similarly, for the inverse power method and the shift-and-invert method, we obtain as contraction factors

$$\text{InvPower:} \quad \left(\frac{1.5\times 10^{-8}}{4\times 10^{-1}}\right)^{1/4} = 0.0139 \approx 0.0195 = \left(\frac{0.2548}{1.8227}\right)^2,$$

$$\text{ShiftInvert:} \quad \left(\frac{5\times 10^{-9}}{4\times 10^{-2}}\right)^{1/4} = 0.0188 \approx 0.0227 = \left|\frac{\mu-3.1773}{\mu-1.8227}\right|^2.$$

Note that the ratios are squared in all cases because A is symmetric.

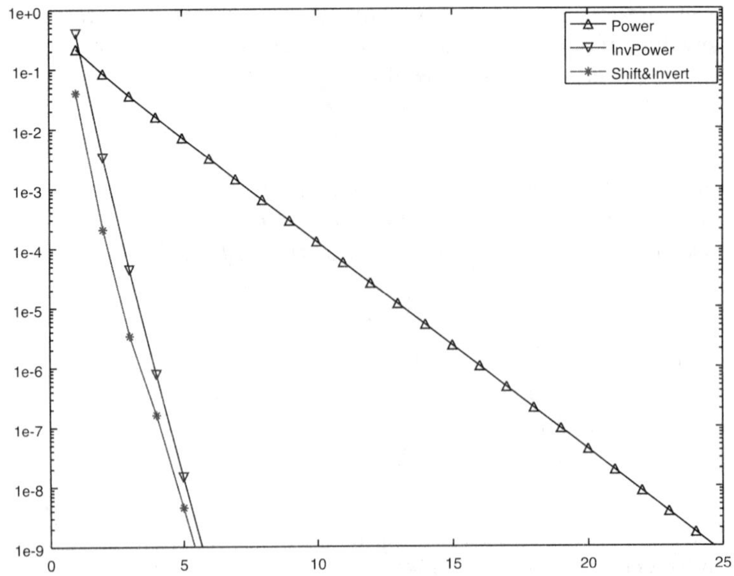

FIGURE 10. Convergence of the power method and its variants for the matrix in Example 4.2.

4.2. Orthogonal iterations and the QR method. If one wishes to calculate several eigenvalues and eigenvectors at a time, one needs to modify the power method to start with several initial vectors. However, in order to prevent those vectors from converging to the same eigenvector, we need to ensure that they remain linearly independent, which is accomplished by *orthogonalization*. This leads to the so-called *orthogonal iteration*: in the case of two starting vectors, we define the $n \times 2$ matrix $U_0 = [\mathbf{u}_1^{(0)}, \mathbf{u}_2^{(0)}]$ and perform the following iteration for $k = 0, 1, 2, \ldots$:

1. Calculate $Z_k = AU_k$;
2. Factor Z_k into $Z_k = U_{k+1}Y_{k+1}$, where U_{k+1} has orthonormal columns and Y_{k+1} is upper triangular (QR decomposition).

The factorization in step 2 can be computed e.g. by the Gram-Schmidt process. The eigenvalues are approximated by the diagonal entries of

$$T_k = U_k^T A U_k = \begin{bmatrix} (\mathbf{u}_1^{(k)})^T A\mathbf{u}_1^{(k)} & (\mathbf{u}_1^{(k)})^T A\mathbf{u}_2^{(k)} \\ (\mathbf{u}_2^{(k)})^T A\mathbf{u}_1^{(k)} & (\mathbf{u}_2^{(k)})^T A\mathbf{u}_2^{(k)} \end{bmatrix}.$$

Note that the first column $\mathbf{u}_1^{(k)}$ of U_k is identical to the normalized vector $\mathbf{x}^{(k)}$ in the power method, so $(\mathbf{u}_1^{(k)})^T A\mathbf{u}_1^{(k)}$ converges to the largest eigenvalue, assuming that $\mathbf{u}_1$ has a non-zero component along the corresponding eigenvector. The second column, however, cannot converge to same eigenvector, since it is always kept orthogonal from the first column. In fact, $(\mathbf{u}_2^{(k)})^T A\mathbf{u}_2^{(k)}$ converges to the second largest eigenvalue, under some mild assumptions on U_0 and A. Moreover, the off-diagonal entries of T_k can be shown to converge to zero. The method can be extended to any number of starting vectors, as long as they are linearly independent; in this case, it can be shown under certain assumptions (see [**2**, Theorem

10.6.1] for the n-vector case) that T_k tends towards a diagonal matrix containing the eigenvalues of A.

For the special case of $U_0 = I$ (i.e., n vectors), it is possible to derive a recurrence based on T_k only, leading to the well-known *QR iteration*. Letting $T_0 = A$, we have the very simple two-step iteration for $k = 0, 1, 2, \ldots$:

1. Factor T_k into $T_k = Q_k R_k$, where Q_k is orthogonal and R_k is upper triangular (QR factorization);
2. Compute $T_{k+1} = R_k Q_k$.

We claim that this is equivalent to the orthogonal iteration with $T_k = U_k^T A U_k$. Indeed, since $U_0 = I$, this is satisfied for $k = 0$. We now proceed inductively by assuming the result for k and proving it for $k+1$. Note that T_k can be written as

$$T_k = U_k^T Z_k = U_k^T U_{k+1} Y_{k+1}.$$

Since $U_k^T U_{k+1}$ is orthogonal and Y_{k+1} is upper triangular, this gives a valid QR factorization of T_k into $Q_k = U_k^T U_{k+1}$ and $R_k = Y_{k+1}$. So the recurrence for T_{k+1} leads to

$$T_{k+1} = R_k Q_k = Y_{k+1} U_k^T U_{k+1} = (U_{k+1}^T Z_k) U_k^T U_{k+1} = U_{k+1}^T A U_{k+1},$$

which completes the induction.

The basic QR method inherits the basic properties of the power method, meaning that the convergence of eigenvalues is linear, with the rate dependent on the gap between eigenvalues, see Example 4.2. Moreover, each iteration requires a QR factorization, which requires $O(n^3)$ arithmetic operations. In order to make the method more efficient and converge faster, practical QR implementations typically use the following tricks:

- A preprocessing step reduces a general (symmetric) matrix A into tridiagonal form, i.e., one finds an orthogonal matrix Q_0 such that $T = Q_0^T A Q_0$ is tridiagonal. This reduces the cost of each QR step to $O(n)$ instead of $O(n^3)$. Note that T and A have the same eigenvalues, and eigenvectors can be recovered from multiplication by Q_0.
- A shift of the form $A - \mu I$ with μ close to an eigenvalue is often applied explicitly or implicitly in order to speed up convergence. This works because the QR step decouples A into diagonal blocks when A is singular. Once the blocks are decoupled, a divide and conquer strategy can be used to solve the smaller subproblems that arise.

Reduction to tridiagonal form. To reduce a symmetric matrix A into tridiagonal form, we use *Householder transformations*, which are orthogonal matrices of the form

$$H = I - 2\mathbf{v}\mathbf{v}^T$$

with $\|\mathbf{v}\|_2 = 1$. The vector $\mathbf{v}$ is called a *Householder vector*. The first step in the tridiagonal reduction is to find a Householder vector of the form

$$\mathbf{v} = (0, v_2, \ldots, v_n)^T,$$

such that H applied to the first column $\mathbf{a}_1$ of A is non-zero only in its first two entries. In other words, we seek H such that $H\mathbf{a}_1 = a_{11}\mathbf{e}_1 + \alpha\mathbf{e}_2$, where $\mathbf{e}_1$ and $\mathbf{e}_2$ are the first two columns of the identity matrix, respectively. (Note that the first coefficient of $H\mathbf{a}_1$ must be a_{11} because $\langle \mathbf{e}_1 \rangle$ and $\langle \mathbf{e}_1 \rangle^\perp$ are both invariant subspaces of H.) Using the fact that $H^2 = I$, we calculate

$$\mathbf{a}_1 = H(H\mathbf{a}_1) = (I - 2\mathbf{v}\mathbf{v}^T)(a_{11}\mathbf{e}_1 + \alpha\mathbf{e}_2) = a_{11}\mathbf{e}_1 + \alpha(\mathbf{e}_2 - 2v_2\mathbf{v}).$$

Since H preserves 2-norms, we conclude that $\alpha = \pm\|\mathbf{a}_1 - a_{11}\mathbf{e}_1\|_2$, so that $\mathbf{v}$ can be calculated by

$$\mathbf{w} = \mathbf{e}_2 \pm \frac{\mathbf{a}_1 - a_{11}\mathbf{e}_1}{\|\mathbf{a}_1 - a_{11}\mathbf{e}_1\|_2}, \qquad \mathbf{v} = \frac{\mathbf{w}}{\|\mathbf{w}\|_2},$$

where the plus or minus is chosen so that $w_2 > 1$, which ensures numerical stability. Applying H to the whole matrix A in the sequence $A \to HA \to HAH^T$ leads to the transformation

$$\begin{bmatrix} X & X & X & X & X \\ X & X & X & X & X \\ X & X & X & X & X \\ X & X & X & X & X \\ X & X & X & X & X \end{bmatrix} \to \begin{bmatrix} X & X & X & X & X \\ Y & Y & Y & Y & Y \\ 0 & Y & Y & Y & Y \\ 0 & Y & Y & Y & Y \\ 0 & Y & Y & Y & Y \end{bmatrix} \to \begin{bmatrix} X & Y & 0 & 0 & 0 \\ Y & Z & Z & Z & Z \\ 0 & Z & Z & Z & Z \\ 0 & Z & Z & Z & Z \\ 0 & Z & Z & Z & Z \end{bmatrix},$$

where we use a different letter each time an entry is modified by a transformation. Note that the zeros in the first column appear because of the choice of $\mathbf{v}$, and the zeros in the first row appear because of symmetry. We can now apply the same reduction recursively to the lower part (the 'Z' part) of the matrix, and after $n-2$ steps, we obtain a matrix in tridiagonal form.

QR with shift. To accelerate convergence of the QR method, we apply a shift μ_k that is close to an eigenvalue of A, often chosen to be the bottom rightmost entry of latest iterate T_k. The modified algorithm is to perform, for $k = 0, 1, 2, \ldots$, the following steps:

1. Factor $T_k - \mu_k I = Q_k R_k$ (QR factorization);
2. Compute $T_{k+1} = R_k Q_k + \mu_k I$.

EXAMPLE 4.3. Consider the 2×2 matrix

$$T_0 = A = \begin{bmatrix} 2 & \epsilon \\ \epsilon & 1 \end{bmatrix}.$$

One step of the QR with no shift gives

$$Q = \frac{1}{\sqrt{4+\epsilon^2}} \begin{bmatrix} 2 & -\epsilon \\ \epsilon & 2 \end{bmatrix}, \quad T_1 = Q^T A Q = \begin{bmatrix} * & \text{sym.} \\ \frac{\epsilon(2-\epsilon^2)}{4+\epsilon^2} & * \end{bmatrix}.$$

We see that the off-diagonal entry is multiplied approximately by $1/2$ when ϵ is small, so the method converges approximately linearly with factor $1/2$. On the other hand, if we use a shift of $\mu_0 = 1$, we get

$$Q = \frac{1}{\sqrt{1+\epsilon^2}} \begin{bmatrix} 1 & -\epsilon \\ \epsilon & 1 \end{bmatrix}, \quad T_1 = Q^T A Q = \begin{bmatrix} * & \text{sym.} \\ -\frac{\epsilon^3}{1+\epsilon^2} & * \end{bmatrix}.$$

The off-diagonal entry is now $O(\epsilon^3)$, which indicates *cubic convergence*. Thus, if ϵ is small (say ≈ 0.1), the shifted version converges much faster than the basic version. The same observation can be made for larger matrices, see Figure 11 for the convergence of off-diagonal entries for the 4×4 matrix in Example 4.2.

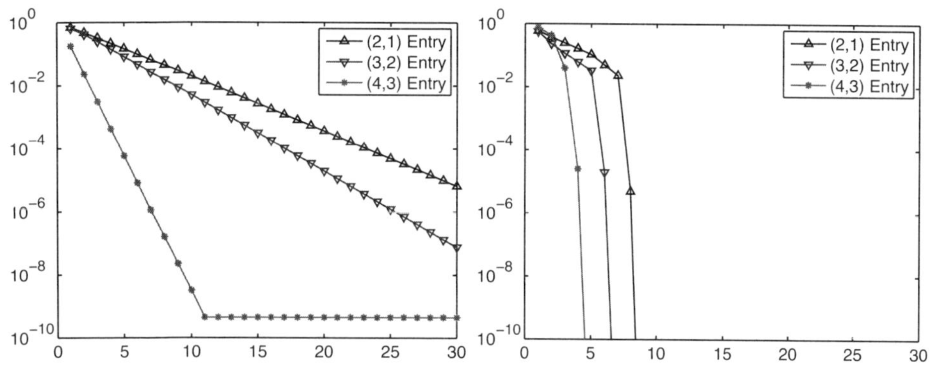

FIGURE 11. Convergence of off-diagonal entries for the 4×4 matrix in Example 4.2. Left: Basic (unshifted) QR iteration. Right: QR method with shift.

Remark. The QR iteration (or, more precisely, its practical variants) is the workhorse for small and/or dense eigenvalue problems, with robust and efficient implementations widely available, e.g. in LAPACK. For more details, see [**21**, **23**, **32**].

4.3. Bisection. In this section, we introduce the *bisecton* method, which gives a different way of computing eigenvalues of a symmetric matrix in tridiagonal form. In addition to approximating specific eigenvalues, this method has the interesting property of revealing the number of eigenvalues that lie within a given interval (λ_-, λ_+); this can be useful in engineering design, for instance, where one would need to ensure a structure has no resonant frequencies within a given range.

We start by considering the $n \times n$ tridiagonal matrix

$$T_n = \begin{bmatrix} \alpha_1 & \beta_1 & & & \\ \beta_1 & \alpha_2 & \ddots & & 0 \\ & \ddots & \ddots & \ddots & \\ 0 & & \ddots & \ddots & \beta_{n-1} \\ & & & \beta_{n-1} & \alpha_n \end{bmatrix}.$$

as well as its *principal submatrices* $T_j = T_n(1:j, 1:j)$, $j = 1, \ldots, n-1$. Such a family of tridiagonal matrices enjoy the following properties:

1. The characteristic polynomials $p_j(\lambda)$ of T_j satisfy the recurrence

$$p_j(\lambda) = (\alpha_j - \lambda)p_{j-1}(\lambda) - \beta_{j-1}^2 p_{j-2}(\lambda), \quad p_0(\lambda) = 1. \tag{11}$$

Proof: Direct expansion of $\det(T_n - \lambda I_n)$*, starting with the bottom row.*

2. If $\lambda_1^{(j)} \le \cdots \le \lambda_j^{(j)}$ are the eigenvalues of T_j for $j = 1, \ldots, n$, then we have the following *interlacing property*:

$$\lambda_1^{(j+1)} \le \lambda_1^{(j)} \le \lambda_2^{(j+1)} \le \cdots \le \lambda_k^{(j)} \le \lambda_{k+1}^{(j+1)}.$$

Proof: Use the minimax principle.

3. T_n has no repeated eigenvalues if it is *unreduced*, i.e., if $\beta_i \neq 0$ for all i. *Proof: If repeated eigenvalues exist, then p_n and p_{n-1} have a common linear factor because of the interlacing property, which means p_{n-1} and p_{n-2} have a common linear factor, etc., leading to a contradiction because $p_0 \equiv 1$ has no linear factors.*

Suppose we wish to calculate $\lambda_k^{(n)}$, the k-th eigenvalue of the original matrix T_n. The following theorem will be useful for determining where this occurs in the spectrum.

THEOREM 4.1 (Sturm sequence property). *Let T_n be a symmetric, unreduced tridiagonal matrix, and let $\omega(\lambda)$ be the number of sign changes in the Sturm sequence $(p_0(\lambda), p_1(\lambda), \dots, p_n(\lambda))$. Then $\omega(\lambda)$ equals the number of eigenvalues of T_n less than λ.*

The proof uses the recurrence (11) and a counting argument, and can be found in [**45**, pp. 300–301]. Since the Sturm sequence is easy to evaluate for a given λ using the recurrence (11), an approximation of $\lambda_k^{(n)}$ can be computed as follows:

Bisection method for finding $\lambda_k^{(n)}$:

1. Find y and z such that $y < \lambda_k^{(n)} < z$ using e.g. Gershgorin's theorem, cf. Section 7.3, Exercise 3.
2. While $|z - y| > tol$, do
   - Compute $x = (y + z)/2$.
   - If $\omega(x) \geq k$, set $z := x$; else set $y := x$.

Remarks.

1. In order to avoid overflow when n is large, it is sometimes preferable to adapt (11) to compute $\varphi_i(\lambda) := p_i(\lambda)/p_{i-1}(\lambda)$ rather than the $p_i(\lambda)$ themselves. In that case, one would count the number of negative values of $\varphi_i(\lambda)$ instead of the number of sign changes.
2. The bisection method converges approximately linearly with a factor $1/2$, since it halves the interval that containing $\lambda_k^{(n)}$ at every step. Once the interval is small enough to contain a single eigenvalue, it is possible to switch to another method (e.g., shift-and-invert) to obtain faster convergence to $\lambda_k^{(n)}$.

4.4. Lanczos method. For large sparse matrices, the Householder tridiagonalization procedure described in Section 4.2 requires too many operations and becomes impractical. A different way of generating a tridiagonal matrix from which eigenvalues can be calculated is based on the idea of Krylov subspaces. Given a matrix A and an initial vector $\mathbf{q}_1$, the *Krylov subspace* $\mathcal{K}_r(A; \mathbf{q}_1)$ is defined as

$$\mathcal{K}_r(A; \mathbf{q}_1) = \text{Span}\{\mathbf{q}_1, A\mathbf{q}_1, \dots, A^r\mathbf{q}_1\}.$$

From this definition, we see that every element of $\mathcal{K}_r(A; \mathbf{q}_1)$ can be written as $P_r(A)\mathbf{q}_1$ for some polynomial $P_r(x)$ of degree at most r. Moreover, we have $\mathcal{K}_r(A; \mathbf{q}_1) \subset \mathcal{K}_{r+1}(A; \mathbf{q}_1)$. If they are equal, then $\mathcal{K}_r(A; \mathbf{q}_1)$ is an *invariant subspace* of A.

As we have seen in the power method, repeated multiplication makes $A^k\mathbf{q}_1$ closer and closer to the largest eigenvector of A, and making the set of vectors more and more linearly dependent. Thus, one often creates an *orthonormal* basis for $\mathcal{K}_r(A;\mathbf{q}_1)$ obtained from the so-called *Arnoldi process*, which uses the Gram-Schmidt orthogonalization procedure:[6] for $k = 1, 2, \ldots$, compute

$$\mathbf{r}_k = A\mathbf{q}_k - \sum_{j=1}^{k}(\mathbf{q}_j^T A\mathbf{q}_k)\mathbf{q}_j, \qquad \mathbf{q}_{k+1} = \frac{\mathbf{r}_k}{\|\mathbf{r}_k\|_2} \quad \text{if } \mathbf{v}_k \neq 0. \tag{12}$$

It is clear that $\{\mathbf{q}_1, \ldots, \mathbf{q}_{r+1}\}$ also spans $\mathcal{K}_r(A;\mathbf{q}_1)$, as long as none of the $\mathbf{r}_k$ vanish. If $\mathbf{r}_k = 0$ for some k, then $\mathcal{K}_{k-1}(A;\mathbf{q}_1) = \mathcal{K}_k(A;\mathbf{q}_1)$, so we have found an invariant subspace of A. For sparse matrices A, the subspace $\mathcal{K}_r(A;\mathbf{q}_1)$ is easy to generate, since it suffices to multiply A by different vectors repeatedly. In fact, it is not even necessary to store the matrix A itself, as long as a routine exists for performing the required matrix-vector multiplication; such procedures are known as *matrix-free methods*.

The key observation in deriving the Lanczos method is that if A is symmetric, then $A\mathbf{q}_k$ is automatically orthogonal to $\mathbf{q}_1, \ldots, \mathbf{q}_{k-2}$ (prove it!). Thus, the sum in (12) runs only from $j = k-1$ to k. Letting $\beta_k = \|\mathbf{r}_k\|_2$, we can rewrite (12) as the three-term recurrence

$$A\mathbf{q}_k = \beta_{k-1}\mathbf{q}_{k-1} + \alpha_k\mathbf{q}_k + \beta_k\mathbf{q}_{k+1}, \tag{13}$$

where $\alpha_k = \mathbf{q}_k^T A\mathbf{q}_k$. Thus, we have derived the *Lanczos process*: given a starting vector $\mathbf{q}_1$ (and assuming $\mathbf{q}_0 = 0$ and $\beta_0 = 0$), do for $k = 1, 2, \ldots$:

1. Compute $\alpha_k = \mathbf{q}_k^T A\mathbf{q}_k$, $\mathbf{r}_k = (A - \alpha_k I)\mathbf{q}_k - \beta_{k-1}\mathbf{q}_{k-1}$;
2. If $\mathbf{r}_k \neq 0$, normalize to get $\mathbf{q}_{k+1} = \mathbf{r}_k/\|\mathbf{r}_k\|$, $\beta_k = \|\mathbf{r}_k\|$.

The Lanczos process is said to *break down* if $\mathbf{r}_k = 0$, or equivalently, if $\beta_k = 0$.

Another way of writing (13) after k steps of Lanczos is the matrix relation

$$AQ_k = Q_kT_k + \beta_{k+1}\mathbf{q}_{k+1}\mathbf{e}_k^T, \tag{14}$$

where $Q_k = [\mathbf{q}_1, \ldots, \mathbf{q}_k]$ has orthonormal columns, $\mathbf{e}_k$ is the k-th column of the identity matrix, and

$$T_k = \begin{bmatrix} \alpha_1 & \beta_1 & & & \\ \beta_1 & \alpha_2 & \ddots & & 0 \\ & \ddots & \ddots & \ddots & \\ 0 & & \ddots & \ddots & \beta_{k-1} \\ & & & \beta_{k-1} & \alpha_k \end{bmatrix}.$$

Observe that by multiplying (14) on the left by Q_k^T and using the fact that $\mathbf{q}_{k+1}$ is orthogonal to all previous $\mathbf{q}_j$, we get

$$Q_k^T A Q_k = T_k. \tag{15}$$

A crucial question is how well the eigenvalues of T_k, also known as the "Ritz values" of A, approximate the eigenvalues of A. Indeed, in the case of breakdown at step k, we have the exact relation

$$AQ_k = Q_kT_k,$$

so Q_k generates an invariant subspace of A, and the spectrum of T_k is a subset of that of A. If $\mathbf{r}_k \neq 0$, then there are many results describing the relationship

[6]In fact, the formula (12) suffers from loss of orthogonality when A is ill-conditioned. In that case, a modified Gram-Schmidt or Householder orthogonalization is preferred, see [**38**].

between the eigenvalues of T_k and A, collectively known as *Kaniel–Page Theory*. Below we attempt to give a flavour of this type of analysis.

Let us estimate the convergence of *extremal* Ritz values, i.e., the smallest and largest eigenvalues of T_k, towards the extremal eigenvalues of A as a function of k, the number of Lanczos steps. Let $\lambda_1 \leq \lambda_2 \leq \cdots \leq \lambda_n$ be the eigenvalues of A, and $\tilde{\lambda}_1 \leq \tilde{\lambda}_2 \leq \cdots \leq \tilde{\lambda}_k$ be those of $T_k = Q_k^T A Q_k$. Then from the minimax principle, we know that

$$\tilde{\lambda}_1 = \min_{\substack{\mathbf{x}\in\mathbb{R}^k \\ \mathbf{x}\neq 0}} \frac{\mathbf{x}^T Q_k^T A Q_k \mathbf{x}}{\mathbf{x}^T\mathbf{x}} = \min_{\substack{\mathbf{w}\in\mathcal{K}_{k-1}(A,\mathbf{q}_1) \\ \mathbf{w}\neq 0}} \frac{\mathbf{w}^T A\mathbf{w}}{\mathbf{w}^T\mathbf{w}} \geq \lambda_1.$$

In order to estimate this minimum, observe that for any $\mathbf{w} \in \mathcal{K}_{k-1}(A,\mathbf{q}_1)$, there exists a polynomial $P_{k-1}(x)$ of degree $k-1$ or lower such that $\mathbf{w} = P_{k-1}(A)\mathbf{q}_1$. Because $\tilde{\lambda}_1$ minimizes the Rayleigh quotient over all such polynomials, we in fact have

$$\tilde{\lambda}_1 \leq \frac{\sum_{j=1}^n \alpha_j^2 \lambda_j P_{k-1}^2(\lambda_j)}{\sum_{j=1}^n \alpha_j^2 P_{k-1}^2(\lambda_j)} = \lambda_1 + \frac{\sum_{j=2}^n (\lambda_j - \lambda_1)\alpha_j^2 P_{k-1}^2(\lambda_j)}{\sum_{j=1}^n \alpha_j^2 P_{k-1}^2(\lambda_j)},$$

where $\mathbf{q}_1 = \sum_{j=1}^n \alpha_j \mathbf{v}_j$ is the eigen-decomposition of $\mathbf{q}_1$. This is true for any polynomial $P_{k-1}(x)$, but we would like to choose a specific polynomial so that the second term on the right-hand side is small. To do so, we choose P_{k-1} to be a *shifted Chebyshev polynomial* that is small on the interval $[\lambda_2, \lambda_n]$ but large at λ_1. Recall that Chebyshev polynomials are defined by the recurrence

$$f_0(x) = 1, \qquad f_1(x) = x, \qquad f_{n+1}(x) = 2x f_n(x) - f_{n-1}(x), \quad n \geq 1$$

and satisfy

$$f_n(x) = \begin{cases} \cos(n \arccos(x)), & |x| \leq 1, \\ (\operatorname{sgn}(x))^n \cosh(n \operatorname{arcosh}|x|), & |x| > 1. \end{cases}$$

In other words, we have $|f_n(x)| \leq 1$ for $x \in [-1,1]$, and $|f_n(x)|$ grows like $2^{n-1}|x|^n$ outside this interval. Thus, the shifted Chebyshev polynomial

$$P_{k-1}(\lambda) = f_{k-1}\left(\frac{2\lambda - (\lambda_2 + \lambda_n)}{2(\lambda_n - \lambda_2)}\right)$$

is small on $[\lambda_2, \lambda_n]$, but large at λ_1, provided that it is far enough from λ_2. Using the same reasoning for the largest eigenvalue λ_n, we obtain the following convergence result.

THEOREM 4.2 (Convergence of Lanczos). *In exact arithmetic, the extremal Ritz values $\tilde{\lambda}_1$ and $\tilde{\lambda}_k$ after k iterations of Lanczos satisfy the estimate*

$$\lambda_1 \leq \tilde{\lambda}_1 \leq \lambda_1 + C_1/\cosh^2((k-1)\gamma_1), \tag{16}$$

$$\lambda_n \geq \tilde{\lambda}_k \geq \lambda_n - C_2/\cosh^2((k-1)\gamma_2), \tag{17}$$

where

$$\gamma_1 = \operatorname{arcosh}\left(\frac{\lambda_2 + \lambda_n - 2\lambda_1}{\lambda_n - \lambda_2}\right), \quad \gamma_2 = \operatorname{arcosh}\left(\frac{2\lambda_n - \lambda_1 - \lambda_{n-1}}{\lambda_{n-1} - \lambda_1}\right).$$

In Figure 12, we plot the error in the largest and smallest Ritz values and the theoretical bound given in Theorem 4.2 for the finite-difference discrete Laplacian for an L-shaped domain. We see that this simple bound is not the sharpest, but nonetheless tracks the general slope of the error curves reasonably well.

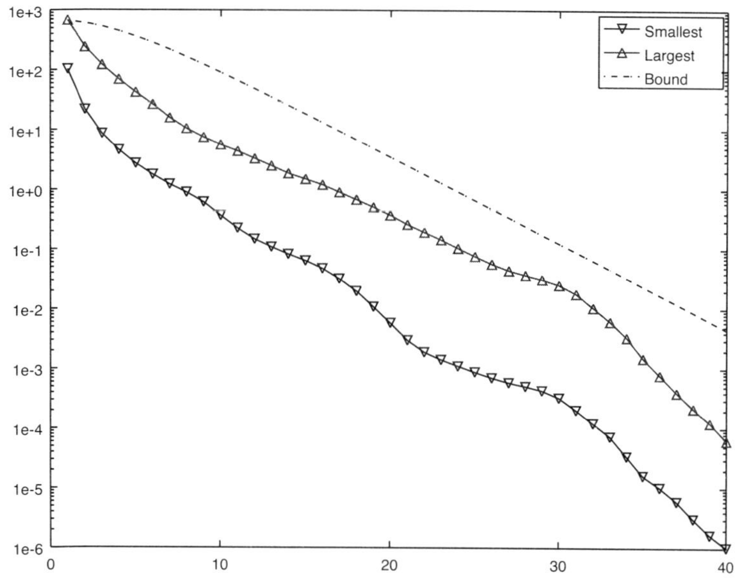

FIGURE 12. Lanczos method applied to the finite-difference Laplacian for an L-shaped domain, $h = 0.1$

Similar techniques can be used to estimate interior eigenvalues, although such bounds may contain constants that become quite large. For instance, we have the estimate

$$\lambda_i \leq \tilde{\lambda}_i \leq \lambda_i + C\kappa_i / \cosh^2((k-i)\gamma_i),$$

where $\gamma_i = \operatorname{arcosh}\left(\frac{\lambda_{i+1}+\lambda_n-2\lambda_i}{\lambda_n-\lambda_{i+1}}\right)$, $\kappa_i = \prod_{j=1}^{i-1}\left(\frac{\lambda_n-\tilde{\lambda}_j}{\lambda_i-\tilde{\lambda}_j}\right)^2$.

Remarks.

1. Just as for power-type methods, the *extremal* eigenvalues (largest and smallest) are among the first to converge in the Lanczos method.
2. Because the short recurrence (13) does not enforce orthogonality explicitly with all previous $\mathbf{q}_i$, orthogonality may be lost gradually due to round-off errors in finite precision arithmetic. When this happens, the subsequent $\mathbf{q}_l$ may be (nearly) linearly dependent on the previous ones, giving rise to repeated *ghost eigenvalues*. To prevent this, one can enforce orthogonality explicitly, either by complete re-orthogonalization (expensive), or by selective re-orthogonalization against "converged" Ritz vectors only. For more details, see [**23**].
3. Lanczos is intimately related to the celebrated Conjugate Gradient method for solving $A\mathbf{x} = \mathbf{b}$. See [**23, 38**] for more details.

5. Application: Vibrating plates

Chladni figures are patterns that were discovered by Ernst Florence Friedrich Chladni (1756–1827) in a series of experiments on vibrating plates. In one of these experiments, he sprinkles fine sand onto a square plate before setting it into vibration using a violin bow. As the plate vibrates, the sand gathers at points that remain stationary and traces out beautiful patterns, such as the ones shown

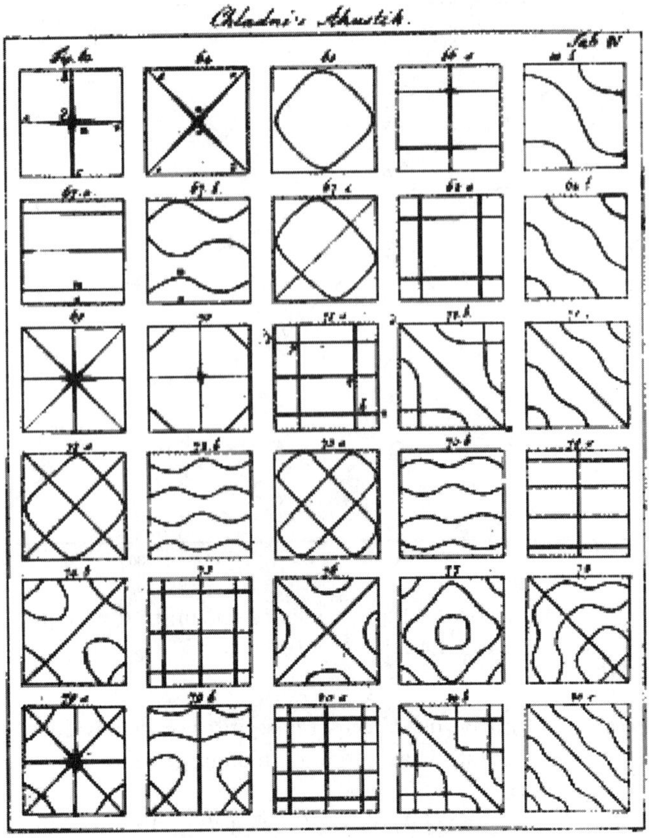

FIGURE 13. Chladni figures as recorded by Chladni in [16].

in Figure 13. It turns out such patterns are zero sets of eigenfunctions of the biharmonic operator, with special boundary conditions that can be derived from a variational formulation. In this section, we will show how this eigenvalue problem can be derived, discretized and calculated using techniques we have seen. The presentation of this section follows [20] closely.

5.1. Vibrating plate model. In order to derive the vibrating plate model, we start just like we did in Section 1, with Newton's second law:

$$\frac{\partial^2 z}{\partial t^2} = -\mathscr{L} z.$$

We saw that such motions can be analyzed using standing wave solutions of the form $z(x, y, t) = T(t)u(x, y)$ is a standing wave, where $u(x, y)$ is a solution of the eigenvalue problem

$$\mathscr{L} u = \lambda u.$$

In the absence of external forces, $\mathscr{L} u$ is simply the restoring force due to the shape of the vibrating plate $u = u(x, y)$. However, instead of $-\mathscr{L}$ being the Laplacian operator, as was the case for vibrating membranes, we will deduce the form of $\mathscr{L}$ based on energy considerations. Gustav Kirchhoff (1824–1887) proposed a model

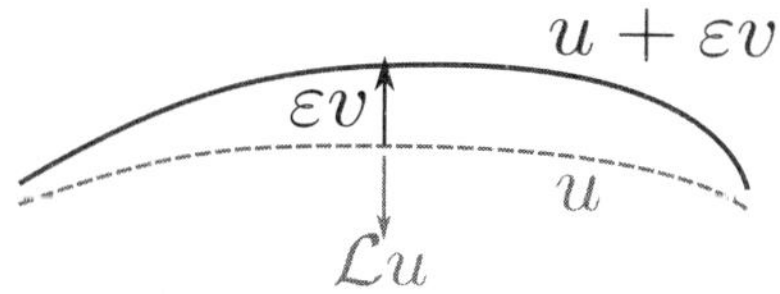

FIGURE 14. Forces involved in deforming of a thin plate from configuration u to a nearby configuration $u + \varepsilon v$.

that describes the *potential energy* stored in a deformed thin square plate with shape $u : \Omega \to \mathbb{R}$, given by

$$J[u(x,y)] = \frac{1}{2} \iint_\Omega (u_{xx} + u_{yy})^2 - 2(1-\mu)(u_{xx}u_{yy} - u_{xy}^2)\, dx\, dy, \tag{18}$$

where $0 < \mu < 1$ is a material constant. This is only one of many models for the bending plate, and it may not be adequate for certain types of problems; see the plate paradox in [**7**], further explored in [**6**]. Other possibilities include the Reissner-Mindlin model or the full 3D model, as described in [**6**]. However, for simplicity, we will only consider the Kirchhoff model (18) in these notes.

To see how this energy functional (18) is related to the operator $\mathscr{L}$, we resort to arguments that are standard in the calculus of variations, see [**22**, Chapter 7]: suppose we wish to change the shape of the plate from an initial configuration $u(x,y)$ to a slighly different shape $u(x,y) + \varepsilon v(x,y)$, where ε is small. Then the change in potential energy is precisely the work done on the system, which is the distance travelled by each particle on the plate times the force needed to counteract the restoring force $\mathscr{L}u(x,y)$, see Figure 14. In other words, we have

$$J[u+\varepsilon v] - J[u] = \iint_\Omega (\mathscr{L}u)(\varepsilon v)\, dx\, dy + O(\varepsilon^2).$$

Dividing both sides by ε and taking the limit as $\varepsilon \to 0$, we obtain

$$\frac{d}{d\varepsilon} J[u+\varepsilon v]\Big|_{\varepsilon=0} = \lim_{\varepsilon \to 0} \frac{J[u+\varepsilon v] - J[u]}{\varepsilon} = \iint_\Omega (\mathscr{L}u) v\, dx\, dy.$$

Substituting the above into the definition (18) and differentiating yields

$$\iint_\Omega (\mathscr{L}u) v\, dx\, dy = \iint_\Omega (u_{xx} + u_{yy})(v_{xx} + v_{yy}) \\ - (1-\mu)(u_{xx}v_{yy} + u_{yy}v_{xx} - 2u_{xy}v_{xy})\, dx\, dy.$$

This is precisely the expression we need to define the *weak form* of the eigenvalue problem

$$\iint_\Omega (\mathscr{L}u) v\, dx\, dy = \iint_\Omega \lambda u v\, dx\, dy.$$

We will use this form in Section 5.3 to derive a *spectral approximation* of the eigenvalue problem, as was done by Walther Ritz [**36**].

For the moment, we continue by deriving the strong form of the problem for the special case of a rectangular plate, where $\Omega = (-L, L) \times (-H, H)$. We integrate by parts to eliminate derivatives of v:

$$\begin{aligned}\iint_\Omega (\mathscr{L}u)v\,dx\,dy &= \iint_\Omega (u_{xxxx} + 2u_{xxyy} + u_{yyyy})v\,dx\,dy \\ &\quad + \int_{x=\pm L} \left([u_{xx} + \mu u_{yy}] \frac{\partial v}{\partial n} + [u_{xxx} + (2-\mu)u_{xyy}]v \right) dy \\ &\quad + \int_{y=\pm H} \left([\mu u_{xx} + u_{yy}] \frac{\partial v}{\partial n} + [u_{yyy} + (2-\mu)u_{xxy}]v \right) dx \\ &\quad + 2(1-\mu)[u_{xy}(L,H)v(L,H) - u_{xy}(-L,H)v(-L,H) \\ &\qquad - u_{xy}(L,-H)v(L,-H) + u_{xy}(-L,-H)v(-L,-H)]\end{aligned}$$

By choosing arbitrary functions v that vanish on the boundary of Ω, we obtain the PDE

$$u_{xxxx} + 2u_{xxyy} + u_{yyyy} = \lambda u, \tag{19}$$

which must hold at every point in the interior of Ω. We can also obtain boundary conditions by choosing arbitrary variations along different parts of the boundary. For instance, by considering the integrals along $x = \pm L$, we obtain the boundary conditions

$$u_{xx} + \mu u_{yy} = 0, \qquad u_{xxx} + (2-\mu)u_{xyy} = 0$$

there. We can combine the above with the similar-looking conditions along the horizontal boundaries $y = \pm H$ by writing

$$u_{nn} + \mu u_{\tau\tau} = 0, \qquad u_{nnn} + (2-\mu)u_{n\tau\tau} = 0, \tag{20}$$

where the subscripts n and τ denote derivatives in the outward normal and tangential directions, respectively. Finally, by taking arbitrary variations at the four corners, we obtain the corner conditions

$$u_{xy}(\pm L, \pm H) = 0. \tag{21}$$

Equations (20) and (21) are known as *free boundary conditions* for the thin plate PDE (19). Another way of writing (19) is $\Delta^2 u = \lambda u$, where Δ is the usual Laplacian operator; Δ^2 is also known as the *biharmonic operator*. Note that the material constant μ does not enter into the PDE itself, but instead appears in the boundary conditions. The PDE (19) has already appeared in a series of articles by Sophie Germain in 1811–21; however, the edge conditions (20) only appears in 1850 in the work by Kirchhoff, and the corner conditions (21) finally emerge in an article by Lamb in 1889, almost 80 years after the discovery of the strong form PDE.

5.2. Finite difference discretization. In order to discretize the strong form (19)–(21), we use the finite difference and finite volume techniques introduced in Section 2. We show this for the case $L = H = 1$, i.e., for the square $\Omega = (-1,1)^2$. Let $(u_{ij})_{i,j=0}^N$ be the grid points distributed uniformly at $(x_i, y_j) = (-1+ih, -1+jh)$, $h = 2/N$. Since the biharmonic operator can be regarded as the Laplacian operator composed with itself, it is natural to discretize it by composing the discrete five-point Laplacian stencil with itself. If we define the auxiliary grid function w using the five-point stencil as shown in the top left panel of Figure 15, i.e.,

$$w_{ij} := \frac{1}{h^2}(4u_{ij} - u_{i-1,j} - u_{i+1,j} - u_{i,j-1} - u_{i,j+1}) \approx -\Delta u,$$

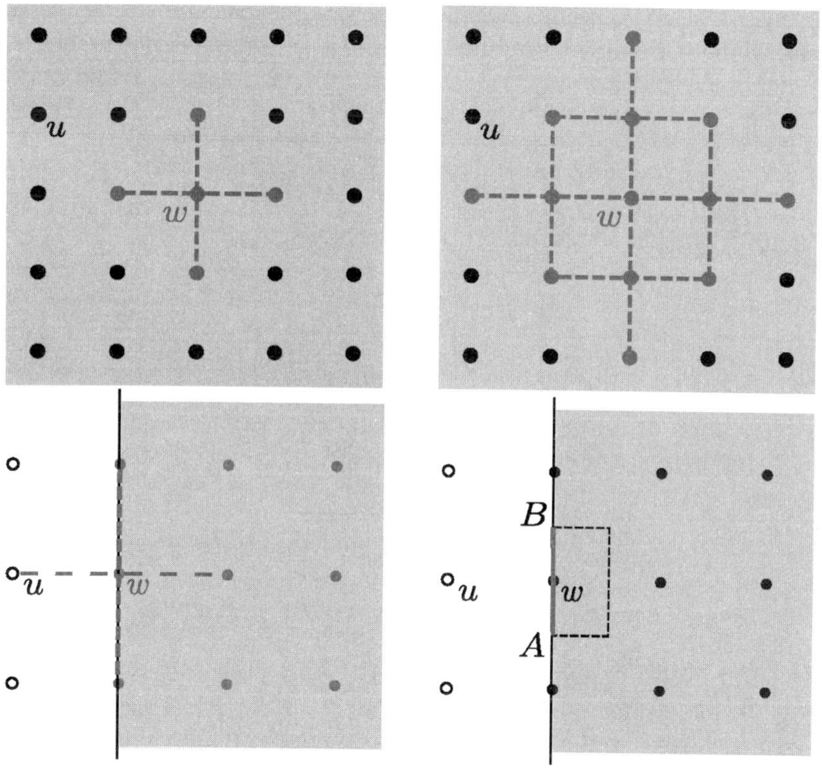

FIGURE 15. Finite difference/finite volume discretization of the square plate problem. Top left: definition of w. Top right: full stencil of the biharmonic operator after composing the five-point stencil with itself. Bottom left: elimination of ghost points using second order boundary conditions. Bottom right: treatment of third order boundary conditions.

then (19) can be discretized i as

$$\frac{1}{h^2}(4w_{ij} - w_{i-1,j} - w_{i+1,j} - w_{i,j-1} - w_{i,j+1}) = \lambda u_{ij}.$$

This leads to the wide stencil shown in the top right panel of Figure 15, which can be written for the interior points $1 \leq i, j \leq N-1$.

Note that if $i = 1$ or $N-1$, the stencil involves the unknown $u_{-1,j}$ or $u_{N+1,j}$, which falls outside Ω; this is a *ghost point* that needs to be eliminated. To do so, we make use of the second order boundary condition in (20): along $x = -1$, the condition reads $u_{xx} + \mu u_{yy} = 0$, which can be discretized as

$$2(1+\mu)u_{0,j} - u_{-1,j} - u_{1,j} - \mu(u_{0,j-1} + u_{0,j+1}) = 0.$$

This allows us to represent the ghost value as

$$u_{-1,j} = 2(1+\mu)u_{0,j} - u_{-1,j} - u_{1,j} - \mu(u_{0,j-1} + u_{0,j+1}),$$

as shown in the bottom left panel of Figure 15. Other ghost points can be treated similarly.

Finally, we need equations for the edge variables themselves, which we will derive using the finite volume method. Consider the edge unknown shown in the bottom right panel of Figure 15. We integrate (19) over the half volume V indicated in dotted lines. Noting that $\Delta^2 u = -\Delta w$, we have

$$-\int_V \Delta w\, dx\, dy = \lambda \int_V u\, dx\, dy \approx \frac{h^2\lambda}{2} u_{ij}.$$

Applying the divergence theorem on the left hand side yields

$$-\int_{\partial V} \frac{\partial w}{\partial n}\, ds \approx \frac{h^2\lambda}{2} u_{ij}.$$

The integral along three of the four edges can be approximated using finite differences. For instance, the integral along the top edge can be approximated by $\frac{1}{2}(w_{0,j} - w_{0,j+1})$. To deal with the edge along the domain boundary, we use the third order boundary condition to eliminate $\frac{\partial w}{\partial n}$: since $w_x = -(u_{xxx} + u_{xyy})$, we have by (20)

$$w_x = -\underbrace{(u_{xxx} + (2-\mu)u_{xyy})}_{=0} + (1-\mu)u_{xyy}.$$

Thus, the integral along the boundary portion of ∂V is given by

$$-\int_A^B \frac{\partial w}{\partial n}\, dy = \int_A^B w_x\, dy = (1-\mu)\int_A^B u_{xyy}\, dy = (1-\mu)(u_{xy}(B) - u_{xy}(A)).$$

The mixed derivative at B can now be approximated by a finite difference involving $u_{-1,j+1}$, $u_{-1,j}$, $u_{1,j+1}$ and $u_{1,j}$, and similarly for the derivative at A. Finally, for corner points, we use the same technique over the quarter volume, with one of the mixed derivatives vanishing directly because of the corner condition (21).

The above discretization leads to a large, sparse *generalized* eigenvalue problem of the form

$$Au = \lambda Bu,$$

where B is a diagonal matrix containing the areas of the control volumes, and A having at most 13 non-zero entries per row. This matrix is easy to set up in MATLAB, requiring only about 80 lines of code, see [**20**] for details. The generalized eigenvalue problem can then be solved using a modified version of the Lanczos method, see Section 7.3. If we plot the nodal lines of these eigenfunctions, we obtain the Chladni figures shown in Figure 16.

5.3. Ritz approximation. Another way of approximating the eigenvalues of the free vibrating plate, developed by Walther Ritz [**36**], bears much similarity to the finite element method introduced in Section 3.[7] Consider once again the eigenvalue problem in weak form, where we seek $u \in H^2(\Omega)$ such that

$$a(u,v) = \lambda(u,v) \quad \text{for all } v \in H^2(\Omega), \tag{22}$$

where

$$a(u,v) = \iint_\Omega (u_{xx} + u_{yy})(v_{xx} + v_{yy}) - (1-\mu)(u_{xx}v_{yy} + u_{yy}v_{xx} - 2u_{xy}v_{xy})\, dx\, dy,$$

[7] In fact, the finite element method has its origin in Ritz's method.

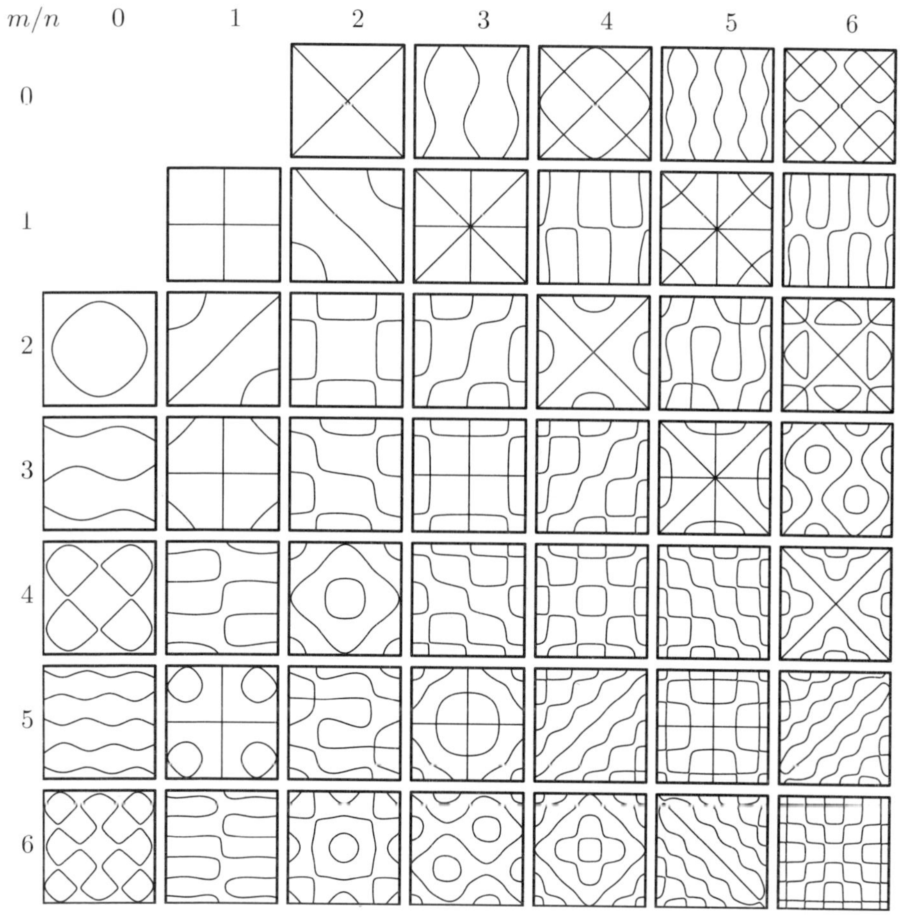

FIGURE 16. Chladni figures obtained by Matlab using the finite difference method.

and $(\cdot,\cdot)$ denotes the L^2 inner product on Ω. Instead of using a basis of hat functions, Ritz's idea is to use a product of one-dimensional eigenfunctions. Suppose $u_m(x)$ satisfies the one-dimensional eigenvalue problem

$$\frac{d^4 u_m}{dx^4} = k_m^4 u_m, \qquad \frac{d^2 u_m}{dx^2} = \frac{d^3 u_m}{dx^3} = 0 \text{ at } x = \pm 1.$$

This two point boundary value problem can be solved easily, and we find

(23)

$$u_m(x) = \begin{cases} \dfrac{\cosh k_m \cos k_m x + \cos k_m \cosh k_m x}{\sqrt{\cosh^2 k_m + \cos^2 k_m}}, & \tan k_m + \tanh k_m = 0, \quad m \text{ even}, \\[2ex] \dfrac{\sinh k_m \sin k_m x + \sin k_m \sinh k_m x}{\sqrt{\sinh^2 k_m - \sin^2 k_m}}, & \tan k_m - \tanh k_m = 0, \quad m \text{ odd}. \end{cases}$$

We now let the finite dimensional subspace V_h to be spanned by functions of the form $u_m(x)u_n(y)$; in other words, for a given $N > 0$, we seek an approximate

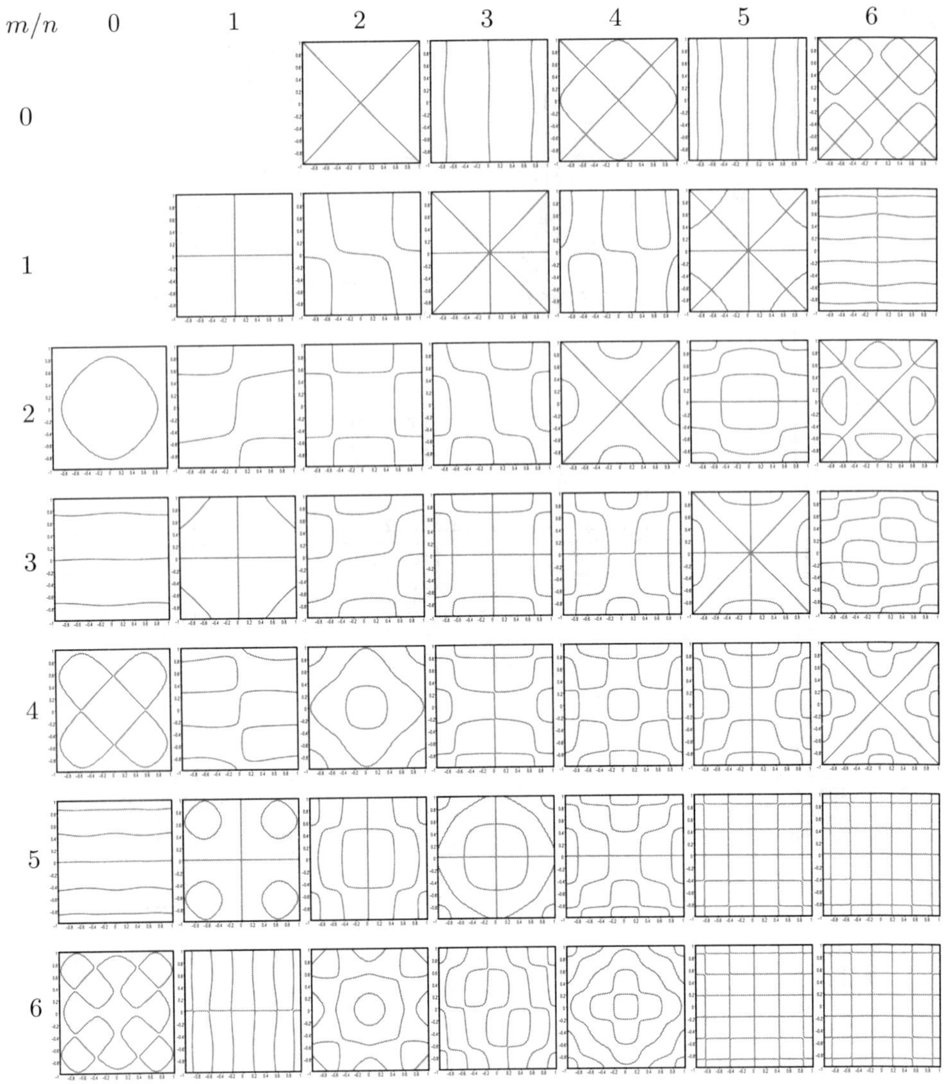

FIGURE 17. Chladni figures obtained by Ritz's method for $N = 6$, i.e., using 49 basis functions.

eigenfunction of the form

$$u(x, y) = \sum_{m=0}^{N} \sum_{n=0}^{N} c_{mn} u_m(x) u_n(y) \tag{24}$$

such that (22) is satisfied for all $v(x, y) = u_p(x)u_q(y)$, $0 \le p, q \le N$. In matrix form, this gives the generalized eigenvalue problem

$$A\mathbf{c}_k = \tilde{\lambda}_k B\mathbf{c}_k,$$

where A and B are symmetric $(N + 1)^2 \times (N + 1)^2$ matrices, with rows indexed by (p, q) and columns indexed by (m, n), corresponding to the basis functions

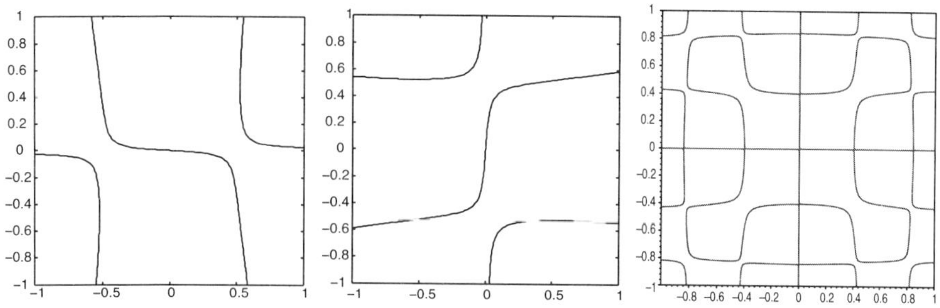

FIGURE 18. Left and middle: nodal lines for linear combinations of the eigenfunctions calculated by Matlab. Right: A better approximation of the eigenfunction obtained from Maple for $m = 5$ and $n = 5$ with 10 basis functions per direction ($0 \le m, n \le 9$).

$u_{pq}(x,y) := u_p(x)u_q(y)$ and $u_{mn}(x,y) := u_m(x)u_n(y)$ respectively. The corresponding entries are

$$a_{mn}^{pq} = a(u_{mn}, u_{pq}), \qquad b_{mn}^{pq} = (u_{mn}, u_{pq}).$$

Note that the matrices are dense; the calculation of the entries, while cumbersome, can be automated by software such as Maple; see [**20**] for details. Once A and B are evaluated numerically, the QR method can be used to obtain the eigenvalues $\tilde{\lambda}_k$, and the eigenvectors $\mathbf{c}_k$ can be substituted back into (24) to yield the approximate eigenfunctions. The nodal lines of the latter are shown in Figure 17 for the case with $N = 6$, i.e., when the matrices are 49×49. We see that even for this small number of degrees of freedom, the approximation is already very good for the low frequency modes. Thus, thanks to a clever choice of basis functions, this method converges much faster to the exact eigenvalues in terms of number of degrees of freedom than the finite difference method, at the price of dealing with dense rather than sparse matrices.

Comparing the Chladni figures in 17 with those obtained by Matlab in 16, we observe that there are substantial qualitative differences between them in two cases:

- When $m+n$ is odd. This is because these correspond to eigenvalues with multiplicity 2: the (m,n) and (n,m) modes together span a two-dimensional subspace with two linearly independent eigenfunctions, so the Chladni figures are not uniquely determined. If we take linear combinations of the eigenfunctions obtained by Matlab, we obtain the ones found by Maple. We show an example in the left and middle panels of Figure 18 for the case of $(m,n) = (1,2)$ and $(2,1)$: denoting by $u^{(12)}$ and $u^{(21)}$ the eigenfunctions shown in positions (1,2) and (2,1) in Figure 16, the left panel shows the the linear combination $u = -1.6u^{(12)} + u^{(21)}$, whereas the middle panel shows $v = u^{(12)} + 1.6u^{(21)}$. These look qualitatively similar to the ones found by Maple in positions (1,2) and (2,1) in Figure 17.
- When m and n are large relative to the number of basis functions used, for example when $(m,n) = (5,5)$. When we increase the number of basis functions, e.g., with $N = 9$, we get a much better approximation that matches the Chladni figure found by Matlab, see the right panel of Figure 18.

TABLE 2. First 10 eigenvalues obtained by the finite difference and Ritz methods, versus experimental values found by Chladni. Up/down arrows mean that the observed notes are higher/lower than the indicated nominal values.

Model			Experiment	
Fin. Diff.	Ritz	Gap	Note	Value
12.5	12.5	0.08%	G	12.4
26.0	26.1	0.46%	D$\downarrow$	26.4
35.6	35.8	0.36%	E$\uparrow$	36.2
80.9	81.2	0.36%	B	77.5
235.4	236.3	0.39%	G$\sharp\downarrow$	215
269.3	270.0	0.24%	A$\uparrow$	260
320.7	322.4	0.53%	B	310
375.2	377.2	0.53%	C$\uparrow$	364
730.0	732.4	0.33%	F$\sharp$	698
876.1	879.8	0.42%	G$\uparrow$	819

From the Minimax principle, we know the Ritz values $\tilde{\lambda}_k$ are overestimates of the exact eigenvalues. In Table 2, we compare the first 10 eigenvalues obtained by the finite difference method with 400 grid points per side with those from the Ritz method using $N = 10$ modes per direction. We see that the Ritz values are indeed larger than the finite difference eigenvalues, but they are already very accurate, with a gap no larger than about half a percent between the two. In the same table, we also show the frequencies observed experimentally by Chladni himself in [**16**], as well as the corresponding eigenvalues. (Recall from Section 1 that the square root of the eigenvalue is the frequency of oscillation, which produces a sound that Chladni heard and recorded as a musical note.) We see that the experimental results also match the model very well, but they tend to be underestimates. This is possibly due to the internal friction present in physical systems, which causes energy to dissipate and leads to lower frequencies when compared with ideal models.

5.4. Other boundary conditions. The vibrating plate model given by the energy functional (18) can be used to obtain other boundary conditions. For example, one is said to apply *clamped boundary conditions* when the edge is not allowed to move, and the plate may not pivot around the edge. In the language of the calculus of variation, this is the same as specifying that perturbations of the form $u + \varepsilon v$ are only allowed if

$$v = \frac{\partial v}{\partial n} = 0.$$

This leads to strong form boundary conditions

$$u = \frac{\partial u}{\partial n} = 0$$

in the corresponding eigenvalue problem. Note that this is the analogue of essential boundary conditions for the Laplacian operator, in the sense that such boundary conditions must be built into the test and trial spaces in the variational problem.

Another common type of boundary conditions is the *simply supported* boundary conditions, where the edge of the plate is not allowed to move, but the plate may pivot around the edge. Thus, any perturbations of the form $u + \varepsilon v$ must satisfy

$$v = 0.$$

This leads to strong form boundary conditions

$$u = 0, \quad \frac{\partial^2 u}{\partial n^2} = 0.$$

Whereas the first condition is rather obvious, the second-order condition is less so – it can be derived by integration by parts and a variational argument, in the same way we deduced the edge and corner conditions for the free boundary conditions. The reader is invited to calculate the eigenvalues and nodal lines of the biharmonic operator for these types of boundary conditions.

6. Further reading

There is a large body of literature on the numerical solution of PDE eigenvalue problems. Instead of attempting to give a (necessarily non-exhaustive) list of references, we will just give pointers to a few classical and recent works on the subject, through which the interested reader can continue to explore via the references therein.

The approximation of eigenvalue problems was first done by finite difference methods, with much important work appearing before the 1970s, such as [**25**, **26**, **42**, **43**]. Such methods can be used to derive lower as well as upper bounds on the exact eigenvalues, but because a minimax type argument is not readily available, the convergence theory often relies on clever tricks that only work for specific stencils and/or domain types. The classical paper by Kuttler [**29**] develops a fairly general convergence theory for this class of methods for the Laplace operator. A list of related work can also be found in Section 9 in the review paper by Kuttler and Sigillito [**30**]. Because of the simplicity of the discretization, finite difference methods are also used for higher order elliptic operators, see [**33**, **34**].

For the finite element method, the survey paper by Boffi [**11**] elegantly describes the essentials of the convergence theory of FEM for eigenvalue problems. The paper deals mainly with symmetric problems, with some comments and references to the non-symmetric case. It also contains many references to relevant work up to around 2010. For more on non-self adjoint problems, see Bramble and Osborn [**12**]. For problems that contain eigenfunctions with less regularity, e.g., the L shpaed domain, an adaptive mesh refinement approach may be preferred, see Dai, Xu and Zhou [**17**] and also Carstensen and Gedicke [**14**], where their methods also provide guaranteed lower bounds for the eigenvalues concerned. For implementation, FreeFem++ ([**24**], available at `www.freefem.org`) is a domain-specific language based on C++ that makes the tasks of gridding and stiffness matrix calculations much easier. The documentation available on the web site contains examples on PDE eigenvalue problems. For more on FEM for eigenvalue problems, we recommend the very readable and comprehensive treatise by Babuška and Osborn [**5**].

There are other methods for solving PDE eigenvalue problems than the ones examined in this manuscript. There is the Method of Particular Solutions, which assumes that the eigenfunction can be written as a linear combination of eigenfunctions with certain symmetries, and the coefficients are chosen so that the boundary conditions are satisfied. This method was used by Fox, Henrici and Moler to calculate the eigenfunctions of the L-shaped domain in [**19**]. The method was subsequently improved by [**9**]. A different possibility is to use a boundary integral method, in which the eigenvalue problem is reformulated in terms of unknown traces (function values or normal derivatives) along the boundary. This has the advantage of reducing the number of unknowns dramatically and can be beneficial when searching for large eigenvalues, where the highly oscillatory eigenfunctions would require a very fine mesh in the interior of the domain. For an easy-to-understand algorithmic description of the method, see Bäcker [**8**]. More recent work that considers a problem with both Dirichlet and Neumann boundaries can be found in [**1**].

The solution of eigenvalue problems for dense matrices is now a fairly mature technology, with highly efficient implementations available in libraries such as LAPACK ([**3**], `www.netlib.org/lapack`) and its parallel version, ScaLAPACK ([**10**], `www.netlib.org/scalapack`). Nonetheless, new shift-and-deflate strategies have been proposed that lead to even faster convergence, see [**28**]. For sparse matrices, considerations such as eigenvalue distribution and sparsity preservation become important. In addition to the Lanczos method, another popular method is the Jacobi–Davidson method, which is described in [**39**]. Some numerical experiments suggest that Jacobi–Davidson does a better job in approximating interior eigenvalues than methods such as Lanczos; for more details, see the relevant references contained in [**39**]. We also recommend the excellent book by Saad [**37**], which contains very thorough discussions of the theory and algorithms related to large, sparse matrix eigenvalue problems.

Finally, we mention the book by Chatelin [**15**], which develops the theory and numerical methods for spectral approximation not only for differential operators, but also for other types of linear operators, e.g., integral operators.

Acknowledgments

The author would like to thank Martin J. Gander, Walter Gander, Ronald D. Haynes and Nilima Nigam for their careful reading of the manuscript and for their helpful suggestions, which significantly improved the quality of these notes. We also thank Catherine Bénéteau, Alexandre Girouard, Dmitry Khavinson, Javad Mashreghi and Thomas J. Ransford for organizing the 2016 CRM Summer School on Spectral Theory and Applications, at which the author benefited from many insightful and stimulating discussions. Finally, we thank the anonymous referees for their many suggestions that led to significant improvements in the manuscript.

7. Exercises

This section contains three problem sets handed out during the 2016 CRM Summer School on Spectral Theory and Applications. They are computational in nature, with the aim of helping participants gain insight into the theory by implementing the numerical methods themselves and experimenting with them. Some additional code was provided to the participants, in particular the finite element modules; those can be downloaded at `http://www.math.hkbu.edu.hk/~felix_kwok/crm/`.

7.1. Problem Set #1: Finite difference methods.

Exercise 1: Grid setup and visualization

The goal of this exercise is to learn how to set up a rectangular grid in MATLAB, as well as visualize functions defined on a grid.

1. Open a new script file in the editor and type the following commands:

```
a = 0; b = 1;
N = 10;            % Number of mesh widths in the grid
h = (b-a)/N;       % Mesh size
t = (a:h:b)'
```

   The last command generates a vector of $N+1$ points from `a` to `b` with increment `h`. Save the script and run it, either by typing its name in the command window, or by clicking the 'Run' button. We will change or include additional commands in this file in the subsequent exercises.

2. Generate a grid of x-values by creating an $(N+1) \times (N+1)$ array with $N+1$ copies of the column vector `t` using the command

```
xx = repmat(t, 1, N+1)
```

3. Generate a grid of y-values by creating an $(N+1) \times (N+1)$ array with $N+1$ copies of the row vector `t'`, the transpose of `t`.
4. To visualize the function $f(x, y) = xy^2$, we first generate a **grid function** u by evaluating f at our grid points `xx` and `yy`:

```
f = @(x,y) x.*(y.^2);
u = f(xx,yy);
```

   Now we can plot the function using the `surf` or the `mesh` command:

```
surf(xx,yy,u)
xlabel('x')
ylabel('y')
```

   The last two commands label the axes. The `mesh` command has an identical syntax to `surf`, but plots the surface a bit differently.

5. To add more resolution to the plot, change `N` in part 1 to `N = 50` and rerun your script.

Exercise 2: Laplacian on a square

We now set up the finite difference matrix for the square grid in Exercise 1.

1. First, we number the interior points using the commands

```
G = zeros(size(xx));
G(xx > a & xx < b & yy > a & yy < b) = (1:(N-1)^2);
```

   Explain what this does by displaying the variable `G` (it may be easier to visualize for $N = 10$).

2. Create an empty $(N-1)^2 \times (N-1)^2$ sparse matrix A using the `sparse` command. Next, fill in the entries by completing the following loop:

```
for i=1:size(G,1),
    for j=1:size(G,2),
        if (G(i,j)>0),
            A(G(i,j),G(i,j)) = 4/h^2;
            if (G(i+1,j)>0), A(G(i,j),G(i+1,j)) = -1/h^2; end;
            ... COMPLETE HERE ...
        end;
    end;
end;
```

3. Compute the five smallest eigenvalues and their associated eigenvectors using the command `eigs`. Do not forget the 's' at the end – `eig` is a different command! If `v` is an eigenvector of `A`, we can visualize it as an eigenfunction on the unit square by running

```
U = G;
U(G>0) = v;
surf(xx,yy,U);
```

4. Compute the **smallest** eigenvalue for different mesh sizes by letting $N = 8$, 16, 32, 64, 128. Plot the difference between these approximations and the exact value $\lambda = 2\pi^2$ in a log-log plot, using the command `loglog`. Also plot on the same graph the curves $y(N) = N^{-\alpha}$ for $\alpha = 1, 2, 3$. At what rate does the smallest eigenvalue converge to the exact value?

Exercise 3: L-shaped domain

The MATLAB commands `numgrid` and `delsq` provide a quick way of setting up the discrete Laplacian matrix. To set up the Laplacian for a square, use the pair of commands

```
G = numgrid('S',N+1)
A = delsq(G);
```

1. Check that `numgrid` and `delsq` give the same grid and matrix as the ones generated in Exercise 2, provided that the appropriate mesh size h is used.
2. Generate a grid for the L-shaped domain shown in the lecture slides, i.e., the region $[-1,1]\times[-1,1]$ with the second quadrant removed. Calculate the smallest eigenvalue and the associated eigenfunction for $N = 16, 32, 64, \ldots, 512$.
3. If we have a slowly converging sequence, such as our sequence of eigenvalues, we can transform it into a new, faster-converging sequence with the same limit using the ϵ-algorithm by P. Wynn (cf. [**21**, §5.2.4]). The function `EpsilonAlgorithm` provided returns the last element of this new sequence. Apply this function to the sequence of eigenvalues obtained in part 2. We will consider the answer to be our best approximation to the exact eigenvalue.
4. Study the convergence of the first eigenvalue as a function of $h = 1/N$. Does the error behave like $O(h^2)$? What is the actual convergence rate?
5. Repeat the above computation for the next four smallest eigenvalues. Why do some eigenvalues converge faster than others?

Exercise 4: Neumann conditions

In this exercise, we will solve numerically for the eigenvalues of the Laplacian of the unit square with Neumann boundary conditions

$$-\Delta u = \lambda u \quad \text{on } \Omega = (0,1)^2, \quad \frac{\partial u}{\partial n} = 0 \quad \text{on } \partial\Omega.$$

Recall that the finite volume method gives the *generalized* eigenvalue problem

$$A\mathbf{u} = \lambda B\mathbf{u},$$

where A has the same structure as the finite difference matrix at interior points, and B is a diagonal matrix.

1. Modify your code to set up the numbered grid `G`, the stiffness matrix `A` and the mass matrix `B`. Points to note:
   - You may set up the matrix A either from scratch like in Exercise 2, or by using `delsq` and dividing appropriate parts of the matrix by 2. Either way, remember to scale the entries correctly along boundary nodes!
   - Be careful with the **size** of these matrices: nodes on the boundary are now part of the degrees of freedom in $\mathbf{u}$.

   For $N = 16$, the first five eigenvalues should be 0.0000, 9.8379, 9.8379, 19.6759, 38.9737.
2. Notice that the first computed eigenvalue is always zero up to machine precision. Can you prove it?
3. Verify numerically that the error of the computed eigenvalues is $O(h^2)$. Can you prove this analytically?

7.2. Problem Set #2: Finite element methods.

Exercise 1: Getting to know a finite element code

In this exercise, you will familiarize yourself with the different parts of a finite element code for solving eigenvalue problems. The following routines are included in the sample code:

NewMesh.m: sets up the data structures for the initial triangular mesh;
PlotMesh.m: displays the mesh stored in the data structures
RefineMesh.m: refines a given mesh by a factor of two;
SmoothMesh.m: moves the nodes of a mesh to get a more uniform triangulation;
ComputeElementStiffnessMatrix.m: computes the element stiffness matrix for a given triangle;
ComputeElementMassMatrix.m: computes the element mass matrix for a given triangle;
FEeig.m: computes the assembled stiffness and mass matrices for a given mesh;
FEeig_fast.m: same as `FEeig.m`, but optimized to exploit the built-in MATLAB routines for manipulating sparse matrices.
PlotSolution.m: displays the finite element function $u(\mathbf{x}) = \sum_i u_i\varphi_i(\mathbf{x})$ given its vector of degrees of freedom $\mathbf{u} = (u_i)_{i=1}^N$.

The following commands show how to run the code from start to finish for the unit square. To run all of them sequentially, save them into an M-file (say `eig_example.m`), then click the 'Run' button or type the name of the file (without the `.m`) on the command line.

1. Change the argument to `NewMesh` to see what other domains are available.
2. Refine the mesh for the unit square until there are at least 9 degrees of freedom. Compute the five smallest eigenvalues of $-\Delta u = \lambda u$, $u|_{\partial\Omega} = 0$, and display their corresponding eigenfunctions. Do the eigenfunctions look the way you expect them to?
3. Continue refining the grid and calculate the first five eigenvalues for each level of refinement. Plot the error as a function of the mesh parameter h. Does the error behave like $O(h^2)$? (**Hint:** If the code becomes too slow, replace `FEeig` by `FEeig_fast` in your code.)

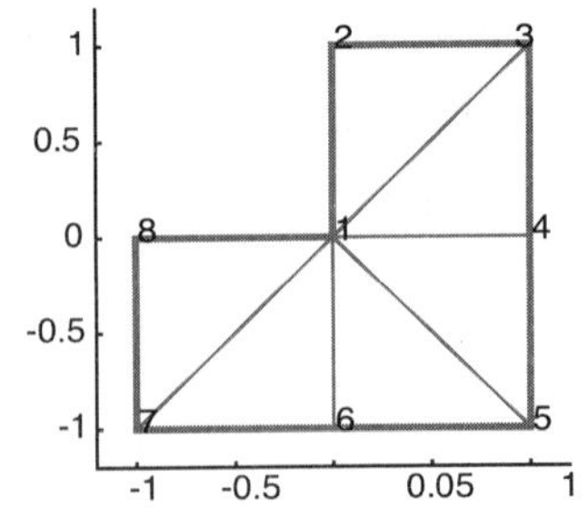

FIGURE 19

Exercise 2: L-shaped domain

In this exercise, you will create an initial mesh for the L-shaped domain shown in Figure 19 and study the convergence of its eigenfunctions.

1. Create the array `N` containing in each row i the x and y coordinates of the ith node. The nodes can appear in any order.
2. Create the array `T` containing a list of triangles defined by the nodes. Each row of `T`, say `T(i,:)`, contains six entries according to the format

   `[v1, v2, v3, e1, e2, e3]`

   where `v1`, `v2` and `v3` are the three nodes of the ith triangular element **in counterclockwise order**. The fourth entry `e1` is 1 if the edge (v_1, v_2) is part of the physical boundary, and zero otherwise. The fifth and sixth entries are similar, except they describe the edges (v_2, v_3) and (v_3, v_1). For example, one of the triangles in the mesh above is described by the row `[1 3 2 0 1 1]`. The triangles can appear in any order.
3. Verify whether your mesh is set up correctly by visualizing it using `PlotMesh`. Once everything is correct, add this new mesh to your copy of `NewMesh.m`.
4. Study the convergence of the first few eigenvalues of $-\Delta u = 0$, $u|_{\partial\Omega} = 0$ for the L-shaped domain under mesh refinement. Do you observe the same phenomenon as in the finite difference case?
5. Suppose we want to study the eigenvalues for the Neumann problem, i.e., we want $\nabla u \cdot n|_{\partial\Omega} = 0$. How would you modify your mesh generation code to handle this? Compute the first few eigenvalues and eigenfunctions for the Neumann problem for different shapes.

Exercise 3: Robin boundary conditions

In this exercise, we consider the eigenvalue problem with Robin boundary conditions

$$-\Delta u = \lambda u \quad \text{on } \Omega, \qquad \nabla u \cdot n + pu = 0 \quad \text{on } \partial\Omega,$$

where $p > 0$ is a constant for simplicity. Recall that in this case, the energy bilinear form contains the extra boundary integral $p\int_{\partial\Omega} uv\, dS(x)$. We will compute the contribution of this term edge by edge along the physical boundary.

1. Given an edge $\mathcal{E}$ defined by the nodes (x_1, y_1), (x_2, y_2), calculate (with pencil and paper) the 2×2 **edge** mass matrix corresponding to the integrals $\int_{\mathcal{E}} \varphi_i \varphi_j\, dS(x)$, $i, j = 1, 2$.
2. Using the function `ComputeElementMassMatrix` as a model, write a function `ComputeEdgeMassMatrix` that implements your calculation in part 1.

3. Modify the assembly routine `FEeig` to incorporate the Robin boundary term. Save your result under a different name, e.g., `FEeig_Robin.m`. Check that your results are correct by calculating the first eigenvalue and eigenfunction for the unit square for $p = 1$. As you refine the grid, this value should approach $\lambda_1 \approx 3.4141\ldots$

7.3. Problem Set #3: Matrix eigenvalue problems.

Exercise 1: Sparse vs. dense matrices

In MATLAB, a matrix can be stored in *dense* or *sparse* format.

- A **dense** matrix is stored in a contiguous block of memory for n^2 real numbers, and zero entries are stored explicitly as the real number '0'.
- A **sparse** matrix only allocates enough memory for storing the non-zero entries in the matrix, plus some supporting data structures for locating these non-zero entries within the matrix. Zeros are not explicitly stored.

MATLAB provides two commands, `eig` and `eigs`, for computing the eigenvalues of dense and sparse matrices respectively. You have already used `eigs` in previous problem sets for calculating a few eigenvalues of sparse matrices arising from PDE discretizations. The aim of this exercise is to compare the two routines.

1. Generate the matrix of a discrete Laplacian using `numgrid` and `delsq` for $N = 4, 8, 16, 32, 64$. Use `eig` to compute **all** eigenvalues of the matrix. Note how much time the computation takes by enclosing your `eig` command within a tic-toc pair:

   ```
   tic; lambdas = eig(A); toc
   ```

   (Depending on the version, you may need to convert A to dense format first by calling `eig(full(A))`.) How does the running time behave as the size of the matrix increases?
2. Again for $N = 4, 8, 16, 32, 64$, use `eigs` to calculate (i) the 5 smallest eigenvalues, and (ii) the $m/2$ smallest eigenvalues, where m is the size of the matrix. Round to the nearest integer if m is odd. Note again the running time for each case.
3. Under what circumstances is `eigs` more efficient than `eig`, and vice versa?

Exercise 2: Power iteration

1. Write a function `PowerMethod` that implements the power method. Use the following header:

   ```
   function [lambda, y] = PowerMethod(A, y0)
   % POWERMETHOD calculates the largest eigenvalue in magnitude using the
   % power method
   % [lambda, y]=PowerMethod(A,y0) uses the power method to approximate
   % the largest eigenvalue lambda of A, using y0 as the initial vector.
   % The return value y is an approximate eigenvector associated with
   % lambda.
   ```

   You may use the function `ShiftInvert` as a reference when writing this routine.
2. Let A be a symmetric positive definite matrix, so that all its eigenvalues are positive. There are two ways of finding its smallest eigenvalue:
   - Use the inverse power method (i.e., shift and invert with a zero shift),
   - Apply the power method to the matrix $\gamma I - A$, where γ is an upper bound on the largest eigenvalue of A, e.g., its maximum row sum $\gamma = \|A\|_\infty$.

Which method is faster in terms of (i) number of iterations and (ii) running time? For the running time, you need to consider the cost of solving the linear system $A\mathbf{x} = \mathbf{b}$ versus multiplying A by a vector $\mathbf{x}$. Try to construct matrices for which one method is faster than the other, and vice versa.

Exercise 3: Tridiagonal matrices

Write a function `Sturm` that calculates the kth eigenvalue of a symmetric tridiagonal matrix. Use the following header:

```
% STURM calculates the kth eigenvalue of a symmetric tridiagonal matrix
% [lambda, y] = sturm(alpha, beta, k) calculates the kth eigenvalue lambda
% of the symmetric tridiagonal matrix A with [beta(i-1), alpha(i), beta(i)]
% in its i-th row. beta is assumed to be non-zero everywhere. The method
% uses Sturm sequences to find a good enough approximation to the k-th
% eigenvalue, then shift-and-invert is used to find a better approximation
% as well as the corresponding eigenvector y.
```

To write such a function, proceed as follows:

1. First, use Gershgorin's theorem to estimate the interval $[y, z]$ containing all the eigenvalues:

   Gershgorin's theorem. If λ is an eigenvalue of a matrix A, then there exists an index i such that $|\lambda - a_{ii}| \leq \sum_{j\neq i} |a_{ij}|$.

   In other words, all the eigenvalues must be contained in intervals of the form $[a_{ii} - \rho_i, a_{ii} + \rho_i]$, where $\rho_i = \sum_{j\neq i} |a_{ij}|$. One can thus obtain y and z by taking some minima and maxima.
2. Next, bisect the interval until $\omega(y) = k-1$ and $\omega(z) = k$. To do so, you will need to evaluate $p_0(\lambda), \dots, p_n(\lambda)$ at the midpoint $x = (y+z)/2$ using the recurrence shown on the slide, and then count the number of sign changes in the sequence.
3. Bisect the interval two more times so that the midpoint obtained is much closer to the kth eigenvalue than to any other eigenvalue.
4. Finally, call `ShiftInvert` to obtain accurate values for the eigenpair.

Test your function on the matrix

$$A = \begin{bmatrix} 2 & -1 & & & \\ -1 & 2 & \ddots & & \\ & \ddots & \ddots & \ddots & \\ & & \ddots & \ddots & -1 \\ & & & -1 & 2 \end{bmatrix},$$

whose kth eigenvalue is given by $4\sin^2\left(\frac{k\pi}{2(n+1)}\right)$. As a further test, generate a symmetric matrix A (e.g. the Laplacian for an L-shaped domain), then run the following:

```
T = hess(full(A));
[lambda, y] = sturm(diag(T),diag(T,-1),7);
```

The first command calculates the tridiagonal form of A having the same eigenvalues, and the `diag` calls inside `sturm` extract the diagonal and subdiagonal of T before passing it to `sturm`, which then computes the seventh smallest eigenvalue. Compare your results with that obtained by `eig` or `eigs`.

Exercise 4: Lanczos method

1. Derive the Lanczos method for the *generalized* eigenvalue problem

$$A\mathbf{x} = \lambda B\mathbf{x},$$

where A and B are symmetric and B is positive definite. Make sure your method can be expressed entirely in terms of the following operations:
   – Scalar operations;
   – Vector manipulations of the form $\mathbf{z} \leftarrow \alpha\mathbf{x} + \mathbf{y}$;
   – Dot products between two vectors;
   – The matrix vector product $A\mathbf{x}$,
   – Linear solves involving B, i.e., solve $B\mathbf{x} = \mathbf{y}$.
2. Using `myLanczos` as a model, implement the Lanczos method for the generalized eigenvalue problem. Use your routine to calculate the lowest eigenmode for the Laplacian problem discretized by the finite element method. Taking the result from `eigs` as the exact solution, plot the error of the smallest eigenvalue as a function of the number of Lanczos iterations.
3. Compare the convergence of Lanczos with that of the power method. Which method converges faster?

References

[1] Eldar Akhmetgaliyev, Oscar P. Bruno, and Nilima Nigam, *A boundary integral algorithm for the Laplace Dirichlet-Neumann mixed eigenvalue problem*, J. Comput. Phys. **298** (2015), 1–28, DOI 10.1016/j.jcp.2015.05.016. MR3374535

[2] Grégoire Allaire and Sidi Mahmoud Kaber, *Numerical linear algebra*, Texts in Applied Mathematics, vol. 55, Springer, New York, 2008. Translated from the 2002 French original by Karim Trabelsi. MR2365296

[3] Edward Anderson, Zhaojun Bai, Christian Bischof, L. Susan Blackford, James Demmel, Jack Dongarra, Jeremy Du Croz, Anne Greenbaum, Sven Hammarling, Alan McKenney, and Danny C. Sorensen. *LAPACK Users' Guide.* SIAM, 1999.

[4] Jared L. Aurentz, Thomas Mach, Raf Vandebril, and David S. Watkins, *Fast and backward stable computation of roots of polynomials*, SIAM J. Matrix Anal. Appl. **36** (2015), no. 3, 942–973, DOI 10.1137/140983434. MR3366913

[5] I. Babuška and J. Osborn, *Eigenvalue problems*, Handbook of numerical analysis, Vol. II, Handb. Numer. Anal., II, North-Holland, Amsterdam, 1991, pp. 641–787. MR1115240

[6] I. Babuška and J. Pitkäranta, *The plate paradox for hard and soft simple support*, SIAM J. Math. Anal. **21** (1990), no. 3, 551–576, DOI 10.1137/0521030. MR1046789

[7] I. Babuška. Stability of the domain under perturbation of the boundary in fundamental problems in the theory of partial differential equations principally in connection with the theory of elasticity, parts I and II. *Czechoslovak Math. J*, 11:75–105, 1961.

[8] Arnd Bäcker, *Numerical aspects of eigenvalue and eigenfunction computations for chaotic quantum systems*, The mathematical aspects of quantum maps, Lecture Notes in Phys., vol. 618, Springer, Berlin, 2003, pp. 91–144, DOI 10.1007/3-540-37045-5_4. MR2159324

[9] Timo Betcke and Lloyd N. Trefethen, *Reviving the method of particular solutions*, SIAM Rev. **47** (2005), no. 3, 469–491, DOI 10.1137/S0036144503437336. MR2178637

[10] L. Susan Blackford, Jaeyoung Choi, Andy Cleary, Eduardo D'Azevedo, James Demmel, Inderjit Dhillon, Jack Dongarra, Sven Hammarling, Greg Henry, Antoine Petitet, K. Stanley, David W. Walker, and R. Clint Whaley. *ScaLAPACK Users' Guide.* SIAM, 1997.

[11] Daniele Boffi, *Finite element approximation of eigenvalue problems*, Acta Numer. **19** (2010), 1–120, DOI 10.1017/S0962492910000012. MR2652780

[12] J. H. Bramble and J. E. Osborn, *Rate of convergence estimates for nonselfadjoint eigenvalue approximations*, Math. Comp. **27** (1973), 525–549, DOI 10.2307/2005658. MR0366029

[13] Susanne C. Brenner and L. Ridgway Scott, *The mathematical theory of finite element methods*, 3rd ed., Texts in Applied Mathematics, vol. 15, Springer, New York, 2008. MR2373954

[14] Carsten Carstensen and Joscha Gedicke, *Guaranteed lower bounds for eigenvalues*, Math. Comp. **83** (2014), no. 290, 2605–2629, DOI 10.1090/S0025-5718-2014-02833-0. MR3246802

[15] Françoise Chatelin, *Spectral approximation of linear operators*, Classics in Applied Mathematics, vol. 65, Society for Industrial and Applied Mathematics (SIAM), Philadelphia, PA, 2011. With a foreword by P. Henrici; With solutions to exercises by Mario Ahués; Reprint of the 1983 original [MR0716134]. MR3405533

[16] Ernst Florens Friedrich Chladni. Neue Beiträge zur Akustik. In *Entdeckungen über die Theorie des Klanges*. Leipzig, 1821.

[17] Xiaoying Dai, Jinchao Xu, and Aihui Zhou, *Convergence and optimal complexity of adaptive finite element eigenvalue computations*, Numer. Math. **110** (2008), no. 3, 313–355, DOI 10.1007/s00211-008-0169-3. MR2430983

[18] Lawrence C. Evans, *Partial differential equations*, Graduate Studies in Mathematics, vol. 19, American Mathematical Society, Providence, RI, 1998. MR1625845

[19] L. Fox, P. Henrici, and C. Moler, *Approximations and bounds for eigenvalues of elliptic operators*, SIAM J. Numer. Anal. **4** (1967), 89–102, DOI 10.1137/0704008. MR0215542

[20] Martin J. Gander and Felix Kwok, *Chladni figures and the Tacoma bridge: motivating PDE eigenvalue problems via vibrating plates*, SIAM Rev. **54** (2012), no. 3, 573–596, DOI 10.1137/10081931X. MR2966726

[21] Walter Gander, Martin J. Gander, and Felix Kwok, *Scientific computing*, Texts in Computational Science and Engineering, vol. 11, Springer, Cham, 2014. An introduction using Maple and MATLAB. MR3287477

[22] I. M. Gelfand and S. V. Fomin, *Calculus of variations*, Revised English edition translated and edited by Richard A. Silverman, Prentice-Hall, Inc., Englewood Cliffs, N.J., 1963. MR0160139

[23] Gene H. Golub and Charles F. Van Loan, *Matrix computations*, 3rd ed., Johns Hopkins Studies in the Mathematical Sciences, Johns Hopkins University Press, Baltimore, MD, 1996. MR1417720

[24] F. Hecht, *New development in freefem++*, J. Numer. Math. **20** (2012), no. 3-4, 251–265. MR3043640

[25] Bert Hubbard, *Bounds for eigenvalues of the free and fixed membrane by finite difference methods*, Pacific J. Math. **11** (1961), 559–590. MR0141223

[26] Bert Hubbard, *Eigenvalues of the non-homogeneous rectangular membrane by finite difference methods*, Arch. Rational Mech. Anal. **9** (1962), 121–133, DOI 10.1007/BF00253339. MR0141221

[27] Claes Johnson. *Numerical solution of partial differential equations by the finite element method.* Courier Corporation, 2012.

[28] Bo Kågström and Daniel Kressner, *Multishift variants of the QZ algorithm with aggressive early deflation*, SIAM J. Matrix Anal. Appl. **29** (2006/07), no. 1, 199–227, DOI 10.1137/05064521X. MR2288022

[29] James R. Kuttler, *Finite difference approximations for eigenvalues of uniformly elliptic operators*, SIAM J. Numer. Anal. **7** (1970), 206–232, DOI 10.1137/0707014. MR0273841

[30] J. R. Kuttler and V. G. Sigillito, *Eigenvalues of the Laplacian in two dimensions*, SIAM Rev. **26** (1984), no. 2, 163–193, DOI 10.1137/1026033. MR738929

[31] Richard S. Laugesen. Spectral Theory of Partial Differential Equations. Available at `https://wiki.cites.illinois.edu/wiki/display/MATH595STP/Math+595+STP`, April 2017.

[32] Beresford N. Parlett, *The symmetric eigenvalue problem*, Classics in Applied Mathematics, vol. 20, Society for Industrial and Applied Mathematics (SIAM), Philadelphia, PA, 1998. Corrected reprint of the 1980 original. MR1490034

[33] V. G. Prikazchikov. The finite difference eigenvalue problem for fourth-order elliptic operator. *U.S.S.R. Comput. Maths Math. Phys.*, 17:89–99, 1978.

[34] V. G. Prikazchikov and A. N. Khimich. The eigenvalue difference problem for the fourth order elliptic operator with mixed boundary conditions. *U.S.S.R. Comput. Maths Math. Phys.*, 25:137–144, 1985.

[35] P.-A. Raviart and J.-M. Thomas, *Introduction à l'analyse numérique des équations aux dérivées partielles* (French), Collection Mathématiques Appliquées pour la Maîtrise. [Collection of Applied Mathematics for the Master's Degree], Masson, Paris, 1983. MR773854

[36] Walther Ritz. Theorie der Transversalschwingungen einer quadratischen Platte mit freien Rändern. *Annalen der Physik*, 18(4):737–807, 1909.

[37] Yousef Saad, *Numerical methods for large eigenvalue problems*, Classics in Applied Mathematics, vol. 66, Society for Industrial and Applied Mathematics (SIAM), Philadelphia, PA, 2011. Revised edition of the 1992 original [1177405]. MR3396212

[38] Yousef Saad, *Iterative methods for sparse linear systems*, 2nd ed., Society for Industrial and Applied Mathematics, Philadelphia, PA, 2003. MR1990645

[39] Gerard L. G. Sleijpen and Henk A. Van der Vorst, *A Jacobi-Davidson iteration method for linear eigenvalue problems*, SIAM Rev. **42** (2000), no. 2, 267–293, DOI 10.1137/S0036144599363084. MR1778354

[40] Walter A. Strauss, *Partial differential equations*, 2nd ed., John Wiley & Sons, Ltd., Chichester, 2008. An introduction. MR2398759

[41] Andrea Toselli and Olof Widlund, *Domain decomposition methods—algorithms and theory*, Springer Series in Computational Mathematics, vol. 34, Springer-Verlag, Berlin, 2005. MR2104179

[42] H. F. Weinberger, *Upper and lower bounds for eigenvalues by finite difference methods*, Comm. Pure Appl. Math. **9** (1956), 613–623, DOI 10.1002/cpa.3160090329. MR0084185

[43] H. F. Weinberger, *Lower bounds for higher eigenvalues by finite difference methods*, Pacific J. Math. **8** (1958), 339–368; erratum, 941. MR0107372

[44] J. H. Wilkinson, *The evaluation of the zeros of ill-conditioned polynomials. I, II*, Numer. Math. **1** (1959), 150–180, DOI 10.1007/BF01386381. MR0109435

[45] J. H. Wilkinson, *The algebraic eigenvalue problem*, Clarendon Press, Oxford, 1965. MR0184422

Department of Mathematics, Hong Kong Baptist University, Kowloon Tong, Hong Kong

Email address: felix_kwok@hkbu.edu.hk

Contemporary Mathematics
Volume **720**, 2018
https://doi.org/10.1090/conm/720/14524

Spectral geometry

Yaiza Canzani

ABSTRACT. I wrote these lectures using the material in set of notes "Analysis on manifolds via the Laplacian" available at `http://www.math.harvard.edu/~canzani/docs/Laplacian.pdf`. The material there corresponds to a year long course, so it contains more topics and many more details. The following references were important sources for these notes:

- *Eigenvalues in Riemannian geometry.* By I. Chavel.
- *Old and new aspects in Spectral Geometry.* By M. Craiveanu, M. Puta and T. Rassias.
- *The Laplacian on a Riemannian manifold.* By S. Rosenberg.
- *Local and global analysis of eigenfunctions on Riemannian manifolds.* By S. Zelditch.

Enjoy!!

CONTENTS

1. What makes the Laplacian special?

In this chapter we motivate the study of the Laplace operator. To simplify exposition, we do this by concentrating on planar domains.

1.1. Daily life problems. Let $\Omega \subset \mathbb{R}^n$ be a connected domain and consider the operator Δ acting on $C^\infty(\Omega)$ that simply differentiates a function $\varphi \in C^\infty(\Omega)$ two times with respect to each position variable:

$$\Delta\varphi = \sum_{i=1}^{n} \frac{\partial^2 \varphi}{\partial x_i^2}.$$

2010 *Mathematics Subject Classification.* Primary 35P15, 35P20, 58J32, 58J50.

Pierre-Simon de Laplace
(Credit: Wikimedia Commons)

Here are some examples where the Laplacian plays a key role:

Heat diffusion. If you are interested in understanding how would heat propagate along $\Omega \subset \mathbb{R}^n$ then you should solve the *Heat Equation*

$$\Delta u(x,t) = \frac{1}{c}\frac{\partial}{\partial t}u(x,t)$$

where c is the conductivity of the material of which Ω is made of, and $u(x,t)$ is the temperature at the point $x \in \Omega$ at time t.

You could also think you have an insulated region Ω (it could be a wire, a ball, etc.) and apply certain given temperatures on the edge $\partial\Omega$. If you want to know what the temperature will be after a long enough period of time (that is, the steady state temperature distribution), then you need to find a solution of the heat equation that be independent of time. The steady state temperature solution will be a function $u(x_1, \ldots, x_n, t)$ such that

$$\Delta u = 0.$$

Wave propagation. Now, instead of applying heat to the surface suppose you cover it with a thin layer of some fluid and you wish to describe the motion of the surface of the fluid. Then you will need to solve the *Wave equation*

$$\Delta u(x,t) = \frac{1}{c}\frac{\partial^2}{\partial t^2}u(x,t)$$

where $\sqrt{c}$ is the speed of sound in your fluid, and $u(x,t)$ denotes the height of the wave above the point x at time t.

You could also think of your domain Ω as the membrane of a drum, in which case its boundary $\partial\Omega$ would be attached to the rim of the drum. Suppose you want to study what will happen with the vibration you would generate if you hit it. Then, you should also solve the wave equation $\Delta u(x,t) = \frac{\partial^2}{\partial^2 t}u(x,t)$ for your drum, but this time you want to make sure that you take into account that the border of the membrane is fixed. Thus, you should also ask your solution to satisfy $u(x,t) = 0$ for all points $x \in \partial\Omega$.

Quantum particles. If you are a bit more eccentric and wish to see how a quantum particle moves inside Ω (under the assumption that there are no external forces) then you need to solve the *Schrödinger Equation*

$$-\frac{\hbar^2}{2m}\Delta u(x,t) = i\hbar\frac{\partial}{\partial t}u(x,t)$$

where $\hbar$ is Planck's constant and m is the mass of the free particle. Normalizing u so that $\|u(\cdot,t)\|_{L^2(\Omega)} = 1$ one interprets $u(x,t)$ as a probability density. That is, if $A \subset \Omega$ then the probability that your quantum particle be inside A at time t is given by $\int_A |u(x,t)|^2 dx$.

Why not another operator? The Laplacian on $\mathbb{R}^n$ commutes with translations and rotations. That is, if T is a translation or rotation then $\Delta(\varphi \circ T) = (\Delta\varphi) \circ T$. Something more striking occurs, if S is any operator that commutes with translations and rotations then there exist coefficients $a_1, \ldots, a_m$ making $S = \sum_{j=1}^m a_j \Delta^j$. Therefore, it is not surprising that the Laplacian will be a main star in any process whose underlying physics are independent of position and direction such as heat diffusion and wave propagation in $\mathbb{R}^n$.

1.2. You want to solve the Helmholtz equation. There are of course many more problems involving the Laplacian, but we will focus on these ones to stress the importance of solving the eigenvalue problem (also known as Helmholtz equation)

$$-\Delta\varphi - \lambda\varphi.$$

It is clear that if one wants to study harmonic functions then one needs to solve the equation

$$-\Delta\varphi = \lambda\varphi \qquad \text{with} \quad \lambda = 0.$$

So the need for understanding solutions of the Helmholtz equation for problems such as the static electric field or the steady-state fluid flow is straightforward. In order to attack the heat diffusion, wave propagation and Schrödinger problems described above, a standard method (inspired by Stone-Weierstrass Theorem) is to look for solutions $u(x,t)$ of the form $u(x,t) = \alpha(t)\varphi(x)$. For instance if you do this and look at the Heat equation then you must have

$$\frac{\Delta\varphi(x)}{\varphi(x)} = \frac{\alpha'(t)}{\alpha(t)} \qquad x \in \Omega, \quad t > 0.$$

This shows that there must exist a $\lambda \in \mathbb{R}$ such that

$$\alpha' = -\lambda\alpha \quad \text{and} \quad -\Delta\varphi = \lambda\varphi.$$

Therefore φ must be an eigenfunction of the Laplacian with eigenvalue λ and $\alpha(t) = e^{-\lambda t}$. Once you have these particular solutions $u_k = e^{-\lambda_k t}\varphi_k$ you use the superposition principle to write a general solution

$$u(x,t) - \sum_k a_k e^{-\lambda_k t}\varphi_k(x)$$

where the coefficients a_k are chosen depending on the initial conditions. You could do the same with the wave equation (we do it in detail for a guitar string in

Section 1.3.1) or with the Schrödinger equation and you will also find particular solutions of the form $u_k(x,t) = \alpha_k(t)\varphi_k(x)$ with

$$-\Delta\varphi_k = \lambda_k\varphi_k \qquad \text{and} \qquad \alpha_k(t) = \begin{cases} e^{-\lambda_k t} & \text{Heat eqn,} \\ e^{i\sqrt{\lambda_k}t} & \text{Wave eqn,} \\ e^{i\lambda_k t} & \text{Schrödinger eqn.} \end{cases}$$

1.3. Inverse problem: Can you hear the shape of a drum?

Inverse problem: If I know (more or less) the Laplace eigenvalues of a domain, what can I deduce of its geometry?

Suppose you have perfect pitch. Could you derive the shape of a drum from the music you hear from it? More generally, can you determine the structural soundness of an object by listening to its vibrations? This question was first posed by Schuster in 1882. As Berger says in his book *A panoramic view of Riemannian Geometry*,

> "Already in the middle ages bell makers knew how to detect invisible cracks by sounding a bell on the ground before lifting it up to the belfry. How can one test the resistance to vibrations of large modern structures by non-destructive essays?... A small crack will not only change the boundary shape of our domain, one side of the crack will strike the other during vibrations invalidating our use of the simple linear wave equation. On the other hand, heat will presumably not leak out of a thin crack very quickly, so perhaps the heat equation will still provide a reasonable approximation for a short time..."

An infinite sequence of numbers determines via Fourier analysis an integrable function. It wouldn't be that crazy if an infinite sequences of eigenvalues would determine the shape of the domain. Unfortunately, the answer to the question *can you hear the shape of a drum?* is no. This was proved in 1992 by Gordon, Web and Wolpert. Nowadays many planar domains are known to have different shapes but exactly the same spectrum. Not all is lost. One can still derive a lot of information of a domain by knowing its eigenvalues. Using the heat kernel, one can prove that as $t \to 0$,

$$\sum_{k=1}^{\infty} e^{-\lambda_k t} \sim \frac{1}{4\pi t}\left(\text{area}(\Omega) - \sqrt{4\pi t}\,\text{length}(\partial\Omega) + \frac{2\pi t}{3}(1-\gamma(\Omega))\right)$$

where $\gamma(\Omega)$ is the number of holes of Ω and Ω is a smooth, bounded domain. The eigenvalues λ_k are the ones corresponding to the Laplacian on Ω enforcing $\varphi_k|_{\partial\Omega} = 0$. The first term was proved by Weyl in 1911. The second term was proved in 1954 by Pleijel. It shows that you can hear wether a drum is circular or not (because of the isoperimetric inequality $(\text{length}(\partial\Omega))^2 \geq 4\pi \cdot \text{area}(\Omega)$ whose equality is only attained by circles). The first term was found by Mark Kac in 1966 and it was rigorously justified by McKean and Singer in 1967.

This means that if you know the full sequence of eigenvalues of your favorite domain Ω then you can deduce its area, its perimeter and the number of holes in it!!

Mark Kac
(Credit: Mathematisches Forschungsinstitut Oberwolfach gGmbH (MFO), used under the terms of the Creative Commons License Attribution-Share Alike 2.0 Germany.)

1.3.1. *Hearing the length of a guitar string.* Consider an interval $[0, \ell]$ with Dirichlet boundary conditions $\varphi(0) = \varphi(\ell) = 0$.

The eigenfunctions are

$$\varphi_k(x) = \sin\left(\frac{k\pi}{\ell}x\right) \qquad \text{for } k \geq 1$$

with eigenvalues $\lambda_k = \left(\frac{k\pi}{\ell}\right)^2$ for $k \geq 1$.

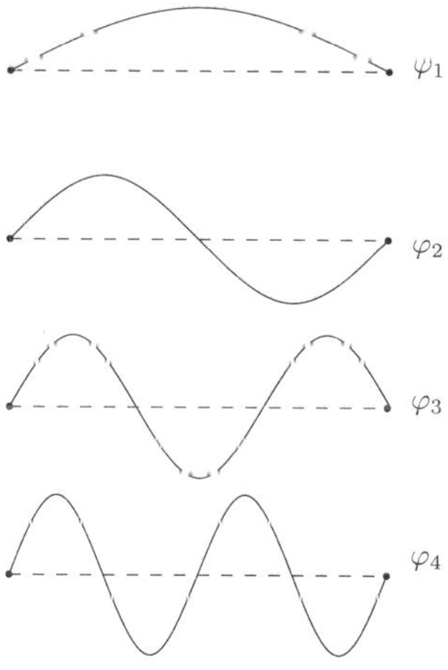

Jean-Baptiste d'Alembert
(Credit: Wikimedia Commons)

Daniel Bernoulli

After the first half of the 18th century mathematicians such as d'Alembert and Bernoulli developed the theory of a vibrational string. As one should expect, the vibrations of a string will depend on many factors such us its length, mass and tension.

To simplify our exposition consider a guitar string of length ℓ which we model as the interval $[0,\ell]$. Assume further that the density mass and the tension are constant and equal to 1. Today it comes as no surprise that the behavior of a vibrating string is described by the wave equation. That is, if we write x for a point in the string $[0,\ell]$ and t for the time variable, then the height $u(x,t)$ of the string above the point x after a time t should satisfy the wave equation

$$\Delta u(x,t) = \frac{\partial^2}{\partial t^2} u(x,t).$$

There are infinitely many solutions to this problem. But we already know that there are constraints to this problem that we should take into account since the endpoints of the string are fixed and so $u(x,t)$ must satisfy $u(0,t) = 0 = u(\ell,t)$ for all time t. In addition having a unique solution to our problem depends upon specifying the initial shape of the string $f(x) = u(x,0)$ and its initial velocity $g(x) = \partial_t u(x,0)$. All in all, we are solving the system

$$\begin{cases} \frac{\partial^2}{\partial x^2} u(x,t) = \frac{\partial^2}{\partial t^2} u(x,t) & x \in [0,\ell],\ t > 0, \\ u(0,t) = 0 = u(\ell,t) & t > 0, \\ u(x,0) = f(x) & x \in [0,\ell], \\ \partial_t u(x,0) = g(x) & x \in [0,\ell]. \end{cases}$$

A general solution of this problem has the form

$$u(x,t) = \sum_{k=1}^{\infty} \alpha_k(t) \varphi_k(x), \tag{1}$$

where

$$\alpha_k(t) = a_k \cos\left(\frac{k\pi}{\ell} t\right) + b_k \sin\left(\frac{k\pi}{\ell} t\right).$$

The coefficients are $a_k = \langle f, \varphi_k \rangle$ and $b_k = \langle g, \varphi_k \rangle$.

The functions φ_k are called harmonic modes for the string $[0, \ell]$. Since $\lambda_k = \left(\frac{k\pi}{\ell}\right)^2$ is the eigenvalue of the wave $\varphi_k(x) = \sin\left(\frac{k\pi}{\ell}x\right)$, the connection between the eigenvalues λ_k and the frequencies f_k of the harmonic modes of the string is obvious:

$$f_k = \frac{\sqrt{\lambda_k}}{2\pi}.$$

Therefore, the higher the eigenvalue, the higher the frequency is.

Consider the Fourier transform of a function φ as

$$\mathcal{F}(\varphi)(\xi) = \int_{-\infty}^{\infty} \varphi(x)e^{-ix\xi}dx.$$

The Fourier transform of the function $\varphi_k(x) = \sin(\sqrt{\lambda_k}x)$ is

$$\mathcal{F}(\varphi_k)(\xi) = \frac{\pi}{i}\Big(\delta(\xi - \sqrt{\lambda_k}) - \delta(\xi + \sqrt{\lambda_k})\Big).$$

This is because

$$\sin(\sqrt{\lambda_k}x) = \frac{e^{i\sqrt{\lambda_k}x} - e^{-i\sqrt{\lambda_k}x}}{2i} \qquad \text{and} \qquad \mathcal{F}(e^{i\sqrt{\lambda_k}x})(\xi) = 2\pi\,\delta(\xi - \sqrt{\lambda_k}).$$

If you pluck a guitar string then you obtain a wave of the form $u(x,t) = \sum_{k=1}^{\infty} \alpha_k(t)\varphi_k(x)$, and by applying the Fourier transform to it you get

$$\mathcal{F}(u(\cdot,t))(\xi) = \frac{\pi}{i}\sum_{k=1}^{\infty} \alpha_k(t)\Big(\delta\Big(\xi - \sqrt{\lambda_k}\Big) - \delta\Big(\xi + \sqrt{\lambda_k}\Big)\Big).$$

So you recover all the relevant frequencies and hence all the eigenvalues. In the picture below we have the graphs of $s_1(x) = \sin(x)$, $s_2(x) = \frac{1}{3}\sin(2x)$ and $s_3(x) = \frac{1}{8}\sin(4x)$.

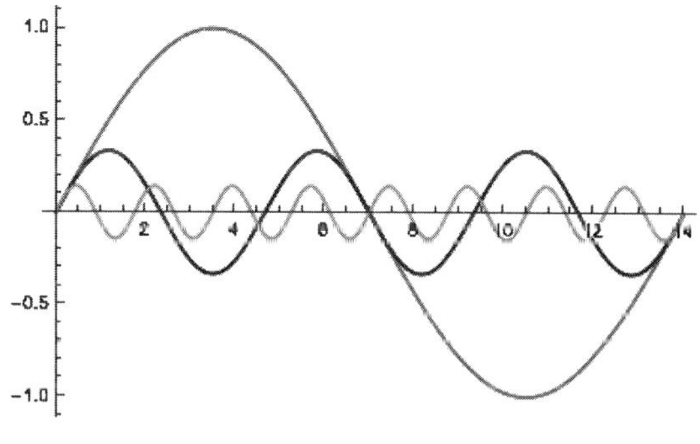

In the first picture we show the graph of the wave $s_1 + s_2 + s_3$, while in the second picture we show the Fourier transform of the function $s_1 + s_2 + s_3$.

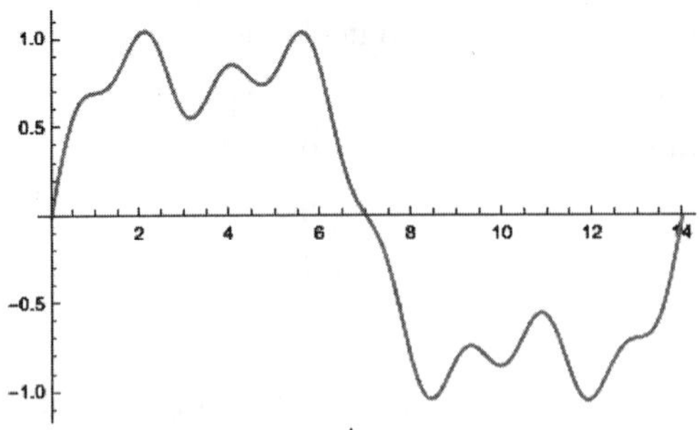

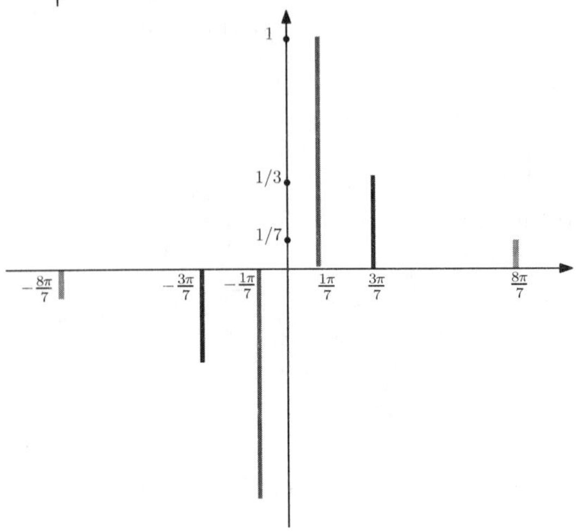

1.4. Direct problems.

Direct problem: If I know (more or less) the shape of a domain, what can I deduce of its Laplace eigenvalues?

1.4.1. *The first eigenvalue: Rayleigh Conjecture.* The first eigenvalue λ_1 of the Laplacian on an interval or a region of the plane is called the *fundamental tone.* This is because either on a vibrating guitar string or drum membrane the first eigenvalue corresponds to the leading frequency of oscillation and it is therefore the leading tone you hear when you play one of these instruments.

Seen from a heat-diffusion point of view, since the solutions of the heat equation are of the form $u(x,t) = \sum_k a_k e^{-\lambda_k t}\varphi_k(x)$, it is clear that (λ_1, φ_1) give the dominant information because $e^{-\lambda_1 t}\varphi_1(x)$ is the mode that decays with slowest rate as time passes by. From this last point of view it is natural to expect that the geometry of Ω should be reflected on λ_1 to some extent. For instance the largest the boundary $\partial\Omega$ is, the more quickly the heat should wear off. That is, if we consider a domain Ω and a ball B of same area as Ω, then we expect the heat on Ω to diffuse more quickly than that of B. Therefore, we should have

Faber-Krahn inequality:

$$\lambda_1(\Omega) \geq \lambda_1(B).$$

This result was proved by Faber and Krahn in 1923. As expected, it extends to any dimension.

Lord Rayleigh
(Credit: Wikimedia Commons)

1.4.2. *Counting eigenvalues: Weyl's law.* Jeans asked once what is the energy corresponding to an infinitesimal frequency interval. In 1966 Mark Kac told this story in a very illustrating manner:

> ...At the end of October of 1910 the great Dutch physicist H. A. Lorentz was invited to Götingen to deliver a Wolfskehl lecture... Lorentz gave five lectures under the overall title "Old and new problems of physics" and at the end of the fourth lecture he spoke as follows (in free translation from the original German):
>
> *In conclusion, there is a mathematical problem which perhaps will arouse the interest of mathematicians who are present. It originates in the radiation theory of Jeans.*
>
> *In an enclosure with a perfectly reflecting surface, there can form standing electromagnetic waves analogous to tones over an organ pipe: we shall confine our attention to very high overtones. Jeans asks for the energy in the frequency interval $d\nu$. To this end, he calculates the number of overtones which lie between frequencies ν and $\nu + d\nu$, and multiplies this number by the energy which belongs to the frequency ν, and which according to a theorem of statistical mechanics, is the same for all frequencies.*
>
> *It is here that there arises the mathematical problem to prove that the number of sufficiently high overtones which lie between ν and $\nu + d\nu$ is independent of the shape of the enclosure, and is simply proportional to its volume. For many shapes for which calculations can be carried out, this theorem has been verified in a Leiden dissertation. There is no doubt that it holds in general even for multiply connected regions. Similar theorems for other vibrating structures, like membranes, air masses, etc., should also hold.*

If we express the Lorentz conjecture in a vibrating membrane Ω, it becomes of the following form: Let $\lambda_1 \leq \lambda_2 \leq \dots$ be the Laplace eigenvalues corresponding to the problem

$$-\Delta\varphi_k = \lambda_k\varphi_k \qquad \varphi_k|_{\partial\Omega} = 0.$$

Then

$$N(\lambda) = \#\{\lambda_k : \lambda_k < \lambda\} \sim \frac{\text{area}(\Omega)}{4\pi}\lambda \qquad \text{as } \lambda \to \infty.$$

D. Hilbert was attending these lectures and predicted as follows: "This theorem would not be proved in my life time." But, in fact, Hermann Weyl, a graduate student at that time, was also attending these lectures. Weyl proved this conjecture four months later in February of 1911.

Hermann Weyl
(Credit: Mathematisches Forschungsinstitut Oberwolfach gGmbH (MFO), used under the terms of the Creative Commons License Attribution-Share Alike 2.0 Germany.)

We will prove this in specific examples such as rectangles and the torus. Later on we will prove the analogue result for compact Riemannian manifolds (M, g). Let $\lambda_0 \leq \lambda_1 \leq \dots$ be the Laplace eigenvalues repeated according to its multiplicity. Then

$$N(\lambda) \sim \frac{\omega_n}{(2\pi)^n} vol_g(M)\lambda^{n/2}, \qquad \lambda \to \infty$$

where ω_n is the volume of the unit ball in $\mathbb{R}^n$.

In particular,

$$\lambda_j \sim \frac{(2\pi)^2}{(\omega_n vol_g(M))^{2/n}} j^{2/n}, \qquad j \to \infty.$$

1.4.3. *Weyl's law for an interval* $[0, \ell]$. Note that if we scale our domain by a factor $a > 0$ we get $\lambda_k\big[(0, a\ell)\big] = \frac{1}{a^2}\lambda_k\big[(0, \ell)\big]$. Intuitively, the eigenvalue λ

must balance $\frac{d^2}{dx^2}$, and so $\lambda \sim (\text{length scale})^{-2}$. We also note that we have the asymptotics

$$\lambda_k \sim C\, k^2$$

where C is a constant independent of k.

For $\lambda > 0$ consider the eigenvalue counting function

$$N(\lambda) = \#\{\text{eigenvalues} < \lambda\}.$$

PROPOSITION 1. *(Weyl's law for intervals) Write λ_j for the Dirichlet or Neumann eigenvalues of the Laplacian on the interval $\Omega = [0, \ell]$. Then,*

$$N(\lambda) \sim \frac{\mathit{length}(\Omega)}{\pi}\sqrt{\lambda}.$$

PROOF.

$$N(\lambda) = \max\{k :\ \lambda_k < \lambda\} = \max\left\{k :\ \frac{k^2\pi^2}{\ell^2} < \lambda\right\} \sim \frac{\ell}{\pi}\sqrt{\lambda}.$$

Note that $\omega_1 = 2$. □

1.4.4. *Weyl's law for a rectangle (Exercise 1).* Consider a rectangle $\Omega = [0, \ell] \times [0, m]$ with Dirichlet boundary conditions. In this exercise you will prove that

$$N(\lambda) \sim \frac{\text{area}(\Omega)}{4\pi}\lambda.$$

(1) Prove that the eigenvalues are

$$\lambda_{jk} = \left(\frac{j\pi}{\ell}\right)^2 + \left(\frac{k\pi}{m}\right)^2 \qquad \text{for } j, k \geq 1.$$

(2) Prove that

$$\#\{(j, k) :\ \lambda_{jk} < \lambda\} \sim \frac{\text{area}(\Omega)}{4\pi}\lambda \qquad \text{as } \lambda \to \infty.$$

Hint: use the two figures below, where E_λ denotes an ellipse and $\tilde{E}_\lambda$ is the copy of the ellipse translated by $(-1, -1)$.

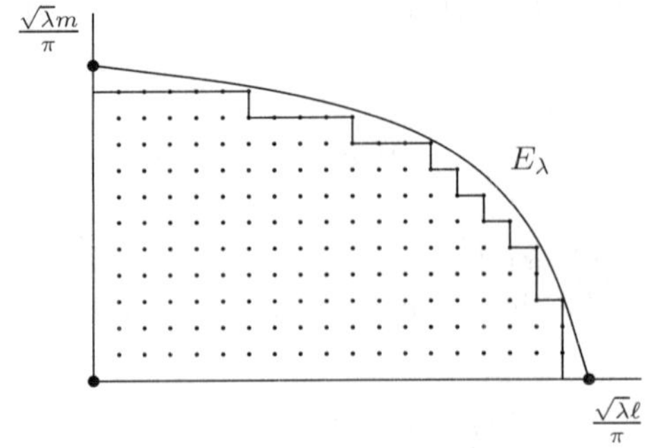

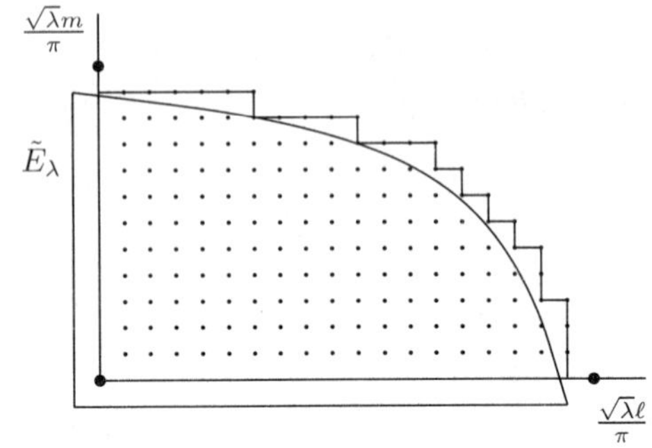

1.5. A hard problem: Understanding the eigenfunctions. As already mentioned, we may interpret the eigenfunction φ_k as the probability density of a quantum particle in the energy state λ_j. That is, the probability that a quantum particle in the state φ_k belongs to the set $A \subset \Omega$ if given by

$$\int_A |\varphi_k(x)|^2 \, dx.$$

High energy eigenfunctions are expected to reflect the dynamics of the geodesic flow. In the energy limit $\lambda \to \infty$ one should be able to recover classical mechanics from quantum mechanics. In the following picture (taken from *Many-body quantum chaos: Recent developments and applications to nuclei*) you can see how the dynamics of the geodesic flow for two different systems is reflected on the eigenfunctions. In the left column a cardioid billiard is represented. In the right column a ball billiard is shown. In the first line the trajectories of the geodesic flow for each system is shown. Then, from the second line to the fifth one, the graph of the functions $|\varphi_j|^2$ are shown for $\lambda_k = 100, 1000, 1500, 2000$. The darker the color, the higher the value of the modulus. One can see how a very chaotic system, such as the cardioid, yields a uniform distribution (chaotic) of the eigenfunctions. On the other hand, a very geometric dynamical system, such as the disc, yields geometric distributions of the eigenfunctions.

A beautiful result about the behavior of eigenfunctions takes place on manifolds with ergodic geodesic flow (like the cardioid above), including all manifolds with negative constant sectional curvature. This result says that in the high energy limit eigenfunctions are equidistributed.

Quantum ergodicity. If Ω is a compact and has ergodic geodesic flow, then there exists a density one subsequence of eigenfunctions $\{\varphi_{j_k}\}_k$ such that for any $A \subset M$

$$\lim_{k\to\infty} \int_\Omega |\varphi_{j_k}(x)|^2\, dx = \frac{\mathrm{vol}(A)}{\mathrm{vol}(\Omega)}.$$

By density one subsequence it is meant that $\inf_m \frac{\#\{k:\, j_k \le m\}}{m} = 1$. This result is due to Schnirelman (1973) finished by Colin de Verdière (1975).

Image processing. One may describe a given surface $M \subset \mathbb{R}^3$ by a function such as the normal vector. That is, to every point $x \in M$ you associate the normal vector at x

$$x \mapsto (n_1(x), n_2(x), n_3(x)).$$

Each function $n_j : M \to \mathbb{R}$ can be rewritten as an infinite series $n_j = \sum_{i=1}^\infty a_i^{(j)} \varphi_i$ by the Sturm-Liouville decomposition Theorem. You may then truncate the series and work with the approximates manifold described by the function

$$x \mapsto \left(\sum_{i=1}^N a_i^{(1)} \varphi_i(x), \sum_{i=1}^N a_i^{(2)} \varphi_i(x), \sum_{i=1}^N a_i^{(3)} \varphi_i(x) \right).$$

This is a way of encoding the geometry of a surface to some extent using as little information as you want (at the risk of having a worse approximation). In practice the way people have of computing eigenfunctions on the dragon is to discretize it and work with a discretized version of the Laplacian and its corresponding eigenfunctions. For instance, the method of approximating a surface by a finite number of eigenfunctions is used to perform a change of the position of some part of the surface. Suppose you have an armadillo standing on two legs and you wish to lift one of its legs reducing as much as possible the amount of computations that need to be carried to get the final result. What people are doing is to compute the first 99 eigenfunctions on the (discretized) armadillo and approximate the armadillo by them. Then, you apply the transformation to the approximate armadillo. Doing this is much cheaper -computation wise- than applying the transformation to the original armadillo. You may then add all the details to the transformed armadillo by another algorithm.

2. The Laplacian on a Riemannian manifold

2.1. Definition of the Laplacian on a manifold.

2.1.1. *Definition of a manifold.* A smooth manifold M of dimension n is a Hausdorff topological space with a countable basis of open sets that has a family of charts $\mathcal{U} = \{(U_\alpha, \varphi_\alpha)\}$, with $\varphi_\alpha : U_\alpha \subset M \to \mathbb{R}^n$ so that

(1) $\bigcup U_\alpha = M$
(2) If $U_\alpha \cap U_\beta \neq \emptyset$ then $\varphi_\beta \circ \varphi_\alpha^{-1} : \varphi_\alpha(U_\alpha \cap U_\beta) \to \varphi_\beta(U_\alpha \cap U_\beta)$ are C^∞ with C^∞ inverse. In this case we say that $(U_\alpha, \varphi_\alpha)$ and (U_β, φ_β) are compatible.
(3) Completness property: If (V, ψ) is a chart which is compatible with every $(U_\alpha, \varphi_\alpha) \in \mathcal{U}$ then $(V, \psi) \in \mathcal{U}$.

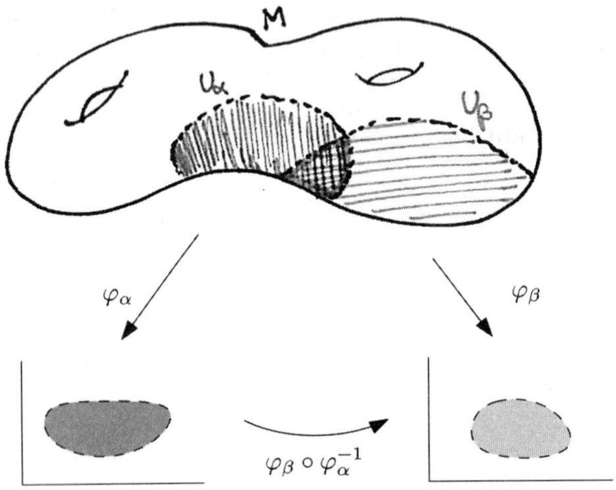

Examples of differential manifolds include $\mathbb{R}^n$, the sphere S^n, the torus $\mathbb{T}^n$. Products of manifolds are manifolds as well.

2.1.2. *Example:* $\mathbb{R}^n$. Only one chart suffices to see that $\mathbb{R}^n$ is a manifold. The chart is the identity:

$$U = \mathbb{R}^n, \qquad \phi : U \to \mathbb{R}^n, \qquad \phi(x) = x.$$

2.1.3. *Example: Circle.* The circle can be thought of as

$$S^1 = \{(\cos(\theta), \sin(\theta)) : \theta \in [0, 2\pi)\}.$$

The circle is a manifold of dimension 1. Let

$$U_1 = \{(\cos(\theta), \sin(\theta)) : \theta \in (0, 2\pi)\}$$

and $\varphi_1 : U_1 \to (0, 2\pi)$ be defined by its inverse $\psi_1 = \varphi_1^{-1}$ by

$$\psi_1(\theta) = (\cos(\theta), \sin(\theta)).$$

Since U_1 doesn't cover the entire circle (we are missing one point!), we need one more chart. For example, we could take $U_2 = \{(\cos(\theta), \sin(\theta)) : \theta \in (-\pi/2, \pi/2)\}$ and $\varphi_2 : U_2 \to (-\pi/2, \pi/2)$ defined by $\psi_2(\theta) = (\cos(\theta), \sin(\theta))$ and $\psi_2 = \varphi_2^{-1}$.

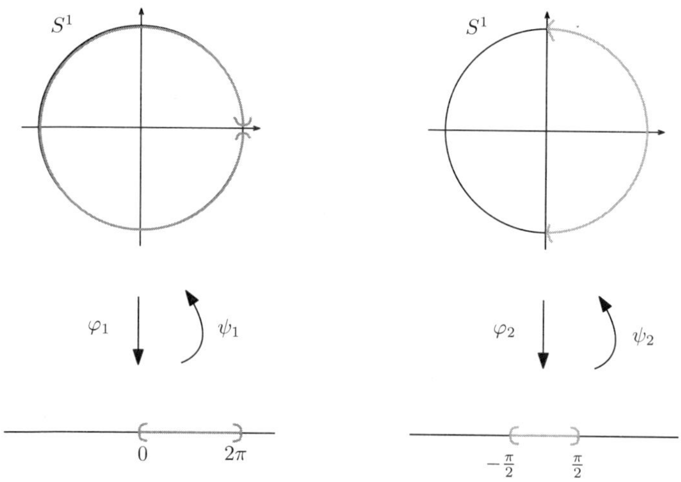

2.1.4. *Example: Sphere.* Here we consider the sphere

$$S^2 = \{(x, y, z) \in \mathbb{R}^3 : x^2 + y^2 + z^2 = 1\}.$$

The sphere is a manifold of dimension 2. Let

$$U_1 = \{(x, y, z) \in \mathbb{R}^3 : x^2 + y^2 + z^2 = 1 \,\text{and}\, x \neq 0\}$$

and $\varphi_1 : U_1 \to (0, \pi) \times (0, 2\pi)$ be defined by its inverse

$$\psi_1(\theta, \phi) = (\sin(\theta)\sin(\phi), \sin(\theta)\cos(\phi), \cos(\theta)).$$

These are called spherical coordinates and are illustrated below.

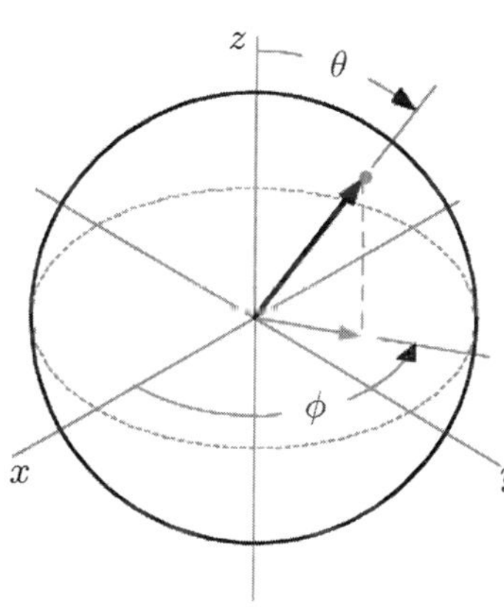

Of course, you should add more charts to include the $\{\theta = \pi, \phi = 2\pi\}$ line. Try it!

2.1.5. *Example: Torus.* The torus is the surface $\mathbb{T}^2 = \mathbb{R}^2/\mathbb{Z}^2$. That is, we put an equivalence relation on $\mathbb{R}^2$ where we say that two points (x_1, y_1) and (x_2, y_2) are the same provided $x_1 - y_1 \in \mathbb{Z}$ and $x_2 - y_2 \in \mathbb{Z}$. You may think of the torus as the surface you get after identifying opposite sides of a unit square, like in the picture below.

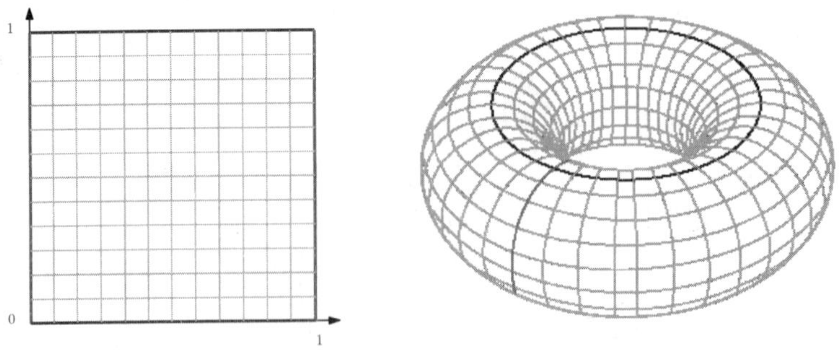

In particular, we will identify the torus $\mathbb{T}^2$ with the set $[0,1) \times [0,1)$. The chart that we will usually choose on the torus for working with coordinates is

$$\phi : (0,1) \times (0,1) \subset \mathbb{T}^2 \to (0,1) \times (0,1) \subset \mathbb{R}^2$$

$$\phi(x,y) = (x,y).$$

Of course, this chart covers all the torus but two lines (the sides of the square). Therefore, to be rigorous, one needs to add more charts.

2.1.6. *Definition of a Riemannian metric.* A metric is an inner product, and to define an inner product we need a space of vectors associated to the manifold. If the manifold lived inside $\mathbb{R}^n$, like the sphere, one could think of the vectors as being the velocities associated to curves on the manifold. The drawback to this is that then one had vectors that live in a space whose dimension is higher than that of the manifold. On the sphere for example, the tangent vectors live in $\mathbb{R}^3$ while the manifold has dimension 2. So how do we define a notion of velocity vectors associated to the manifold? The answer is that if $\gamma : [-1,1] \to M$ is a curve on M, then $\dot{\gamma}(0)$ is going to be defined as an operator. Indeed, $\dot{\gamma}(0)$ is the operator that to a function $f : M \to \mathbb{R}$ associates the derivative of f in the direction of $\dot{\gamma}$ evaluated at $t = 0$. Namely, if we let (U, φ) be a coordinate chart about $x \in M$ induce coordinates $(x_1, \dots, x_n)$, then

$$\dot{\gamma}(0)(f) = \sum_{i=1}^{n} (\varphi \circ \gamma)'(0) \frac{\partial}{\partial x_i}\Big|_{\varphi(\gamma(0))} (f \circ \varphi^{-1}).$$

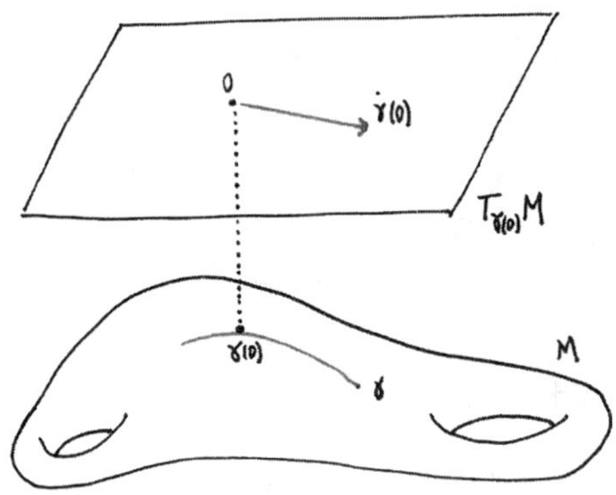

In other words, $\dot{\gamma}(0)$ is the operator

$$\dot{\gamma}(0) = \sum_{i=1}^{n} (\varphi \circ \gamma)'(0) \frac{\partial}{\partial x_i}\Big|_{\gamma(0)},$$

where we have set

$$\frac{\partial}{\partial x_i}\Big|_{\gamma(0)}(f) := \frac{\partial}{\partial x_i}\Big|_{\varphi(\gamma(0))}(f \circ \varphi^{-1}).$$

More generally, let (U, φ) be a coordinate chart about $x \in M$. We define the derivations

$$\left\{ \frac{\partial}{\partial x_i}\Big|_x \ : \ 1 \le i \le n \right\}$$

for $x \in U$, where

$$\frac{\partial}{\partial x_i}\Big|_x f := \frac{\partial f \circ \varphi^{-1}}{\partial x_i}(\varphi(x)).$$

Then, we define the tangent space T_xM at x as the vector space whose basis is $\left\{ \frac{\partial}{\partial x_i}\Big|_x, \ 1 \le i \le n \right\}$. This is our space of velocities!

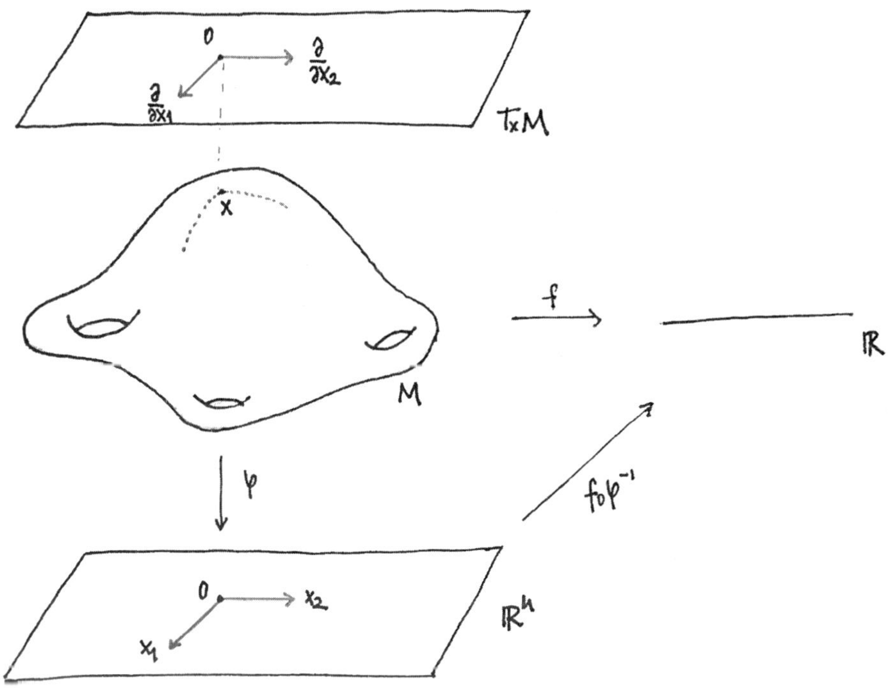

A **Riemannian manifold** is a pair (M, g) where M is a C^∞ manifold and g is a map that assigns to any $x \in M$ a non-degenerate, symmetric, positive definite bilinear form $\langle \cdot, \cdot \rangle_{g(x)} : T_xM \times T_xM \to \mathbb{R}$ such that for all V, W derivations, the map

$$x \mapsto \langle V, W \rangle_{g(x)}$$

is smooth.

Notation. Let (U, φ) be a chart with local coordinates $(x_1, \dots, x_n)$, and consider the corresponding natural basis $\{\frac{\partial}{\partial x_1}\big|_x, \dots, \frac{\partial}{\partial x_n}\big|_x\}$ of T_xM. We adopt the following notation:

$$g_{ij}(x) := \left\langle \frac{\partial}{\partial x_i}\Big|_x, \frac{\partial}{\partial x_j}\Big|_x \right\rangle_{g(x)}.$$

We will also denote by g^{ij} the entries of the inverse matrix of $(g_{ij})_{ij}$.

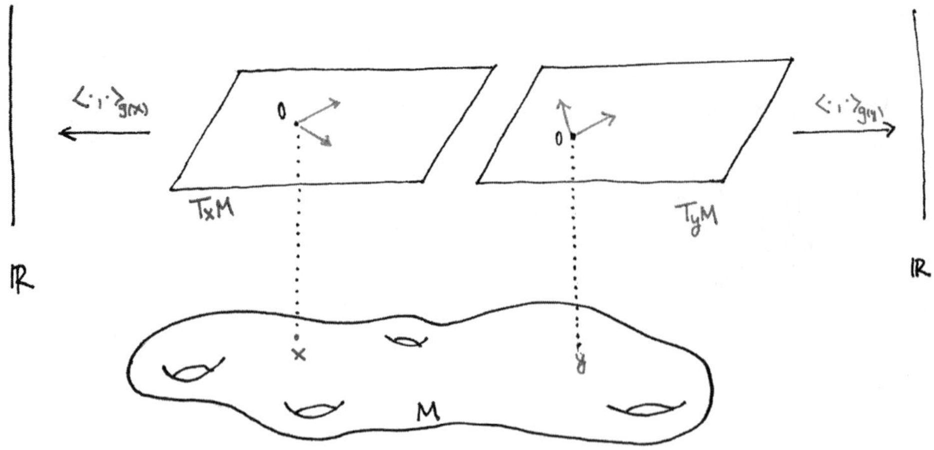

THEOREM 2. *Every manifold carries a Riemannian metric.*

2.1.7. *Example:* $\mathbb{R}^n$. The chart is $(\mathbb{R}^n, \phi)$ where $\phi(x) = x$. The metric is

$$g_{\mathbb{R}^n}(x) = \begin{pmatrix} 1 & 0 & \cdots & 0 & 0 \\ 0 & 1 & \cdots & 0 & 0 \\ \vdots & \vdots & \ddots & \vdots & \vdots \\ 0 & 0 & \cdots & 1 & 0 \\ 0 & 0 & \cdots & 0 & 1 \end{pmatrix}.$$

This means that, using coordinates $(x_1, \ldots, x_n)$, we have

$$\left\langle \frac{\partial}{\partial_{x_i}}\bigg|_x, \frac{\partial}{\partial_{x_j}}\bigg|_x \right\rangle_{g_{\mathbb{R}^n}} = \delta_{ij}.$$

2.1.8. *Example: Circle.* We use the chart φ defined by its inverse $\psi : (0, 2\pi) \to S^1$ with $\psi(\theta) = (\cos(\theta), \sin(\theta))$. Since everything is 1-dimensional, we we define the metric as

$$g_{S^1}(\theta) = 1.$$

That is, we set

$$\left\langle \frac{\partial}{\partial\theta}\bigg|_\theta, \frac{\partial}{\partial\theta}\bigg|_\theta \right\rangle = 1.$$

2.1.9. *Example: Sphere.* Endow the sphere with spherical coordinates

$$\psi : (0, \pi) \times (0, 2\pi) \to S^2 \subset \mathbb{R}^3$$

$$\psi(\theta, \phi) = (\sin\theta\, \cos\phi,\ \sin\theta\, \sin\phi,\ \cos\theta).$$

The round metric g_{S^2} on the 2-sphere is defined as

$$g_{S^2}(\theta, \phi) = \begin{pmatrix} 1 & 0 \\ 0 & \sin^2\theta \end{pmatrix}.$$

2.1.10. *Example: Torus.* Identifying $\mathbb{T}^2$ with $[0,1)\times[0,1)$ and using the coordinates

$$\psi : (0,1)\times(0,1) \subset \mathbb{R}^2 \to \mathbb{T}^2$$
$$\psi(x,y) = (x,y),$$

the flat Riemannian metric on the torus takes the form

$$g_{\mathbb{T}^2}(x,y) = \begin{pmatrix} 1 & 0 \\ 0 & 1 \end{pmatrix}.$$

2.1.11. *Definition of the Laplacian.* Having a metric on a manifold allows us to define geodesics on the manifold. Indeed, given any two points x and y sufficiently close in M there exists a unique path that has the shortest possible length (called geodesic) joining x with y.

Fix a point $x \in M$ and let $\beta_1, \ldots, \beta_n : [-1,1] \to M$ be n geodesics in M with $\gamma_i(0) = x$ for all i. Suppose further that the geodesics are orthogonal at x. That is,

$$\langle \dot{\beta}_i(0)\,,\, \dot{\beta}_j(0)\rangle = 0, \qquad i \neq j.$$

We define the **Laplacian** as the operator

$$\Delta_g : C^\infty(M) \to C^\infty(M)$$

making

$$\Delta_g f(x) = \sum_{i=1}^{n} (f\circ\beta_i)''(0).$$

It is possible to show that if $(x_1, \ldots, x_n)$ are local coordinates in M, then the Laplacian on (M,g)takes the form

$$\Delta_g = \frac{1}{\sqrt{|\det g|}} \sum_{i,j=1}^{n} \frac{\partial}{\partial x_i}\left(g^{ij}\sqrt{|\det g|}\frac{\partial}{\partial x_j}\right).$$

2.1.12. *Example:* $\mathbb{R}^n$. Since $g_{\mathbb{R}^n}(x) = Id$, we have

$$\Delta_{g_{\mathbb{R}^n}} = \frac{\partial^2}{\partial x_1^2} + \cdots + \frac{\partial^2}{\partial x_n^2}.$$

2.1.13. *Example: Circle.* Since $g_{S^1}(\theta) = 1$, we have

$$\Delta_{g_{S^1}} = \frac{\partial^2}{\partial \theta^2}.$$

2.1.14. *Example: Sphere.* Since

$$g_{S^2}(\theta,\phi) = \begin{pmatrix} 1 & 0 \\ 0 & \sin^2\theta \end{pmatrix},$$

$$\begin{aligned}
\Delta_{g_{S^2}} &= \frac{1}{\sqrt{|\det g_{S^2}|}}\left(\frac{\partial}{\partial\theta}\left(g^{\theta\theta}\sqrt{|\det g_{S^2}|}\frac{\partial}{\partial\theta}\right) + \frac{\partial}{\partial\phi}\left(g^{\phi\phi}\sqrt{|\det g_{S^2}|}\frac{\partial}{\partial\phi}\right)\right)\\
&= \frac{1}{\sin\theta}\left(\frac{\partial}{\partial\theta}\left(\sin\theta\frac{\partial}{\partial\theta}\right) + \frac{\partial}{\partial\phi}\left(\frac{1}{\sin\theta}\frac{\partial}{\partial\phi}\right)\right)\\
&= \frac{1}{\sin\theta}\frac{\partial}{\partial\theta}\left(\sin\theta\frac{\partial}{\partial\theta}\right) - \frac{1}{\sin^2\theta}\frac{\partial^2}{\partial\phi^2}.
\end{aligned}$$

And so, in spherical coordinates,

$$\Delta_{g_{S^2}} = \frac{1}{\sin\theta}\frac{\partial}{\partial\theta}\left(\sin\theta\frac{\partial}{\partial\theta}\right) - \frac{1}{\sin^2\theta}\frac{\partial^2}{\partial\phi^2}. \tag{2}$$

2.1.15. *Example: Torus.* It is clear that

$$\Delta_{g_{\mathbb{T}^2}} = \frac{\partial^2}{\partial x^2} + \frac{\partial^2}{\partial y^2}.$$

2.2. Spectral theory for the Laplacian.

Volume form. Having a Riemannian metric g allows us to have a volume form dv_g on M with respect to which we integrate. Given $A \subset M$,

$$\mathrm{vol}(A) = \int_M \mathbf{1}_A(x)\, dv_g(x).$$

Here, $\mathbf{1}_A(x) = 1$ if $x \in A$ and $\mathbf{1}_A(x) = 0$ if $x \notin A$. In local coordinates $\varphi(x) = (x_1, \ldots, x_n)$ the volume form is

$$dv_g(x) = \sqrt{\det(g_{ij}(x))}\, dx_1 \ldots dx_n.$$

Hilbert space. The riemannian metric induces an inner product for functions on M. Given $\phi, \psi : M \to \mathbb{R}$ we set

$$\langle \phi, \psi \rangle_{L^2} := \int_M \phi(x)\, \psi(x) dv_g(x).$$

This yields a Hilbert space

$$L^2(M) := \{f : M \to \mathbb{R}\,\text{function so that } \langle f, f \rangle_{L^2} < \infty\}.$$

Gradient. The metric also allows us to have a notion of gradient ∇_g. The gradient of a function $f \in C^\infty(M)$ measures the direction in which the function grows the most. In local coordinates $\varphi(x) = (x_1, \ldots, x_n)$ the gradient is defined as

$$\nabla_g f(x) = \sum_{i=1}^n g^{ij}(x) \frac{\partial f \circ \varphi^{-1}}{\partial x_i}(x) \frac{\partial}{\partial x_j}\Big|_x.$$

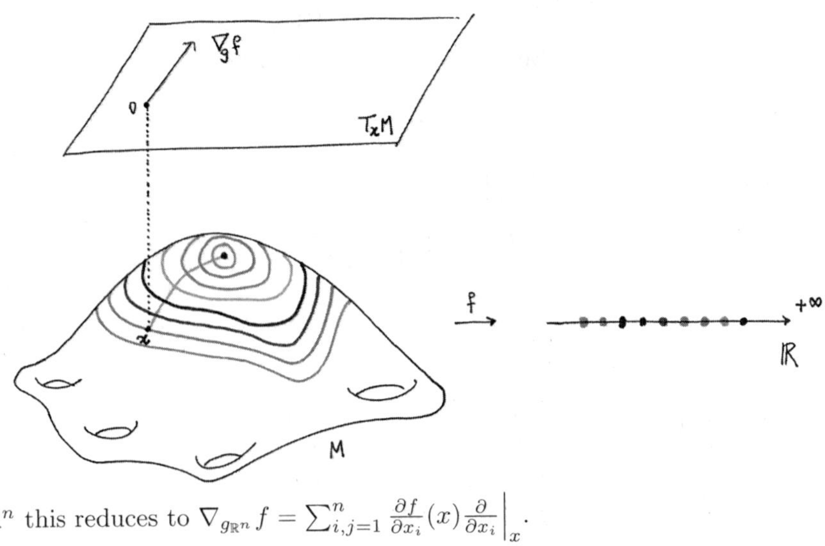

In $\mathbb{R}^n$ this reduces to $\nabla_{g_{\mathbb{R}^n}} f = \sum_{i,j=1}^n \frac{\partial f}{\partial x_i}(x) \frac{\partial}{\partial x_i}\Big|_x$.

Green's identity. Let (M, g) be a compact Riemannian manifold with boundary ∂M. Let $\psi \in C^1(M)$ and $\phi \in C^2(M)$. Then,

$$\int_M \psi \cdot \Delta_g \phi \; dv_g = -\int_M \langle \nabla_g \psi, \nabla_g \phi \rangle_g \, dv_g + \int_{\partial M} \phi \cdot \partial_\nu \phi \, d\sigma_g.$$

In particular, when $\partial M = \emptyset$, we get that the Laplacian is self-adjoint:

$$\langle \Delta_g \phi, \psi \rangle_{L^2} = \langle \phi, \Delta_g \psi \rangle_{L^2}.$$

Sobolev space. As for domains in $\mathbb{R}^n$ one can define the Sobolev inner product between functions $\phi, \psi \in C^\infty(M)$

$$\langle \phi, \psi \rangle_{H^1} = \int_M \langle \nabla_g \phi, \nabla_g \psi \rangle_g + \phi \psi \; dv_g.$$

The Sobolev space denoted by $H^1(M)$ is the completion of $C^\infty(M)$ with respect to the $\| \cdot \|_{H^1}$ norm.

2.2.1. *Spectral Theorem.* The Hilbert space is $L^2(M)$ and the Sobolev space is $H^1(M)$. The sesquilinear form is

$$a(\phi, \psi) := \langle \phi, \psi \rangle_{H^1}.$$

The ellipticity follows from the fact that

$$a(\phi, \phi) = \|\phi\|^2_{H^1} \geq 0.$$

The sesquilinear form a is continuous and symmetric on $H^1(M)$. Then, there exists a basis $\{\varphi_j\}$ of $L^2(M)$ and eigenvalues $\{\gamma_j\}$ staisfying that

$$a(\varphi_j, \psi) = \gamma_j \langle \varphi_j, \psi \rangle_{L^2}$$

for all $\psi \in H^1(M)$. Since $a(\varphi_j, \psi) = \int_M \langle \nabla_g \varphi_j, \nabla_g \psi \rangle_g dv_g + \langle \varphi_j, \psi \rangle_{L^2}$, we get

$$\int_M \langle \nabla_g \varphi_j, \nabla_g \psi \rangle_g dv_g = (\gamma_j + 1) \langle \varphi_j, \psi \rangle_{L^2}$$

for all $\psi \in H^1(M)$. In particular,

$$\langle -\Delta_g \varphi_j, \psi \rangle_{L^2} = (\gamma_j + 1) \langle \varphi_j, \psi \rangle_{L^2}$$

for all $\psi \subset H^1(M)$, and so

$$-\Delta_g \varphi_j = \lambda_j \varphi_j$$

for $\lambda_j = \gamma_j + 1$.

THEOREM 3 (Spectral Theorem). *For (M, g) compact, without boundary, there exists a complete orthonormal basis $\{\varphi_1, \varphi_2, \dots\}$ of $L^2(M)$ consisting of eigenfunctions of Δ_g with φ_j having eigenvalue λ_j satisfying*

$$0 = \lambda_1 \leq \lambda_2 \leq \dots \;\; \to \infty.$$

2.2.2. *Eigenvalue characterization (Exercise 4).* Let (M, g) be a compact Riemannian manifold and write $\lambda_1 \leq \lambda_2 \leq \dots$ for the eigenvalues of the Laplacian repeated according to multiplicity (for any initial problem). Write $\varphi_1, \varphi_2, \dots$ for the corresponding L^2-normalized eigenfunctions. In this exercise you will prove the following result.

THEOREM 4. *For $k \in \mathbb{N}$ and $E_k(g) := \{\varphi_1, \varphi_2, \dots, \varphi_{k-1}\}^\perp$,*

$$\lambda_k = \inf\Big\{ \frac{\|\nabla_g \phi\|_{L^2}^2}{\|\phi\|_{L^2}^2} : \quad \phi \in H_1(M) \cap E_k(g) \Big\}.$$

The infimum is achieved if and only if ϕ is an eigenfunction of eigenvalue λ_k. Here, $\|\nabla_g \phi\|_{L^2}^2 := \int_M \|\nabla_g \phi\|_g^2 dv_g$.

To ease the exposition, define $D_g(\phi, \psi) = \int_M \langle \nabla_g \phi, \nabla_g \psi \rangle_g dv_g$. Fix $\phi \in H^1(M) \cap E_k(g)$ and assume it has expansion $\phi = \sum_{j=1}^\infty a_j \varphi_j$.

(1) Fix $\ell \in \mathbb{N}$ and prove that $D_g(\phi, \phi) \geq \sum_{j=1}^\ell \lambda_j a_j^2$. Hint: compute

$$D_g(\phi - \sum_{j=1}^\ell \lambda_j a_j^2, \phi - \sum_{j=1}^\ell \lambda_j a_j^2).$$

(2) Conclude that $D_g(\phi, \phi) \geq \lambda_k \|\phi\|_{L^2}^2$ for all $\phi \in E_k(g)$.
(3) To prove the second statement in the theorem, assume that the equality is attained. From your previous work deduce that $\sum_{j=k} \lambda_j a_j^2 = \lambda_k \sum_{j=k}^\infty a_j^2$. Conclude that ϕ must be a linear combination of the φ_j's.

2.3. Examples of eigenfunctions and eigenvalues.

2.3.1. *Circle.* The eigenfunctions are

$$1, \quad \cos(\theta), \quad \sin(\theta), \quad \cos(2\theta), \quad \sin(2\theta), \quad \dots, \quad \cos(k\theta), \quad \sin(k\theta), \dots$$

with eigenvalues $0, 1, 1, 4, 4, \dots, k^2, k^2, \dots$ respectively.

2.3.2. *Torus.* We claim that the following are a basis of eigenfunctions on the torus:

$$\varphi_y(x) := e^{2\pi i \langle x, y \rangle} \qquad \text{for } y \in \mathbb{Z}^2.$$

If you want, you may take their real and imaginary part to get the eigenfunctions

$$1, \qquad \sin(2\pi i \langle x, y \rangle), \qquad \cos(2\pi i \langle x, y \rangle), \qquad y \in \mathbb{Z}^2$$

with eigenvalues

$$0, \qquad 4\pi^2 |y|^2, \qquad 4\pi^2 |y|^2, \qquad y \in \mathbb{Z}^2.$$

Let us see that the

$$\{\varphi_y : \ y \in \mathbb{Z}^2\}$$

is a basis of $L^2(\mathbb{T}^2)$.

We first check that these functions are linearly independent by induction: suppose $\varphi_{y_1}, \dots, \varphi_{y_k}$ are linearly independent and suppose also that $\sum_{i=1}^{k+1} a_i \varphi_{y_i} = 0$. Since $\varphi_{y_j} \circ \varphi_{y_i} = \varphi_{y_j + y_i}$ we know that $0 = \sum_{i=1}^{k+1} a_i \varphi_{y_i - y_{k+1}} = a_{k+1} + \sum_{i=1}^k a_i \varphi_{y_i - y_{k+1}}$, after applying the Laplacian we get

$$0 = \sum_{i=1}^k a_i 4\pi^2 |y_i - y_{k+1}|^2 \varphi_{y_i - y_{k+1}} = e^{-2\pi i \langle x, y_{k+1} \rangle} \sum_{i=1}^k a_i 4\pi^2 |y_i - y_{k+1}|^2 \varphi_{y_i}.$$

It then follows that $a_1 = \cdots = a_k = 0$. Therefore $a_{k+1} = 0$.

To see that $\mathcal{B} = \operatorname{span}\{\varphi_y : \ y \in \mathbb{Z}^2\}$ is dense in $L^2(\mathbb{T}^2)$ we use the Stone Weirstrass Theorem. Clearly $\mathcal{B}$ is a subalgebra. Let us see that it separates points. Fix two points $x_1, x_2 \in \mathbb{T}^2$ and assume $x_1 \neq x_2$ (then $x_1 - x_2 \notin \mathbb{Z}^2$). Suppose now that $\varphi_y(x_1) = \varphi_y(x_2)$ for all $y \in \mathbb{Z}^2$. It then follows that $e^{2\pi i \langle x_1 - x_2, y \rangle} = 1$ for all $y \in \mathbb{Z}^2$ and so $\langle x_1 - x_2, y \rangle \in \mathbb{Z}$ for all $y \in \mathbb{Z}^2$ which implies that $x_1 - x_2 \in \mathbb{Z}^2$, and this is a contradiction.

Weyl's law on the torus (Exercise 2). We continue to write $\omega_2 := \mathrm{vol}(B_1(0))$ where $B_1(0)$ is the unit ball in $\mathbb{R}^2$. Denote by $N(\lambda)$ the counting function $N(\lambda) := \#\{j : \lambda_j < \lambda\}$. In this exercise you will prove that

$$N(\lambda) \sim \frac{\omega_2 \ \mathrm{vol}(\mathbb{T}^2)}{(2\pi)^2}\lambda \qquad \lambda \to \infty.$$

(1) Define $N^*(r) := \#\{\mathbb{Z}^2 \cap B_r(0)\}$. Find how $N(\lambda)$ and $N^*(r)$ are related. Restate your goal in terms of $N^*(r)$.
(2) Let $P^*(r)$ denote the number of $\mathbb{Z}^2$-lattice squares inside $B_r(0)$. Show that $P^*(r) \leq N^*(r) \leq P^*(r+d)$.
(3) Find upper and lower bounds for the function $P^*(r)$ in terms of volumes of Euclidean balls centered at 0.

2.3.3. *Eigenfunctions on the sphere (Exercise 3).* Let $\psi(\theta,\phi) = (\psi_1(\theta,\phi), \psi_2(\theta,\phi), \psi_3(\theta,\phi))$ represent the spherical coordinate system for the sphere $S^2 \subset \mathbb{R}^3$. Then any $x \in \mathbb{R}^3$ can be written as $r\psi(\theta,\phi)$ for $r > 0$ and is represented by the coordinates (r,θ,ϕ).

(1) Use that if $(x_1,x_2,x_3) = F(\xi_1,\xi_2,\xi_3)$ is a change of coordinates then $\frac{\partial}{\partial \xi_j} = \sum_{i=1}^3 \frac{\partial(F_i)}{\partial \xi_j}\frac{\partial}{\partial x_i}$, to prove that in spherical coordinates

$$g_{\mathbb{R}^3}(r,\theta,\phi) = \begin{pmatrix} 1 & 0 \\ 0 & r^2\, g_{S^2}(\theta,\phi) \end{pmatrix}.$$

Hint: express $\partial_r, \partial_\theta, \partial_\phi$ in terms of $\partial_{x_1}, \partial_{x_2}, \partial_{x_3}$.
(2) Deduce that in spherical coordinates

$$\Delta_{g_{\mathbb{R}^3}} = \frac{1}{r^2}\frac{\partial}{\partial r}\left(r^2\frac{\partial}{\partial r}\right) + \frac{1}{r^2}\Delta_{g_{S^2}}.$$

(3) Define

$$\begin{aligned} \mathcal{P}_k &= \{\text{homogeneous polynomials of degree } k\}, \\ \mathcal{H}_k &= \{P \in \mathcal{P}_k : \ \Delta_{g_{\mathbb{R}^3}} P = 0\}, \\ H_k &= \{Y = P|_{S^2} : \ P \in \mathcal{H}_k\}. \end{aligned}$$

Prove that the spaces $\mathcal{H}_k$ and H_k are isomorphic. That is, find the inverse for the restriction map $\mathcal{H}_k \to H_k$ given by $P \mapsto P|_{S^2}$.
(4) Prove that

$$H_k \subset \{Y \in C^\infty(S^2) : \quad \Delta_{g_{S^2}} Y = -k(k+1)Y\}.$$

The space H_k is known as the space of Spherical Harmonics of degree k.
(5) Find explicit bases for H_1 and H_2.
(6) Use separation of variables to find a basis for H_k. That is, look for solutions of the form $Y_k(\theta,\phi) = P_k(\theta)\Phi_k(\phi)$. The functions $P_k(\theta)$ should satisfy a second order differential equation (do not try to solve it explicitly!). Such solutions $P_k(\theta)$ are called Legendre polynomials. The solutions you obtain should have the form

$$Y_k^m(\theta,\phi) = e^{im\phi} P_k^m(\cos(\theta)).$$

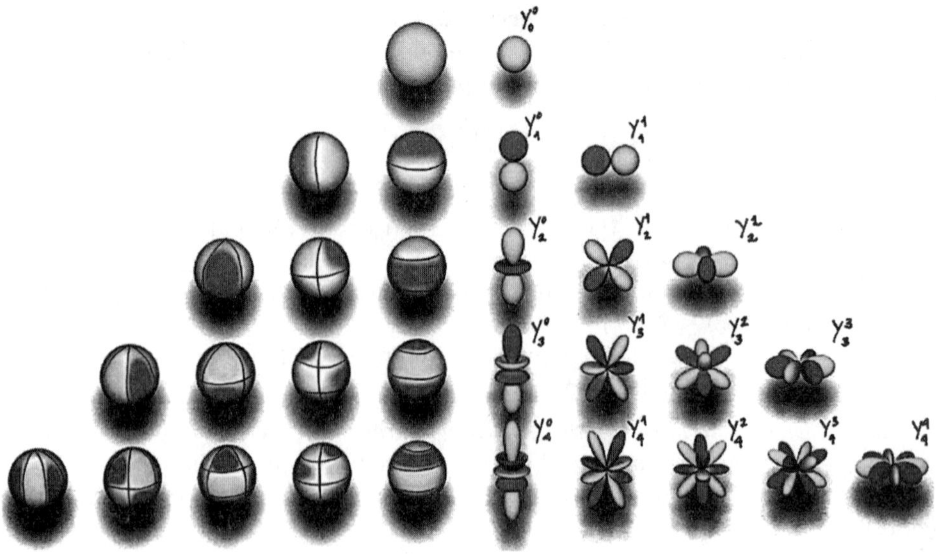

3. Hearing the geometry of a manifold

3.1. Heat equation. Throughout this section we assume that (M, g) is a compact Riemannian manifold without boundary. The **Heat operator**

$$-\Delta_g + \partial_t$$

acts on functions in $C(M \times (0, +\infty))$ that are C^2 on M and C^1 on $(0, \infty)$.

The *homogeneous heat equation* is

$$\begin{cases} (-\Delta_g + \partial_t)u(x,t) = 0 & (x,t) \in M \times (0, +\infty) \\ u(x,0) = f(x) & x \in M \end{cases}.$$

The function $u(x,t)$ represents the temperature at the point x at time t assuming that the initial temperature accross the manifolds was given by the function $f(x)$.

3.1.1. *Solutions to heat equation (Exercise 4).*

(1) Prove that the temperature of the manifold decreases as time evolves. That is, show that $t \mapsto \|u(\cdot, t)\|_{L^2}$ is decreasing with t.
(2) Using the previous part show that the solution to the Heat Equation is unique.

3.1.2. *Definition of the fundamental solution.* We say that a **fundamental solution** of the heat equation is a continuous function $p : M \times M \times (0, \infty) \to \mathbb{R}$ which is C^2 in (x, y), and C^1 with respect to t and such that

$$L_y p = 0 \qquad \text{and} \qquad \lim_{t \to 0} p(\cdot, y, t) = \delta_y.$$

OBSERVATION 1. It can be shown that that the fundamental solution is symmetric: $p(x, y, t) = p(y, x, t)$.

OBSERVATION 2. The function

$$u(x,t) = \int_M p(x,y,t) f(y)\, dv_g(y)$$

solves the homogeneous heat equation.

3.1.3. *Fundamental solution in* $\mathbb{R}^n$. Eventhough we have been working on compact manifolds throughout this chapter, we digress briefly to inspire the form of the fundamental solution on compact manifolds.

PROPOSITION 5. *The function* $p : \mathbb{R}^n \times \mathbb{R}^n \times (0, +\infty) \to [0, +\infty)$

$$p(x, y, t) = \frac{1}{(4\pi t)^{\frac{n}{2}}} e^{-\|y-x\|^2/4t}$$

is a fundamental solution for the heat equation on $\mathbb{R}^n$.

PROOF. It is easy to check that $-\Delta_{g_{\mathbb{R}^n}} p + \partial_t p = 0$. Let us prove that $p(x, y, t) \to \delta_x(y)$ as $t \to 0$. We first need to show that $\int_{\mathbb{R}^n} p(x, y, t)dy = 1$ for all $(x, t) \in \mathbb{R}^n \times (0, +\infty)$. Indeed,

$$\begin{aligned}
\int_{\mathbb{R}^n} p(x, y, t)dy &= \int_0^\infty \int_{S^{n-1}(x)} p(x, x + r\xi, t) r^{n-1} d\xi dr \\
&= \int_0^\infty \int_{S^{n-1}(x)} \frac{1}{(4\pi t)^{\frac{n}{2}}} e^{-\frac{r^2}{4t}} r^{n-1} d\xi dr \\
&= \int_0^\infty \int_{S^{n-1}(x)} \frac{1}{(4\pi t)^{\frac{n}{2}}} e^{-s^2} s^{n-1} (4t)^{\frac{n-1}{2}} (4t)^{\frac{1}{2}} d\xi dr \\
&= \frac{1}{\pi^{\frac{n}{2}}} \int_0^\infty \int_{S^{n-1}(x)} e^{-s^2} s^{n-1} d\xi ds \\
&= \frac{1}{\pi^{\frac{n}{2}}} \int_{\mathbb{R}^n} e^{-\|y\|^2} dy \\
&= 1.
\end{aligned}$$

Let $f : \mathbb{R}^n \to \mathbb{R}$ be a bounded and continuous function.

$$\begin{aligned}
&\left| f(x) - \int_{\mathbb{R}^n} p(x, y, t) f(y) dy \right| = \\
&= \left| \int_{\mathbb{R}^n} p(x, y, t)(f(x) - f(y)) dy \right| \\
&\leq \int_{B_{2\sqrt{t}R}(x)} p(x, y, t)|f(x) - f(y)| dy + \left| \int_{\mathbb{R}^n \setminus B_{2\sqrt{t}R}(x)} p(x, y, t)(f(x) - f(y)) dy \right| \\
&\leq \sup_{y \in B_{2\sqrt{t}R}(x)} |f(x) - f(y)| + \int_{2\sqrt{t}R}^\infty \int_{S^{n-1}(x)} p(x, x + r\xi, t)|f(x) - f(x + r\xi)| r^{n-1} d\xi dr.
\end{aligned}$$

We now need to choose R sufficiently large so that the second term is as small as we wish. Once R is fixed the first term can be chosen to be small since $t \to 0$. To handle the second term note that

$$\begin{aligned}
&\int_{2\sqrt{t}R}^{\infty} \int_{S^{n-1}(x)} p(x, x+r\xi, t)|f(x) - f(x+r\xi)|r^{n-1}\, d\xi dr = \\
&\quad \leq \int_{R}^{\infty} \int_{S^{n-1}(x)} p(x, x+2\sqrt{t}s, t)2\|f\|_\infty (2\sqrt{t}s)^{n-1} 2\sqrt{t}\, d\xi ds \\
&\quad = \int_{R}^{\infty} \int_{S^{n-1}(x)} e^{-s^2} s^{n-1} 2\|f\|_\infty d\xi ds \\
&\quad = 2\mathrm{vol}(S^{n-1})\|f\|_\infty \int_{R}^{\infty} e^{-s^2} s^{n-1} ds.
\end{aligned}$$

□

3.1.4. *Fundamental solution on manifolds.*

THEOREM 6. *There exists $\epsilon > 0$ such that the fundamental solution for the heat equation has expansion*

$$p(x, y, t) = \frac{e^{-d_g^2(x,y)/4t}}{(4\pi t)^{n/2}} \Big(\sum_{j=0}^{k} t^j u_j(x, y) + O(t^{k+1}) \Big),$$

for all $x, y \in M$ with $d_g(x, y) \leq \varepsilon/2$. In addition, $u_0(x, x) = 1$ and $u_1(x, x) = \frac{1}{6} R_g(x)$.

Formally, we have that $(-\Delta + \partial_t)e^{t\Delta} = 0$ and $\lim_{t\to 0} e^{t\Delta} f = f$. So it should be true that

$$e^{t\Delta} f\,(x) = \int_M p(x, y, t) f(y)\, dv_g(y).$$

It can be shown that the operator $e^{t\Delta}$ is well defined and that $p(x, y, t)$ is indeed its kernel.

THEOREM 7. *Let $\{\varphi_j\}_j$ be an orthonormal basis of Laplace eigenfunctions with eigenvalues $\{\lambda_j\}_j$. Then,*

$$p(x, y, t) = \sum_{j=1}^{\infty} e^{-\lambda_j t} \varphi_j(x)\varphi_j(y).$$

PROOF. Since $p(x, y, t) = \sum_{k=0}^{\infty} \langle p(x, \cdot, t), \varphi_k\rangle \varphi_k(y)$ and

$$\langle p(x, \cdot, t), \varphi_k\rangle = e^{t\Delta_g}\varphi_k(x) = e^{-\lambda_k t}\varphi_k(x),$$

it is straight forward that

$$p(x, y, t) = \sum_{k=0}^{\infty} e^{-\lambda_k t}\varphi_k(x)\varphi_k(y).$$

□

3.2. Weyl's law and other high energy asymptotics. Let (M, g) be a compact boundary-less Riemannian manifold. Write

$$0 = \lambda_1 < \lambda_2 \leq \lambda_3 \leq \dots$$

for all the Laplace eigenvalues repeated according to their multiplicity. We begin this section by introducing the **Zeta function** $Z_g : (0, +\infty) \to \mathbb{R}$

$$Z_g(t) = \sum_{j=1}^{\infty} e^{-\lambda_j t}.$$

Since the series is uniformly convergent on intervals of the form $[t_0, +\infty)$ for all $t_0 > 0$ we know that Z_g is continuous. We also have that it is decreasing in t, that $\lim_{t\to 0^+} Z_g(t) = +\infty$, and $\lim_{t\to+\infty} Z_g(t) = 0$.

PROPOSITION 8.

$$Z_g(t) \sim \frac{1}{(4\pi t)^{n/2}}(vol_g(M) + O(t)) \qquad as\ t \to 0^+.$$

PROOF.

$$\begin{aligned} Z_g(t) &= \sum_{j=0}^{\infty} e^{-\lambda_j t} \\ &= \int_M p(x,x,t)\, dv_g(x) \\ &= \frac{1}{(4\pi t)^{n/2}}\left(\sum_{j=0}^{k} t^j \int_M u_j(x,x)\, dv_g(x) + O(t^{k+1})\right) \\ &= \frac{1}{(4\pi t)^{n/2}}\left(\mathrm{vol}_g(M) + O(t)\right) \end{aligned}$$

□

Let us now write

$$0 = \nu_1 < \nu_2 < \nu_3 < \dots$$

for all the distinct eigenvalues. Then, setting m_j for the multiplicity of ν_j we can rewrite

$$Z_g(t) = \sum_{j=1}^{\infty} m_j e^{-\nu_j t}.$$

THEOREM 9. *The function Z_g determines all the eigenvalues and their multiplicities.*

PROOF. Note that for $\mu > 0$ with $\mu \neq 2$,

$$\lim_{t\to\infty} e^{\mu t}(Z_g(t) - 1) = \lim_{t\to\infty} \sum_{j=2}^{\infty} m_j e^{(\mu-\nu_j)t} = \begin{cases} 0 & \text{if } \mu < \nu_2, \\ +\infty & \text{if } \mu > \nu_2 \\ m_2, & \text{if } \mu = \nu_2. \end{cases}$$

It follows that ν_2 is the unique strictly positive real number μ such that the limit $\lim_{t\to\infty} e^{\mu t}(Z_g(t) - 1)$ is a natural number. By induction, ν_k is the unique strictly positive real number μ such that the limit

$$m_k := \lim_{t\to\infty} e^{\mu t}\Big(Z_g(t) - 1 - \sum_{j=2}^{k-1} m_j e^{-\nu_j t}\Big)$$

is a natural number.

$\square$

Note that $N(\lambda) = \#\{j : \lambda_j < \lambda\} = \int_0^\lambda d\mu$ for $\mu = \sum_{j=1}^\infty \delta_{\lambda_j}$. To understand the bahavior of μ we use the following Tauberian Theorem.

THEOREM 10 (Karamata). *Suppose that μ is a positive measure on $[0,\infty)$ and that $\alpha \in (0,\infty)$. Supose that*

$$\int_0^\infty e^{-tx}\, d\mu(x) \sim at^{-\alpha} \qquad t \to 0.$$

Then,

$$\int_0^\lambda d\mu(x) \sim \frac{a}{\Gamma(\alpha+1)}\lambda^\alpha \qquad x \to \infty.$$

Let ω_n be the volume of the unite ball in $\mathbb{R}^n$,

$$\omega_n := \frac{2\pi^{n/2}}{n\Gamma(n/2)}.$$

Our aim is to prove the following Theorem:

THEOREM 11 (Weyl's asymptotic formula). *Let M be a compact Riemannian manifold with eigenvalues $0 = \lambda_1 \le \lambda_2 \le \dots$, each distinct eigenvalue repeated according to its multiplicity.*

Then for $N(\lambda) := \#\{j : \lambda_j \le \lambda\}$, we have

$$N(\lambda) \sim \frac{\omega_n}{(2\pi)^n} vol_g(M)\lambda^{n/2}, \qquad \lambda \to \infty.$$

In particular,

$$\lambda_j \sim \frac{\sqrt{2\pi}}{(\omega_n vol_g(M))^{2/n}} j^{2/n}, \qquad j \to \infty.$$

PROOF. For the measure $\mu = \sum \delta_{\lambda_j}$, Proposition 8 asserts that

$$\int_0^\infty e^{-t\lambda}\, d\mu(\lambda) \sim \frac{1}{(4\pi)^{\frac{n}{2}}}\mathrm{vol}_g(M)t^{-n/2}.$$

Using Karamata's theorem on μ and $\alpha = n/2$ we obtain

$$N(\lambda) = \int_0^\lambda d\mu(\lambda) \sim \frac{\mathrm{vol}_g(M)}{(4\pi)^{n/2}\Gamma(n/2+1)}\lambda^{n/2} = \frac{\omega_n \mathrm{vol}_g(M)}{(2\pi)^n}\lambda^{n/2}.$$

$\square$

3.3. Isospectral manifolds. In this section we prove that if (M, g_M) and (N, g_N) are compact Riemannian manifolds which share the same eigenvalues then they must have the same dimension, same volume and same total curvature.

THEOREM 12. *If (M, g_M) and (N, g_N) are isospectral compact Riemannian manifolds, then*

$$dim\, M = dim\, N, \quad vol_{g_M}(M) = vol_{g_N}(N) \quad and \quad \int_M R_{g_M} dv_{g_M} = \int_N R_{g_N} dv_{g_N}.$$

PROOF. Let $\lambda_0 \leq \lambda_1 \leq \ldots$ be the eigenvalues of both Δ_{g_M} and Δ_{g_N}. Then

$$\sum_{j=0}^{\infty} e^{-\lambda_j t} = \frac{1}{(4\pi t)^{\dim M/2}} \left(\sum_{j=0}^{k} t^j \int_M u_j^{g_M}(x,x)\, dv_{g_M}(x) + O(t^{k+1}) \right)$$

and

$$\sum_{j=0}^{\infty} e^{-\lambda_j t} = \frac{1}{(4\pi t)^{\dim N/2}} \left(\sum_{j=0}^{k} t^j \int_M u_j^{g_N}(x,x)\, dv_{g_N}(x) + O(t^{k+1}) \right).$$

It follows immediately that $dim\, M = dim\, N := n$. Next, note that

$$\frac{1}{(4\pi t)^{n/2}} \left(\int_M u_0^{g_M}(x,x)\, dv_{g_M}(x) - \int_N u_0^{g_N}(x,x)\, dv_{g_N}(x) \right) =$$

$$= \frac{1}{(4\pi t)^{n/2}} \left[\sum_{j=1}^{k} t^j \left(\int_M u_j^{g_M}(x,x)\, dv_{g_M}(x) - \int_N u_j^{g_N}(x,x)\, dv_{g_N}(x) \right) + O(t^{k+1}) \right]$$

yields

$$\int_M u_0^{g_M}(x,x)\, dv_{g_M}(x) = \int_N u_0^{g_N}(x,x)\, dv_{g_N}(x)$$

and since $u_0^{g_M}(x,x) = u_0^{g_N}(x,x) = 1$, we have $vol_{g_M}(M) = vol_{g_N}(N)$. Repeating the same argument it follows that if (M, g_M) and (N, g_N) are isospectral compact Riemannian manifolds, then for all j

$$\int_M u_j^{g_M}(x,x)\, dv_{g_M}(x) = \int_N u_j^{g_N}(x,x)\, dv_{g_N}(x).$$

In particular, since $u_1(x,x) = \frac{1}{6} R_g(x)$ we have that (M, g_M) and (N, g_N) have the same total curvature. □

We next prove that in the case of compact surfaces isospectrality implies a strong result:

COROLLARY 13. *If (M, g_M) and (N, g_N) are isospectral compact Riemannian surfaces, then M and N are diffeomorphic.*

PROOF. By Gauss-Bonnet Theorem

$$\int_M R_{g_M}\, dv_{g_M} = 8\pi(1 - \gamma_M)$$

where γ_M is the genus of M. The same result holds for N. We then use that

$$\int_M R_{g_M}\, dv_{g_M} = \int_N R_{g_N}\, dv_{g_N}$$

which yields $\gamma_M = \gamma_N$. The result follows from the fact that two orientable surfaces with the same genus are diffeomorphic. □

4. Exercises

4.1. Exercise 0: Recovering the eigenvalues of a guitar string.

(1) Show that the vibrations of a guitar string of length ℓ are of the form

$$u(x,t) = \sum_k \left(a_k \sin\left(\frac{k\pi t}{\ell}\right) + b_k \cos\left(\frac{k\pi t}{\ell}\right)\right) \sin\left(\frac{k\pi x}{\ell}\right),$$

where $u(x,t)$ is the height of the wave at the point x at time t. Hint: Look first for solutions of the form $\alpha(t)\varphi(x)$.

(2) Find a_k and b_k if you knew that the initial conditions are $u(x,0) = f(x)$ and $\partial_t u(x,0) = h(x)$.

(3) Prove that the Fourier transform of the sound wave you heard recovers all the frequencies present in the wave. Hint: decompose $\sin(x)$ as a sum of exponentials and compute $\mathcal{F}(\sin)(\xi)$.

4.2. Exercise 1: Weyl's law for a rectangle. Consider a rectangle $\Omega = [0,\ell] \times [0,m]$ with Dirichlet boundary conditions. In this exercise you will prove that

$$N(\lambda) \sim \frac{\text{area}(\Omega)}{4\pi}\lambda.$$

(1) Prove that the eigenvalues are

$$\lambda_{jk} = \left(\frac{j\pi}{\ell}\right)^2 + \left(\frac{k\pi}{m}\right)^2 \qquad \text{for} j,k \geq 1.$$

(2) Prove that

$$\#\{(j,k) : \lambda_{jk} \leq \lambda\} \sim \frac{\text{area}(\Omega)}{4\pi}\lambda \qquad \text{as } \lambda \to \infty.$$

Hint: use the two figures below, where E_λ denotes an ellipse and $\tilde{E}_\lambda$ is the copy of the ellipse translated by $(-1,-1)$.

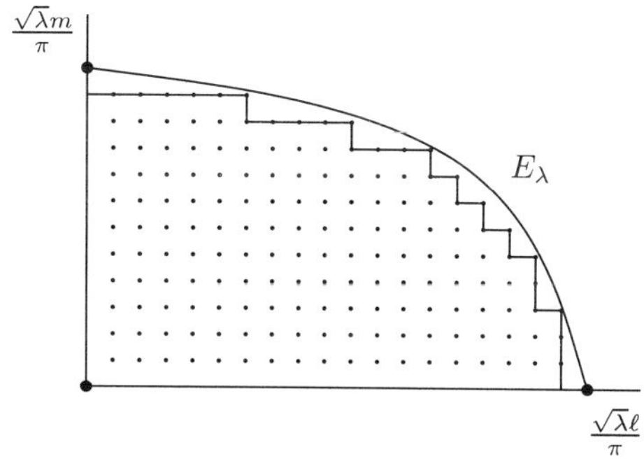

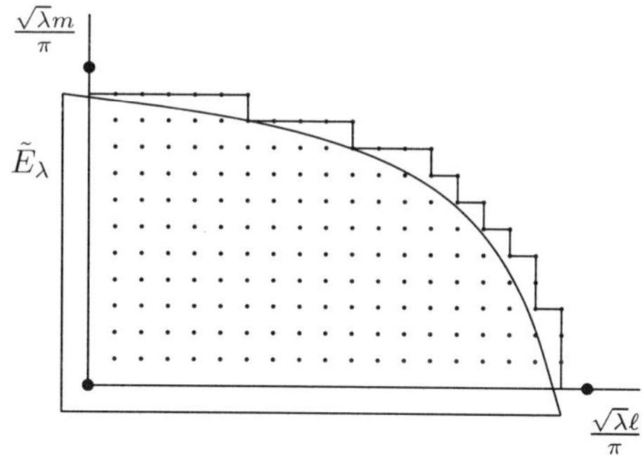

4.3. Exercise 2: Weyl's law on the torus. We continue to write $\omega_2 := \mathrm{vol}(B_1(0))$, where $B_1(0)$ is the unit ball in $\mathbb{R}^2$. Denote by $N(\lambda)$ the counting function $N(\lambda) := \#\{j : \lambda_j < \lambda\}$. In this exercise you will prove that

$$N(\lambda) \sim \frac{\omega_2 \,\mathrm{vol}(\mathbb{T}^2)}{(2\pi)^2}\lambda \qquad \lambda \to \infty.$$

(1) Define $N^*(r) := \#\{\mathbb{Z}^2 \cap B_r(0)\}$. Find how $N(\lambda)$ and $N^*(r)$ are related. Restate your goal in terms of $N^*(r)$.
(2) Let $P^*(r)$ denote the number of $\mathbb{Z}^2$-lattice squares inside $B_r(0)$. Show that $P^*(r) \leq N^*(r) \leq P^*(r+d)$.
(3) Find upper and lower bounds for the function $P^*(r)$ in terms of volumes of Euclidean balls centered at 0.

4.4. Exercise 3: Eigenfunctions on the sphere. Let $\psi(\theta,\phi) = (\psi_1(\theta,\phi), \psi_2(\theta,\phi), \psi_3(\theta,\phi))$ represent the spherical coordinate system for the sphere $S^2 \subset \mathbb{R}^3$. Then any $x \in \mathbb{R}^3$ can be written as $r\psi(\theta,\phi)$ for $r > 0$ and is represented by the coordinates (r,θ,ϕ). Hint: express $\partial_r, \partial_\theta, \partial_\phi$ in terms of $\partial_{x_1}, \partial_{x_2}, \partial_{x_3}$.

(1) Use that if $(x_1,x_2,x_3) = F(\xi_1,\xi_2,\xi_3)$ is a change of coordinates then $\frac{\partial}{\partial \xi_j} = \sum_{i=1}^3 \frac{\partial F_i}{\partial \xi_j}\frac{\partial}{\partial x_i}$, to prove that in spherical coordinates

$$g_{\mathbb{R}^3}(r,\theta,\phi) = \begin{pmatrix} 1 & 0 \\ 0 & r^2\, g_{S^2}(\theta,\phi) \end{pmatrix}.$$

Hint: express $\partial_r, \partial_\theta, \partial_\phi$ in terms of $\partial_{x_1}, \partial_{x_2}, \partial_{x_3}$.
(2) Deduce that in spherical coordinates

$$\Delta_{g_{\mathbb{R}^3}} = \frac{1}{r^2}\frac{\partial}{\partial r}\left(r^2 \frac{\partial}{\partial r}\right) + \frac{1}{r^2}\Delta_{g_{S^2}}.$$

(3) Define

$$\begin{aligned}\mathcal{P}_k &= \{\text{homogeneous polynomials of degree } k\},\\ \mathcal{H}_k &= \{P \in \mathcal{P}_k : \Delta_{g_{\mathbb{R}^3}} P = 0\},\\ H_k &= \{Y = P|_{S^2} : P \in \mathcal{H}_k\}.\end{aligned}$$

Prove that the spaces $\mathcal{H}_k$ and H_k are isomorphic. Hint: find the inverse for the restriction map $\mathcal{H}_k \to H_k$ given by $P \mapsto P|_{S^2}$.

(4) Prove that

$$H_k \subset \{Y \in C^\infty(S^2) : \quad \Delta_{g_{S^2}} Y = -k(k+1)Y\}.$$

The space H_k is known as the space of Spherical Harmonics of degree k.

(5) Find explicit bases for H_1 and H_2.

(6) Use separation of variables to find a basis for H_k. That is, look for solutions of the form $Y_k(\theta, \phi) = P_k(\theta)\Phi_k(\phi)$. The functions $P_k(\theta)$ should satisfy a second order differential equation (don't try to solve it explicitly!). Such solutions $P_k(\theta)$ are called Legendre polynomials. The solutions you obtain should have the form

$$Y_k^m(\theta, \phi) = e^{im\phi} P_k^m(\cos(\theta)).$$

4.5. Exercise 4: Characterization of eigenvalues. Let (M, g) be a compact Riemannian manifold and write $\lambda_1 \leq \lambda_2 \leq \dots$ for the eigenvalues of the Laplacian repeated according to multiplicity (for any initial problem). Write $\varphi_1, \varphi_2, \dots$ for the corresponding L^2-normalized eigenfunctions. In this exercise you will prove the following result.

THEOREM 14. *For $k \in \mathbb{N}$ and $E_k(g) := \{\varphi_1, \varphi_2, \dots, \varphi_{k-1}\}^\perp$,*

$$\lambda_k = \inf\Big\{\frac{\|\nabla_g \phi\|_{L^2}^2}{\|\phi\|_{L^2}^2} : \quad \phi \in H_1(M) \cap E_k(g)\Big\}.$$

The infimum is achieved if and only if ϕ is an eigenfunction of eigenvalue λ_k. Here, $\|\nabla_g \phi\|_{L^2}^2 := \int_M \|\nabla_g \phi\|_g^2 dv_g$.

To ease the exposition, define $D_g(\phi, \psi) = \int_M \langle \nabla_g \phi, \nabla_g \psi\rangle_g dv_g$. Fix $\phi \in H^1(M) \cap E_k(g)$ and assume it has expansion $\phi = \sum_{j=1}^\infty a_j \varphi_j$.

(1) Fix $\ell \in \mathbb{N}$ and prove that $D_g(\phi, \phi) \geq \sum_{j=1}^\ell \lambda_j a_j^2$. Hint: compute

$$D_g(\phi - \sum_{j=1}^{\ell} \lambda_j a_j^2, \phi - \sum_{j=1}^{\ell} \lambda_j a_j^2).$$

(2) Conclude that $D_g(\phi, \phi) \geq \lambda_k \|\phi\|_{L^2}^2$ for all $\phi \in E_k(g)$.

(3) To prove the second statement in the theorem, assume that the equality is attained. From your previous work deduce that $\sum_{j=k} \lambda_j a_j^2 = \lambda_k \sum_{j=k}^\infty a_j^2$. Conclude that ϕ must be a linear combination of the φ_j's.

4.6. Exercise 5: Temperature decreases with time and solutions are unique.

(1) Prove that the temperature of the manifold decreases as time evolves. That is, show that $t \mapsto \|u(\cdot, t)\|_{L^2}$ is decreasing with t.

(2) Using the previous part show that the solution to the Heat Equation is unique.

4.7. Exercise 5: Nodal domains and first eigenfunctions.

(1) Prove the Courant nodal domain Theorem: Write $\lambda_1 \leq \lambda_2 \leq \ldots$ for the eigenvalues of the Laplacian repeated according to multiplicity. Write $\varphi_1, \varphi_2, \ldots$ for the corresponding L^2-normalized eigenfunctions. Then, the number of nodal domains of φ_k is strictly smaller than $k+1$.

Follow the next steps:

(a) Suppose that φ_k has at least $k+1$ nodal domains $D_1, \ldots, D_{k+1}, \ldots$ and define

$$\psi_j = \begin{cases} \varphi_k \mid_{D_j} & \text{on } D_j, \\ 0 & \text{else} \end{cases}.$$

Show that there exists $\phi = \sum_{j=1}^k a_j \psi_j \in H_1(M)$ orthogonal to $\varphi_1, \ldots, \varphi_{k-1}$.

(b) Explain why $\lambda_k \leq \frac{\|\nabla_g \phi\|_g^2}{\|\phi\|_g^2}$.

(c) Using that $\langle \nabla\phi, \nabla\phi\rangle_g = \sum_{ij=1}^k a_i a_j \langle \Delta_g \psi_i, \psi_j\rangle_g$ show that $\lambda_k \geq \frac{\|\nabla_g \phi\|_g^2}{\|\phi\|_g^2}$.

(d) Conclude that ϕ is an eigenfunction of eigenvalue λ_k. The theorem then follows from a result known as "Unique Continuation" which says that eigenfunctions cannot vanish on an open set unless they are equal to zero everywhere.

(2) Conclude that φ_1 always has constant sign.

(3) Conclude that the multiplicity of λ_1 is 1.

(4) Conclude that φ_2 has precisely two nodal domains and φ_k has at least two nodal domains for all $k \geq 2$.

DEPARTMENT OF MATHEMATICS, UNIVERSITY OF NORTH CAROLINA AT CHAPEL HILL, CHAPEL HILL, NORTH CAROLINA 27599

Email address: canzani@email.unc.edu

Contemporary Mathematics
Volume **720**, 2018
https://doi.org/10.1090/conm/720/14525

Quantum graphs via exercises

Ram Band and Sven Gnutzmann

ABSTRACT. Studying the spectral theory of Schrödinger operator on metric graphs (also known as "quantum graphs") is advantageous on its own as well as to demonstrate key concepts of general spectral theory. There are some excellent references for this study such as a book by Berkolaiko and Kuchment (mathematically oriented book), a review by Gnutzmann and Smilansky (review with applications to theoretical physics), and lecture notes by Berkolaiko (elementary lecture notes). Here, we provide a set of questions and exercises which can accompany the reading of these references or an elementary course on quantum graphs. The exercises are taken from courses on quantum graphs which were taught by the authors.

1. Basic spectral theory of quantum graphs

1.1. δ-type vertex conditions.

Background: The most common vertex conditions of a graph are called Dirichlet and Neumann. Note that the Neumann conditions are also sometimes called Kirchhoff, standard or natural. Nevertheless, we stick to the name Neumann in this note.

For the following set of questions we consider a particle in a box described by the Schrödinger equation

$$-\frac{d^2 f(x)}{dx^2} + V(x)f(x) = k^2 f(x) \quad \text{for } x \in (-l_1, l_2), \tag{1}$$

where $l_1 > 0$, $l_2 > 0$ and $V(x)$ is called the electric potential and can be taken for example in $L^2([-l_1, l_2])$. We employ here the Dirichlet boundary conditions $f(-l_1) = 0 = f(l_2)$, which physically mean that the particle is trapped in an infinite well, whose walls are at $x = -l_1$ and $x = l_2$.

Questions:

1. In this question, we choose the potential in (1) to be $V(x) = \alpha\delta(x)$, where $\alpha \in \mathbb{R}$ and $\delta(x)$ is Dirac's delta function. By doing so, we extend the original validity of equation (1), since the potential is now considered as a distribution and not merely a function. In order for the action of the δ distribution to be well-defined, we must require that f is continuous at $x = 0$ with $f_0 := f(0^+) = f(0^-)$

2010 *Mathematics Subject Classification.* Primary 35P05.

The first author was supported by ISF (Grant No. 494/14) and Marie Curie Actions (Grant No. PCIG13-GA-2013-618468).

Show that any eigenfunction f in (1) satisfies the matching condition

$$\alpha f_0 = f'(0^+) - f'(0^-). \tag{2}$$

Comment: Note that the sign of α determines whether the δ-potential is attractive ($\alpha < 0$), repulsive ($\alpha > 0$) or vanishing ($\alpha = 0$).

The relation (2) is called the δ-type vertex condition. Indeed, in our example $x = 0$ may be considered as a graph vertex whose degree is two. If the vertex under consideration is of degree higher than two, then this condition is generalized and takes the form

$$af_0 = \sum_e f|_e'(0), \tag{3}$$

where the sum is taken over all edges adjacent to the considered vertex, and in all the summands the derivative is taken to be directed towards the vertex. In addition, we always require the continuity of the function at the considered vertex, i.e., all adjacent edges agree in the value they obtain at the vertex, and this value equals f_0.

Note that the Dirichlet and Neumann conditions are obtained as special cases of the δ-type vertex condition for some particular values of the parameter ($\alpha = 0$ for Neumann and $\alpha \to \infty$ for Dirichlet).

2. Next, we find the spectrum of the problem above: the interval $[-l_1, l_2]$ with Dirichlet conditions at the end points,

$$f(-l_1) = f(l_2) = 0, \tag{4}$$

and a δ-type vertex condition at $x = 0$,

$$f(0^+) = f(0^-) \tag{5}$$

$$\alpha f_0 = f'(0^+) - f'(0^-). \tag{6}$$

(a) Show that the non-negative eigenvalues k^2 of the one-dimensional Laplacian, $-\frac{\mathrm{d}^2}{\mathrm{dx}^2}$, with the vertex conditions (4),(5),(6) are given as the zeros of the following function

$$\zeta_\alpha(k) = \alpha + k\cot(kl_1) + k\cot(kl_2). \tag{7}$$

We call such a function, whose zeros provide the graph's eigenvalues, a secular function.

(b) For an irrational lengths ratio ($l_1/l_2 \notin \mathbb{Q}$) show that there cannot be any eigenfunctions with $f_0 = 0$.
Conclude that the poles of the secular function, (7), cannot belong to the spectrum.

(c) Show that $\zeta_\alpha(k)$ has only single poles if $l_1/l_2 \notin \mathbb{Q}$.
However, if $l_1/l_2 \in \mathbb{Q}$ there are single and double poles of the secular function. Locate them and show that the double poles belong to the spectrum, while the single poles do not belong to it. This justifies the need of regularization of the secular function

$$\tilde{\zeta}_\alpha(k) = \zeta_\alpha(k)\sin(kl_1)\sin(kl_2). \tag{8}$$

Namely, the zeros of the regularized secular function, $\tilde{\zeta}_\alpha(k)$, correspond to the Laplacian eigenvalues (irrespective to the value of l_1/l_2).

(d) Consider an attractive δ-potential ($\alpha < 0$). Are there negative eigenvalues among the eigenvalues of the Schrödinger operator? Under what conditions and how many?

(e) Show that if $\alpha = 0$ in the problem above, this is equivalent to solving the eigenvalue problem of the Laplacian on the interval, $[-l_1, l_2]$ with Dirichlet conditions at the boundaries, but with no additional condition at $x = 0$.
Conclude that this is the case in general. Namely for any graph, a vertex of degree two with Neumann vertex conditions ($\alpha = 0$) is superfluous, i.e. one can erase the vertex and join the two incident edges to a one single edge such that the lengths add up.
Hint: You can read about this in [**5**, section 2.2]

Comment: Equation (7) introduces a secular function which possess poles. The regularization process provides the secular function (8) with no poles. This latter secular function is the more standard one. There is an explicit formula for it (with no need of regularization), which makes it also more convenient to use in proofs and analysis on quantum graphs. See more about that in Section 1.4.

1.2. From the interval to the star graph.

Background: Consider a star graph which consists of one central vertex which is connected to all other $V - 1$ vertices. We will enumerate the vertices as $i = 0, 1, \ldots, E$ where $E = V - 1$ is the total number of edges. The central vertex is privileged to have the index $i = 0$. The lengths of the edges are $\{l_{01}, \ldots, l_{0E}\}$. The restrictions of a function f to the edges are denoted by $\{f_{01}, \ldots, f_{0E}\}$. The coordinates along the edges are chosen according to the convention that $x_{0i} = 0$ at the central vertex and $x_{0i} = l_{0i}$ at the vertex i. We will assume Dirichlet boundary conditions at the vertices $i = 1, \ldots, E$, and a repulsive vertex potential whose strength is $\alpha_0 \geq 0$ at the central vertex. Namely, the vertex conditions at the boundary vertices are δ-type vertex conditions:

$$\forall\, 1 \leq i \leq E\ ;\ f_{0i}(l_{0i}) = 0,$$

and the vertex conditions at the central vertex are

$$\forall 1 \leq i < j \leq E\ ;\quad f_{0i}(0) = f_{0j}(0) \equiv f_0$$

and

$$\sum_{i=1}^{E} f'_{0i}(0) = \alpha_0 f_0.$$

Note that the problem considered in section 1.1 is a special case of the problem considered here with $E = 2$.

Questions:

3. Show that a possible secular function for the graph is

$$\zeta(k) = \frac{\alpha_0}{k} + \sum_{i=1}^{E} \cot(k l_{0i}). \tag{9}$$

Namely, show that zeros of this function are all eigenvalues of the graph (note that there is not necessarily one to one correspondence – see the next question).

Note that the secular function, (7) in the previous question is a special case of the one above (with $E = 2$).

4. What is the weakest assumption you need to assume in order to have a one to one correspondence between the zeros of the secular function (9) and the graph's eigenvalues?

5. Show that under the assumption you got in the previous question, the graph's spectrum is non degenerate. Namely, that each eigenvalue appears with multiplicity one.

6. Under the same assumption as in questions **4** and **5** show the following interlacing properties:

(a) The Dirichlet spectrum ($\alpha_0 \to \infty$) of the star graph interlaces with the spectrum of the same star graph but with a Neumann vertex condition at the central point ($\alpha_0 = 0$).

(b) For any positive value of α_0 the n-the eigenvalue of the star graph is bounded from below by the n-th eigenvalue for $\alpha_0 = 0$ and from above by the n-th eigenvalue of the Dirichlet spectrum.

(c) Similarly to the previous question, which interlacing property holds for the n-th eigenvalue of the star graph with a negative value of α_0?

Comment: See [**7**, theorem 3.1.8] for a general statement on eigenvalue interlacing for quantum graphs.

7. Assume now that $\alpha = 0$ (Neumann vertex conditions at the center). Consider all star graphs with any number of edges E and any positive edge lengths $\{l_{01}, \ldots, l_{0E}\}$, such that the total length of the edges, $L = \sum_{i=1}^{E} l_{0i}$ is fixed.

(a) What is the supremum of the first eigenvalue among all choices of values for E and $\{l_{01}, \ldots, l_{0E}\}$ (under the constraint above)? Is this supremum attained (i.e., is it a maximum)?

(b) What is the infimum? Is it attained?

Hint: The answer can be found in [**9**].

Comment: Similar questions of eigenvalue optimization on graphs and bounds on the eigenvalues are discussed in the recent works [**1, 2, 6, 14**]

8. *This is a numerical exercise!* Choose any value for the number of edges, E, and any values for the edge lengths $\{l_{01}, \ldots, l_{0E}\}$. Plot the secular function for your choice and find its thirteen first zeros.

1.3. Secular function - first approach.

Question:

9. Consider an arbitrary quantum graph with V vertices and E edges. Assume there is an edge connecting vertices 1 and 2 and write the restriction of a function f to this edge as

$$f_{12}(x_{12}) = A \cos kx_{12} + B \sin kx_{12}.$$

Similarly, write

$$f_{ij}(x_{ij}) = \frac{f_j \sin(kx_{ij}) + f_i \sin(k(l_{ij} - x_{ij}))}{\sin kl_{ij}},$$

for the restriction of f to an edge connecting vertices i, j ($i < j$). Using this, obtain a set of homogeneous equations for the coefficients A, B, f_i ($i = 1, 2, \ldots V$) and derive a secular function that does not have poles at the Dirichlet spectrum of the edge $e = (1, 2)$.

Hint: There is more than one solution to this - one may also reduce the number of equations (and variables) easily to V (the number of vertices) without re-introducing poles.

1.4. Secular function - scattering approach.

Background: One may express the restriction of the eigenfunction to the edge connecting vertices i, j by

$$f_{ij}(x_{ij}) = a_{ij}^{\text{in}} e^{-ikx_{ij}} + a_{ij}^{\text{out}} e^{ikx_{ij}} , \tag{10}$$

where a_{ij}^{in} and a_{ij}^{out} are some coefficients. If the eigenfunction belongs to the eigenvalue k^2, we can always choose values to those coefficients, such that expression (10) holds.

For a given vertex i of degree d_i we have d_i coefficients of the type a_{ij}^{in} and another d_i coefficients of the type a_{ij}^{out}. Let us collect these into d_i-dimensional vectors $\vec{a}^{(i),\text{out}}$, $\vec{a}^{(i),\text{in}} \in \mathbb{C}^{d_i}$. The vertex conditions on the vertex i allow us to express the $\vec{a}^{(i),\text{out}}$ as a linear transform of the $\vec{a}^{(i),\text{in}}$

$$\vec{a}^{(i),\text{out}} = \boldsymbol{\sigma}^{(i)}(k)\vec{a}^{(i),\text{in}}. \tag{11}$$

The components of the vectors $\vec{a}^{(i),\text{out}}$, $\vec{a}^{(i),\text{in}} \in \mathbb{C}^{d_i}$ have the meaning of incoming and outgoing wave amplitudes; $\boldsymbol{\sigma}^{(i)}(k)$ is a unitary matrix of size $d_i \times d_i$ and is called the vertex scattering matrix.

To each quantum graph with E edges one may associate a unitary matrix $U(k)$ of dimension $2E \times 2E$, known as the graph's *quantum map* or *(discrete) quantum evolution operator* that describes the connectivity of the graph, the matching conditions at the vertices and the eigenvalue spectrum of the graph. In Question **12** the quantum map is derived explicitly for a particular graph. The same procedure may be applied to other quantum graphs to find the quantum map.

Questions:

10. Show that the δ-type vertex conditions (i.e., continuity of the eigenfunction at a vertex i of degree d_i and , $\alpha_i f_i = \sum_{j\in V_i} f'_{ij}(0)$) are equivalent to the vertex scattering matrix

$$\sigma_{jj'}^{(i)}(k) = \frac{2}{d_i + i\frac{\alpha_i}{k}} - \delta_{jj'} = \begin{cases} \frac{2}{d_i+i\frac{\alpha_i}{k}} & j \neq j' \\ \frac{2}{d_i+i\frac{\alpha_i}{k}} - 1 & j = j' \end{cases}. \tag{12}$$

11. Show that the vertex scattering matrix

$$\boldsymbol{\sigma}^{(i)} = -\mathbf{1}_{d_i} + \frac{2}{v_i + i\frac{\alpha_i}{k}}\mathbb{E}_{d_i}$$

is unitary. Here, $\mathbb{E}_{d_i}$ is the full $d_i \times d_i$ matrix with all matrix elements equal to one. You may use $\mathbb{E}_{d_i}^2 = d_i\mathbb{E}_{v_i}$ and $\mathbb{E}^* = \mathbb{E}$ where $\mathbb{E}^*$ denotes the hermitian conjugate of $\mathbb{E}$.

12. Consider a star graph which consists of three edges. The central vertex is denoted by 0 and supplied with Neumann vertex conditions. The boundary vertices are denoted by $1, 2, 3$ and are supplied with Neumann, Dirichlet, Dirichlet conditions, correspondingly. These notations and the edge lengths are shown in Fig. 1.

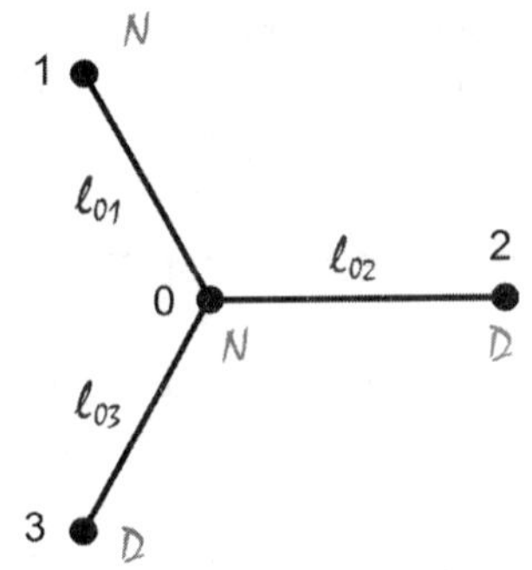

FIGURE 1. A star graph.

In this question you will explicitly build up the quantum map of the graph above by following the given sequence of instructions.

(a) Write explicitly the scattering matrices, $\boldsymbol{\sigma}^{(i)}$, which correspond to each of the vertices $0, 1, 2, 3$. You may compare with equation (12) above.

(b) Write the equation (11) for each of the vertices. Use the explicit matrices which you have found in the previous section and write the components of the vectors $\vec{a}^{(i),\text{in}}$ and $\vec{a}^{(i),\text{out}}$ with explicit indices in each case (i.e., write $a_0^{(2),\text{in}}$, $a_3^{(0),\text{out}}$, etc.).

(c) Write (explicitly again) the "big" scattering matrix $S(k)$ to fit the following set of equations

$$\vec{a}^{\,\text{out}} = S(k)\vec{a}^{\,\text{in}},$$

where

$$\vec{a}^{\,\text{in}} = \begin{pmatrix} a_{10}^{\text{in}} \\ a_{01}^{\text{in}} \\ a_{20}^{\text{in}} \\ a_{02}^{\text{in}} \\ a_{30}^{\text{in}} \\ a_{03}^{\text{in}} \end{pmatrix} \quad \text{and} \quad \vec{a}^{\,\text{out}} = \begin{pmatrix} a_{01}^{\text{out}} \\ a_{10}^{\text{out}} \\ a_{02}^{\text{out}} \\ a_{20}^{\text{out}} \\ a_{03}^{\text{out}} \\ a_{30}^{\text{out}} \end{pmatrix}.$$

Remember that $S(k)$ merely consists of the different components of the single vertex scattering matrices $\boldsymbol{\sigma}^{(i)}(k)$ and zero elements for edges that are not connected to each other. Pay special care to the order of the entries of the vectors above.

(d) Write the matrix $T(k)$ such that it fits into the set of equations

$$\vec{a}^{\,\text{in}} = T(k)\vec{a}^{\,\text{out}}$$

with $\vec{a}^{\,\text{in}}$ and $\vec{a}^{\,\text{out}}$ as given above.
Hint: you may use the fact $f_{ij}(x) = f_{ji}(l_{ji} - x)$

(e) A few tips to check yourself (no need to calculate, just in order to verify your answer).
i. The matrix $S(k)$ should be k-independent.
ii. The matrix $S(k)$should be unitary.
iii. The matrix $T(k)$ should be diagonal.

If you did all the above correctly, the quantum evolution operator is obtained by matrix multiplication of the two matrices, $U(k) = T(k)S(k)$.

13. Consider a star graph with Neumann vertex conditions at the central vertex $i = 0$ and Dirichlet vertex conditions at the boundary vertices $i = 1, 2, \ldots, E$. Derive the quantum evolution map $U(k)$ and show that the secular function can be reduced to the form

$$\zeta(k) = \det\left(\mathbf{1}_{2E} - U(k)\right) = \det\left(\mathbf{1}_E + \widetilde{T}(2k)\boldsymbol{\sigma}^{(0)}\right),$$

where $\boldsymbol{\sigma}^{(0)}$ is the central vertex scattering matrix and $\widetilde{T}(k)$ is a diagonal $E \times E$ matrix, $T(k)_{ee'} = \delta_{ee'}e^{ikl_e}$. You can gain a good intuition for the solution of this question from your solution to the previous question.

14. Show that the following secular function is real

$$\tilde{\zeta}(k) = \sqrt{\det\left(S^*(k)T^*(k)\right)}\det\left(\mathbf{1}_{2E} - U(k)\right). \tag{13}$$

Remember that $U(k) = T(k)S(k)$ and use the unitarity of $T(k)$ and $S(k)$. *Comment*: The secular function $\tilde{\zeta}(k)$ above may even be differentiable in k, if the complex branch of the square root is appropriately chosen.

2. Trace formula and periodic orbits

2.1. The trace formula for the spectrum of a unitary matrix.

Background: Consider an $M \times M$ unitary matrix U with unimodular eigenvalues $e^{i\theta_\ell}$ for $\ell = 1, \ldots, M$. One may extend the spectrum of eigenphases θ_ℓ periodically beyond the interval $0 \le \theta < 2\pi$. The extended spectrum then consists of the numbers

$$\theta_{\ell,n} = \theta_\ell + n2\pi \qquad n \in \mathbb{Z}\,.$$

Assume that $\theta_\ell \neq 0$ and $\theta \neq \theta_{\ell,n}$ is real.

Question:

15. Consider the spectral counting function

$$N(\theta) = \sum_{n=0}^{\infty}\sum_{\ell=1}^{M} \vartheta(\theta - \theta_\ell - n2\pi),$$

where ϑ is the Heaviside step function. Show that one may write it as the following trace formula.

$$N(\theta) = \frac{M\theta}{2\pi} - \frac{1}{\pi}\mathrm{Im}\,\log\,\det\,(1 - U) + \frac{1}{\pi}\mathrm{Im}\,\log\,\det\,(1 - e^{-i\theta}U).$$

Comment: The definition of the spectral counting function may be extended to $\theta = \theta_{n,\ell}$ such that both expressions (the defining expression and the trace formula) remain consistent. For this one replaces the last term in the trace formula by the limit

$$\frac{1}{\pi}\mathrm{Im}\,\log\,\det\,(1 - e^{-i\theta}U) \mapsto \lim_{\epsilon\to 0^+}\frac{1}{\pi}\mathrm{Im}\,\log\,\det\,(1 - e^{-i\theta-\epsilon}U)$$

and sets $\vartheta(0) = 1/2$. Replacing $U \mapsto e^{-\epsilon}U$ and considering the limit $\epsilon \to 0^+$ also helps to regularize certain expansions that may come up in the proof of the trace formula because the trace formula for finite $\epsilon > 0$ does not have any singularities for θ on the real line.

We strongly recommend plotting the regularized expression for $N(k)$ with a (small) positive value for ϵ for a given unitary matrix U (which may be chosen diagonal).

Hint: There are several ways to perform the derivation. One interesting derivation is based on Poisson summation. This method requires ϵ-regularization as mentioned in the comment above.

(a) Write the spectral counting function as

$$N(\theta) = \sum_{n=-\infty}^{\infty} \sum_{\ell=1}^{M} \vartheta(\theta - \theta_\ell - n2\pi)\vartheta(\theta_\ell + n2\pi).$$

(b) The Poisson summation formula for a smooth function $f(x)$ which decays sufficiently fast for $|x| \to \infty$ (so that all sums and integrals converge absolutely) reads

$$\sum_{n=-\infty}^{\infty} f(n) = \sum_{\nu=-\infty}^{\infty} \int_{-\infty}^{\infty} e^{2\pi i \nu x} f(x) dx \ .$$

We want to apply it to $f(x) = \sum_{\ell=1}^{M} \vartheta(\theta - \theta_\ell - 2\pi x)\, \vartheta(\theta_\ell + 2\pi x)$ which is not smooth. In this case the Poisson sum is not absolutely convergent but may be regularized by introducing an additional factor $e^{-\epsilon|\nu|}$ and taking the limit $\epsilon \to 0^+$.

(c) While using the formula above, evaluate separately the $\nu = 0$ term from the other terms.

(d) Compare this term by term to the expansion

$$\log \det\, (1 - e^{-i\theta} U) = \operatorname{tr} \log\, (1 - e^{-i\theta} U) = - \lim_{\epsilon \to 0+} \sum_{n=1}^{\infty} \frac{1}{n} e^{-in\theta - n\epsilon} \operatorname{tr}\, U^n$$

(or the complex conjugate version) in order to perform the sum over ν in the Poisson summation.

Comment: You may compare this to the trace formula of a quantum graph. Do this by showing that the above formula is equivalent to the one obtained in [**11**] for quantum graphs if all the edge lengths of the graph are the same. Note that the derivation of the trace formula in this reference uses a different (more general) method.

2.2. Periodic orbits.

Background: Consider a quantum graph with E edges of lengths l_e ($e = 1, \ldots, E$) with Neumann matching conditions and let $U(k)$ be the unitary, k-dependent $2E \times 2E$ matrix

$$U(k)_{\alpha'\alpha} = e^{ikl_{\alpha'}} S_{\alpha'\alpha}$$

where $S_{\alpha'\alpha}$ is the scattering amplitude from the directed edge α to directed edge α'.

Note that $S_{\alpha'\alpha} = 0$ unless the end vertex of the directed edge α coincides with the start of α' – one then says that α' follows α.

We have

$$\operatorname{tr}\, U(k)^n = \sum_{\alpha_1, \ldots, \alpha_n = 1}^{2E} e^{ikl_{\alpha_1}} S_{\alpha_1 \alpha_n} e^{ikl_{\alpha_n}} S_{\alpha_n \alpha_{n-1}} \cdots e^{ikl_{\alpha_3}} S_{\alpha_3 \alpha_2} e^{ikl_{\alpha_2}} S_{\alpha_2 \alpha_1} \ ,$$

where $\{\alpha_1, \ldots, \alpha_n\}$ is a set of n directed edges of the graph. In the sum above, each term is of the form $A_\gamma e^{ik\ell_\gamma}$

$$A_\gamma = S_{\alpha_1\alpha_n} S_{\alpha_n\alpha_{n-1}} \ldots S_{\alpha_2\alpha_1}$$
$$\ell_\gamma = l_{\alpha_n} + l_{\alpha_{n-1}} + \ldots l_1$$

Here $\gamma = (\alpha_1, \ldots, \alpha_n)$ is a fixed set of summation indices which corresponds to a sequence of directed edges.

Note that $A_\gamma \neq 0$ only if the edge α_{j+1} follows the edge α_j in the graph, for $j = 1, \ldots, n$ (in this context $\alpha_{n+1} \equiv \alpha_1$) – i.e. only if γ is a closed trajectory on the graph. For such a trajectory, l_γ is its total length, i.e., the sum of its edge lengths. By definition, the ***length spectrum*** of the graph is the set of all lengths $\{l_\gamma\}$ of closed trajectories. You can see a few examples of such closed trajectories in Fig. 2.

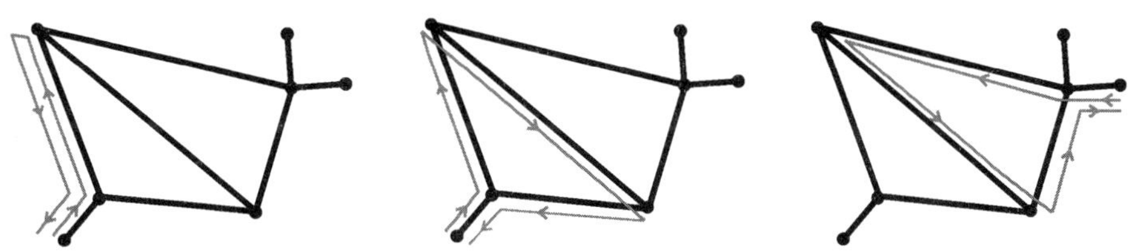

FIGURE 2. Three examples of closed trajectories on a graph.

We now list a few observations and definitions related to periodic orbits.

- Note that any cyclic permutation, γ', of the indices in the trajectory γ gives a different closed trajectory with the same contribution $A_{\gamma'} = A_\gamma$ and $\ell_{\gamma'} = \ell_\gamma$.
- The equivalence class $\overline{\gamma} = \overline{\alpha_1 \ldots \alpha_n}$ that contains all cyclic permutations of a given closed trajectory $\gamma = (\alpha_1, \ldots, \alpha_n)$ is called a ***periodic orbit*** with ***period*** n on the graph.
- The periodic orbit $\gamma = \alpha_1 \ldots \alpha_n$ is a ***primitive periodic orbit*** of primitive period n if the sequence of indices $\alpha_1, \ldots, \alpha_n$ is not a repetition of a shorter sequence.

  All the closed trajectories in Fig. 2 represent primitive periodic orbits.
- If $\overline{\gamma}$ is a periodic orbit with period n, then there exists a unique primitive periodic orbit $\overline{\gamma}_p$ with ***primitive period*** n_p such that $\overline{\gamma}$ is a repetition of $\overline{\gamma}_p$ and $n = rn_p$. Here $r \geq 1$ is the integer ***repetition number*** of $\overline{\gamma}$ and n_p the ***primitive period*** of γ.
- If $r = 1$ then $\overline{\gamma} = \overline{\gamma}_p$ and $\overline{\gamma}$ is primitive.
- We write $\overline{\gamma} = \overline{\gamma}_p{}^r$ for the r-th repetition of the primitive orbit $\overline{\gamma}_p$.

Comment: The trace formula of the spectral counting function may be expressed in terms of an infinite sum over the graph's periodic orbits. To answer the following questions, you should read more about that, for example in [**7**, section 3.7.4] or [**11**, section 5.2].

Questions:

16. Consider the dihedral graphs as given in Fig. 3

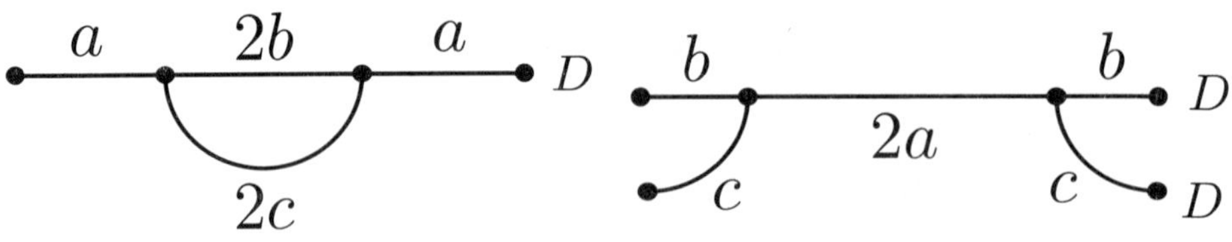

FIGURE 3. Two isospectral quantum graphs. Their edge lengths are indicated by the parameters a, b, c. The vertices marked with 'D' have Dirichlet conditions and all other vertices have Neumann conditions.

It is known that those graphs are isospectral [**3**]. Solve the following 'paradox':

The isospectrality of those graphs means that their spectral counting functions are equal. Hence, the periodic orbit expansions of those counting functions are the same. Therefore, both graphs should have the same set of periodic orbits. Nevertheless, the graph on the left has a periodic orbit of length $2a$, whereas the graph on the right does not have such an orbit. How is this possible?

17. This question concerns the tetrahedron graph, i.e. the complete Neumann graph with $V = 4$ vertices.

(a) *Warm-up:* Choose two periodic orbits on the graph, such that one of them is primitive and the second is some repetition of the first. For each of those periodic orbits, γ, evaluate the following quantities:
  (i) The period n of the orbit.
  (ii) The length ℓ_γ of the orbit (expressed in terms of the graph edge lengths).
  (iii) The coefficient A_γ which corresponds to the orbit in the periodic orbits expansion (write the explicit number).
  (iv) The primitive period n_p and the repetition number r ?

(b) Assume that all bond lengths are incommensurate. Go over all periodic orbits of period $n = 5$ and write their contribution to the length spectrum σ_ℓ (expressed in terms of the graph edge lengths). What are the corresponding quantum amplitudes A_γ of those orbits?

(c) Now assume that all edge lengths are equal. How does your answer to the previous question change?

(d) The connectivity matrix C of a simple graph with V vertices is the real symmetric $V \times V$ matrix with entries

$$C_{i,j} = \begin{cases} 1 & \text{vertices } i \text{ and } j \text{ are connected by an edge,} \\ 0 & \text{else.} \end{cases}$$

Note that for simple graphs there are no loops, i.e. edges that connect a vertex to itself, so $\forall i$, $C_{ii} = 0$. The connectivity matrix may be used to count the number of trajectories that connect the vertices i and j in n steps – this number is just $[C^n]_{ji}$.

Consider the connectivity matrix of the tetrahedron, $C_{ij} = 1 - \delta_{ij}$. Show that as $n \to \infty$, $[C^n]_{ji} \sim ce^{\alpha n}$ and find α.
Hint: diagonalize C.

Comment: The last question shows that there is an exponential growth in the number of orbits on the graph. This holds in particular for periodic orbits and makes it difficult to use the periodic orbit expansion of the trace formula for spectral computations.

18. In the previous question we counted the number of trajectories between two vertices via the (vertex) connectivity matrix. An alternative approach is based on the $2E \times 2E$ edge adjacency matrix $\mathcal{B}$ whose indices correspond to the directed edges where

$$\mathcal{B}_{\alpha\alpha'}. = \begin{cases} 1 & \text{if } \alpha \text{ follows } \alpha', \\ 0 & \text{else.} \end{cases}$$

The main difference is that $[\mathcal{B}^n]_{\alpha\alpha'}$ counts the number of trajectories that start on the directed edge α' and end after n steps on the directed edge α. Both approaches can also be used to count the number of periodic orbits via the traces. We will explore this here for the edge connectivity matrix $\mathcal{B}$.

(a) Show that $\frac{1}{n}\operatorname{tr}\mathcal{B}^n = \sum_{\gamma:n_\gamma=n}\frac{1}{r_\gamma}$ where the sum is over all periodic orbits of period n. Conclude that if n is a prime number then $\frac{1}{n}\operatorname{tr}\mathcal{B}^n$ is the number of periodic orbits of period n.

(b) Derive an expression for the number of periodic orbits of period n in terms of traces of powers of $\mathcal{B}$ for
(*i.*) $n = p^j$ where p is a prime number and $j \geq 2$ an integer, and
(*ii.*) $n = p_1 p_2$ where p_1 and p_2 are prime numbers.
Make an educated guess for the general expression when n has the prime number decomposition $n = \prod_m p_m^{j_m}$ where $j_m \geq 0$ is the multiplicity of the m-th prime.

19. This question demonstrates that given some lengths of periodic orbits of an unknown graph, one can reconstruct the graph.

(a) Find the graph with the following properties:
(*i.*) the total length (sum of all edge lengths) is $1\frac{5}{6}$;
(*ii.*) the lengths of all periodic orbits whose length is not greater than 5 are given by the list

$$\frac{2}{3},\ 1\frac{1}{3},\ 2,\ 2\frac{2}{3},\ 3,\ 3\frac{1}{3},\ 3\frac{2}{3},\ 4,\ 4\frac{1}{3},\ 4\frac{2}{3},\ 5.$$

Draw the graph and indicate the edge lengths on the drawing.

(b) Find the graph with the following properties:
(*i.*) the total length is $5\frac{13}{15}$;
(*ii.*) the lengths of all periodic orbits whose length is not greater than 5 are given by the list

$$2,\ 2\frac{1}{3},\ 2\frac{1}{6},\ 2\frac{2}{5},\ 4,\ 4\frac{1}{6},\ 4\frac{1}{3},\ 4\frac{1}{3},\ 4\frac{2}{5},\ 4\frac{1}{2},\ 4\frac{17}{30},\ 4\frac{2}{3},\ 4\frac{11}{15},\ 4\frac{4}{5},\ 5.$$

Note that there are two different periodic orbits of length $4\frac{1}{3}$. Also, any number which appears only once in the list above indicates that there is exactly one periodic orbit of that length.
What is the graph this time?

(c) Try to think how to construct a general algorithm for finding the graph out of knowing its total metric length and lengths of all of its periodic orbits.
Assume that the graph is simple (no loops and no multiple edges) and that its edge lengths are incommensurate.

Hint: What is the shortest periodic orbit of a graph?
Another hint: The answer can be found in [**12**].

2.3. The constant term of the trace formula.
Background: The spectral counting function of a quantum graph is

$$N(k) := \left|\{\lambda \in \mathbb{R} \text{ is an eigenvalue } : \lambda < k^2\}\right|,$$

where eigenvalues are counted with their multiplicity. One of the forms of the trace formula for the spectral counting function is

$$N(k) = N_0 + \frac{\mathcal{L}}{\pi}k - \lim_{\epsilon\to 0}\frac{1}{\pi}\text{Im}\,\log\tilde{\zeta}(k+i\epsilon), \tag{14}$$

where $\tilde{\zeta}(k)$ is the real secular function given in (13), and N_0 is a constant term. The last two terms in (14) equal to the number of real zeros of $\tilde{\zeta}$ with absolute value smaller than k. Those zeros are in one to one correspondence with the graph eigenvalues (including multiplicity), with the exception of $k = 0$. The value of $\tilde{\zeta}$ at $k = 0$ does not correspond to the multiplicity of the zero eigenvalue and this 'mismatch' is compensated by the constant term N_0 in (14).

The expression for this term was originally derived in [**16**, lemma 1],[**17**]. Other works related to this subject are [**10, 13**]. Further reading in [**7**, section 3.7] and [**11**, section 5] is recommended.

Question:

20. In the following question we derive the value of N_0 for a Neumann graph (all vertex matching conditions are of Neumann aka Kirchhoff type). Initially, assume that the graph has a single connected component. Let $\vec{E}$ be the space of directed edges on the graph (this space is of dimension $2E$, where E is the number of edges).

(a) Let $\omega : \vec{E} \to \mathbb{C}$ such that

$$\forall\,(i,j)\in\vec{E}\quad \omega\,(i,j) = -\omega\,(j,i)$$

and

$$\forall i\quad \sum_{j\sim i}\omega\,(i,j) = 0,$$

where $j \sim i$ means that the vertex j is adjacent to the vertex i. All such functions $\omega : \vec{E} \to \mathbb{C}$ form a vector space (over $\mathbb{C}$). Prove that the dimension of this space is $\beta := E - V + 1$.

Hint: Start by considering a tree graph.

(b) Let $\vec{a}^{in} \in \mathbb{C}^{2E}$ with entries denoted by $a_j^{(i),in}$ (for $i \sim j$), such that the following is satisfied

$$\forall i\sim j\,,\, i\sim k\quad a_j^{(i),in} + a_i^{(j),in} = a_k^{(i),in} + a_i^{(k),in}$$

and

$$\forall i\quad \sum_{j\sim i}\left(-a_j^{(i),in} + a_i^{(j),in}\right) = 0.$$

Prove that the dimension of the vector space which contains all such solutions $\vec{a}^{in} \in \mathbb{C}^{2E}$ is $\beta + 1$.

(c) Note that you have shown $\dim\ker(\mathbf{1}-S) = \beta+1$, which implies $N_0 = \frac{1-\beta}{2}$. Show that for a Neumann graph with C (disjoint) connected components, the constant term of the trace formula (14) is $N_0 = \frac{C-\beta}{2}$. You can use the generalized definition of β, which is $\beta := E - V + C$ (this value can be obtained by summing over all the β's of the different components).

3. Further topics

3.1. Quantum to classical correspondence for quantum graphs.

Background: In this question we study the classical dynamics of a quantum graph. This will help us to understand in what sense the classical dynamics that corresponds to a quantum graph is 'chaotic'. Remember that the quantum evolution map, $U(k)$, contains amplitudes for scattering processes to go from one directed edge to another. We define a corresponding classical map, M, by replacing the amplitudes $U(k)_{\alpha\alpha'}$ by

$$M_{\alpha\alpha'} = |U(k)_{\alpha\alpha'}|^2 = |S_{\alpha\alpha'}|^2 .$$

Hence, M is a matrix of dimensions $2E \times 2E$, which contains the probabilities for the scattering events.

Question:

21. By following the steps below prove that the matrix M defines a Markov process on the set of directed edges with the stated additional properties.

(a) Prove that the matrix M is a *bi-stochastic (doubly stochastic) matrix*. Namely, prove that

$$\sum_{\alpha=1}^{2B} M_{\alpha\alpha'} = \sum_{\alpha'=1}^{2B} M_{\alpha\alpha'} = 1.$$

(b) Use the bi-stochastic property to verify that the following definition of a Markov process on the directed edges of the graph is well-defined. Let $P_\alpha(n)$ be the probability to find a particle on the directed edge α, at some (discrete) time n. We can then define the probabilities to find the particle on the directed edge α, at time $n+1$, by

$$P_\alpha(n+1) = \sum_{\alpha'} M_{\alpha\alpha'} P_{\alpha'}(n)$$

or, in short $P(n+1) = MP(n)$. In particular show that if $P(n)$ satisfies $\sum_\alpha P_\alpha(n) = 1$ and $P(n)_\alpha \geq 0$, then $P(n+1)$ satisfies the same properties. That is if $P(n)$ is a probability vector then $P(n+1)$ is a probability vector.

Further background: We next consider the equilibration properties of the Markov process $P(n+1) = MP(n)$. Let $P^{\text{inv}} = \frac{1}{2E}$ be the equi-distributed probability vector on the graph. For any quantum graph this is an invariant probability vector, i.e.

$$M\, P^{\text{inv}} = P^{\text{inv}}$$

The classical dynamics which corresponds to a quantum graph is chaotic in the following sense: The Markov process on the graph is called *ergodic* if

$$\lim_{n\to\infty} \frac{1}{n} \sum_{m=1}^{n} P(m) = P^{\text{inv}}$$

for every initial probability vector $P(0)$. We call a graph dynamically connected[1], if for any two directed edges α and α' there is an integer $n > 0$ such that $(M^n)_{\alpha\alpha'} \neq 0$ (i.e one can get from one directed edge to another with a non-vanishing probability in a finite number of steps). Every dynamically connected graph is ergodic. Most graphs are also *mixing* which is the stronger property

$$\lim_{n\to\infty} P(n) = P^{\text{inv}}$$

for every initial probability vector $P(0)$.

Question:

22. In this question we will characterize ergodicity and mixing in terms of the eigenvalue spectrum of the bi-stochastic matrix M. The results apply to any bi-stochastic Markov process on a directed graph whether or not there is a corresponding quantum graph such that $M_{\alpha\alpha'} = |U(k)_{\alpha\alpha'}|^2$ in terms of the quantum map.

(a) Prove that all the eigenvalues of M are either on the unit circle or inside it. Namely, if we denote the set of eigenvalues by $\{\lambda_i\}$, then $\forall i;\ |\lambda_i| \le 1$.

(b) We know that M has at least one eigenvalue which equals 1 (the corresponding eigenvector is P^{inv}). Let us denote this eigenvalue by λ_1 ($\lambda_1 = 1$). Prove that if $\min_{2\le i\le 2E}(1 - |\lambda_i|) > 0$ then the graph is mixing.
Hint: It might be useful to prove the convergence property using the vectors L_1-norm.

(c) Using the notation above (λ_1=1), prove that if $\min_{2\le i\le 2E}(|1-\lambda_i|) > 0$ then the graph is ergodic.
Hint: Again, use the L_1-norm .

Comment: Note that the conditions above are consistent with the trivial fact that mixing is a stronger notion than ergodicity (namely, that every mixing system is also ergodic which follows directly from the definition).

The quantity $\Delta := \min_{2\le i\le 2E}(|1 - \lambda_i|)$ is called the spectral gap and it determines the convergence rate of $\lim_{n\to\infty} \frac{1}{n}\sum_{m=1}^{n} P(m)$ (the greater the gap, the quicker is the convergence).

Similarly, $\tilde{\Delta} := \min_{2\le i\le 2E}(1 - |\lambda_i|)$ determines the convergence rate of $\lim_{n\to\infty} P(n)$.

[1] In the literature a matrix M with this property is sometimes called irreducible.

3.2. The quadratic form.

Background: Consider a quantum graph with the edge and vertex sets, $\mathcal{E}$ and $\mathcal{V}$. We take the operator to be the Laplacian with δ-type vertex conditions (see (3)). The quadratic form of this operator is:

$$h[f] := \sum_{e\in\mathcal{E}} \int_0^{l_e} \frac{\mathrm{d}f}{\mathrm{d}x_e}\overline{\frac{\mathrm{d}g}{\mathrm{d}x_e}}\mathrm{d}x_e + \sum_{v\in\mathcal{V}\,;\,\alpha_v\neq\infty} \alpha_v f(v)\overline{g(v)},$$

where α_v is the coupling coefficient of the δ-type vertex condition at vertex v, and $\alpha_v = \infty$ indicates Dirichlet vertex conditions. The length of the edge e is denoted by l_e.

The domain $D(h)$ of this quadratic form consists of all functions f on the metric graph that satisfy the following three conditions:

($i.$) for each edge e the restriction $f|_e$ belongs to the Sobolev space $H^1([0, l_e])$,
($ii.$) f is continuous at each vertex, and
($iii.$) $f(v) = 0$ at each vertex v for which $\alpha_v = \infty$.

The quadratic form is useful for variational characterization of the spectrum. More on this topic is found in [**7**, section 1.4.3].

Questions:

23. Consider a Schrödinger operator with a bounded non-negative potential ($V \geq 0$) on a quantum graph, $H\psi = -\psi'' + V\psi$. Show that the spectrum of this operator is non-negative if all the vertex conditions are of δ-type with non-negative coupling coefficients (i.e., $\forall\, v \in \mathcal{V}\ ;\ \alpha_v \geq 0$). *Hint*: You need to modify the quadratic form given above to fit the case of Schrödinger operator with a potential.

24. Prove the following statements:

(a) Let $\lambda = \lambda(\alpha)$ be a simple eigenvalue of a graph with δ-type vertex condition at a certain vertex v with the coupling coefficient $\alpha \neq \infty$. The operator on this graph is just the Laplacian (no potential). Then

$$\frac{\mathrm{d}\lambda}{\mathrm{d}\alpha} = |f(v)|^2 .$$

(b) Now, re-parameterize the δ-type vertex condition at v as:

$$\zeta \sum_{e\in E_v} \frac{\mathrm{d}f}{\mathrm{d}x_e}(v) = -f(v),$$

with E_v denoting the set of edges adjacent to v. This parametrization allows Dirichlet condition ($\zeta = 0$) and excludes Neumann condition ($\zeta = \infty$). Show that if the simple eigenvalue is now given by $\lambda = \lambda(\zeta)$ then the derivative is

$$\frac{\mathrm{d}\lambda}{\mathrm{d}\zeta} = \left|\sum_{e\in E_v} \frac{\mathrm{d}f}{\mathrm{d}x_e}(v)\right|^2 .$$

Hint: The answer is given in [**7**, proposition 3.1.6]

3.3. From quantum graphs to discrete graphs.

Question:

25. In this question we consider the spectral connection between quantum graphs and discrete graphs.

(a) Consider an arbitrary quantum graph with V vertices and E edges. Assume that Neumann conditions are imposed at all vertices and that all edges are of the same length, l. Use the following representation for an eigenfunction with eigenvalue k^2 on the edge (i,j)

$$f_{ij}(x_{ij}) = \frac{f_j \sin(kx_{ij}) + f_i \sin(k(l - x_{ij}))}{\sin(kl)}.$$

and the Neumann conditions to obtain a set of V homogeneous equations for the variables f_i $(i = 1, 2, \ldots V)$.

(b) Denote by $\vec{f}$ the vector whose entries are all the f_i variables. Assume that $\sin(kl) \neq 0$ and manipulate the linear set of equations you got in the previous section to have the following form

$$A\vec{f} = \cos(kl)\vec{f}.$$

What is the matrix A? Note that this matrix describes the underlying discrete graph and this establishes a spectral connection between the discrete and the quantum graph.

(c) Denote by $\{\lambda_i\}_{i=1}^{V}$ the eigenvalues of A. Express the k-eigenvalues of the quantum graph (remember that there are infinitely many of those) in terms of the eigenvalues of A. Are all the eigenvalues of the quantum graph can be obtained in this way? If so, prove it, or otherwise, point on the eigenvalues which are not obtained in this way.

Comment: Further reading on the spectral connection between discrete and quantum graphs may be found in [**4, 8, 13, 15, 19**]. The most general derivation of this connection, treating electric and magnetic potentials as well as δ-type vertex conditions appears in [**18**].

Acknowledgments

We thank the referee whose critical remarks and constructive suggestions has lead to a profound improvement of the manuscript.

References

[1] S. Ariturk. Eigenvalue estimates on quantum graphs. preprint `arXiv:1609.07471`, 2016.

[2] R. Band and G. Lévy, *Quantum graphs which optimize the spectral gap*, Ann. Henri Poincaré **18** (2017), no. 10, 3269–3323, DOI 10.1007/s00023-017-0601-2. MR3697195

[3] R. Band, T. Shapira, and U. Smilansky, *Nodal domains on isospectral quantum graphs: the resolution of isospectrality?*, J. Phys. A **39** (2006), no. 45, 13999–14014, DOI 10.1088/0305-4470/39/45/009. MR2277370

[4] J. von Below, *A characteristic equation associated to an eigenvalue problem on c^2-networks*, Linear Algebra Appl. **71** (1985), 309–325, DOI 10.1016/0024-3795(85)90258-7. MR813056

[5] G. Berkolaiko, *An elementary introduction to quantum graphs*, Geometric and computational spectral theory, Contemp. Math., vol. 700, Amer. Math. Soc., Providence, RI, 2017, pp. 41–72, DOI 10.1090/conm/700/14182. MR3748521

[6] G. Berkolaiko, J. B. Kennedy, P. Kurasov, and D. Mugnolo, *Edge connectivity and the spectral gap of combinatorial and quantum graphs*, J. Phys. A **50** (2017), no. 36, 365201, 29, DOI 10.1088/1751-8121/aa8125. MR3688110

[7] G. Berkolaiko and P. Kuchment, *Introduction to quantum graphs*, Mathematical Surveys and Monographs, vol. 186, American Mathematical Society, Providence, RI, 2013. MR3013208

[8] C. Cattaneo, *The spectrum of the continuous Laplacian on a graph*, Monatsh. Math. **124** (1997), no. 3, 215–235, DOI 10.1007/BF01298245. MR1476363

[9] L. Friedlander, *Extremal properties of eigenvalues for a metric graph* (English, with English and French summaries), Ann. Inst. Fourier (Grenoble) **55** (2005), no. 1, 199–211. MR2141695

[10] S. A. Fulling, P. Kuchment, and J. H. Wilson, *Index theorems for quantum graphs*, J. Phys. A **40** (2007), no. 47, 14165–14180, DOI 10.1088/1751-8113/40/47/009. MR2438118

[11] S. Gnutzmann and U. Smilansky. Quantum graphs: Applications to quantum chaos and universal spectral statistics. *Adv. Phys.*, 55(5–6):527–625, 2006.

[12] B. Gutkin and U. Smilansky, *Can one hear the shape of a graph?*, J. Phys. A **34** (2001), no. 31, 6061–6068, DOI 10.1088/0305-4470/34/31/301. MR1862642

[13] J. Harrison and T. Weyand, *Relating zeta functions of discrete and quantum graphs*, Lett. Math. Phys. **108** (2018), no. 2, 377–390, DOI 10.1007/s11005-017-1017-0. MR3748369

[14] J. B. Kennedy, P. Kurasov, G. Malenová, and D. Mugnolo, *On the spectral gap of a quantum graph*, Ann. Henri Poincaré **17** (2016), no. 9, 2439–2473, DOI 10.1007/s00023-016-0460-2. MR3535868

[15] P. Kuchment, *Quantum graphs. I. Some basic structures*, Waves Random Media **14** (2004), no. 1, S107–S128, DOI 10.1088/0959-7174/14/1/014. Special section on quantum graphs. MR2042548

[16] P. Kurasov and M. Nowaczyk, *Inverse spectral problem for quantum graphs*, J. Phys. A **38** (2005), no. 22, 4901–4915, DOI 10.1088/0305-4470/38/22/014. MR2148632

[17] P. Kurasov and M. Nowaczyk, *Inverse spectral problem for quantum graphs*, J. Phys. A **38** (2005), no. 22, 4901–4915, DOI 10.1088/0305-4470/38/22/014. MR2148632

[18] K. Pankrashkin, *Spectra of Schrödinger operators on equilateral quantum graphs*, Lett. Math. Phys. **77** (2006), no. 2, 139–154, DOI 10.1007/s11005-006-0088-0. MR2251302

[19] J. von Below and D. Mugnolo, *The spectrum of the Hilbert space valued second derivative with general self-adjoint boundary conditions*, Linear Algebra Appl. **439** (2013), no. 7, 1792–1814, DOI 10.1016/j.laa.2013.05.011. MR3090437

Department of Mathematics, Technion–Israel Institute of Technology, Haifa 32000, Israel

Email address: `ramband@technion.ac.il`

School of Mathematical Sciences, University of Nottingham, Nottingham NG7 2RD, United Kingdom

Email address: `sven.gnutzmann@nottingham.ac.uk`

Contemporary Mathematics
Volume **720**, 2018
https://doi.org/10.1090/conm/720/14526

Spectral properties of classical integral operators and geometry

Dmitry Khavinson

ABSTRACT. These informal notes represent a lecture given as part of the CRM Summer School on Spectral Theory and Applications at Laval University in July 2016.

1. Classical integral operators

The unifying theme of this lecture is the following question. Given a domain in the euclidean space and one of the most popular integral operators in function theory or potential theory, could one derive certain geometric characteristics of the domain from the spectral properties of the operator in question. The theme comes under a very broad umbrella of "Inverse Problems" and " Isospectral Geometry", yet, usually, the operators under consideration are differential operators, e.g., the Laplacian, or pseudo-differential operators. Our goal is to show that a similar program for several well studied classical integral operators is rich with many appealing questions and deserves further attention.

Let's start with a very simple operator, the Cauchy integral operator (cf. [**1**,**2**,**6**] for background and some history).

Let $\Omega \subset \mathbb{R}^2$ be a bounded domain, $\Gamma = \partial\Omega$ is assumed to be sufficiently smooth.

$$L^2(\Omega) = \left\{ f : \Omega \to \mathbb{C}, \|f\|_2^2 = \int_\Omega |f|^2 dA(\xi) < \infty, \, dA = \text{area measure} \right\}.$$

$$(Cf)(z) = \left[-\frac{1}{\pi} \int_\Omega \frac{f(\xi)}{\zeta - z} dA(\xi) \right].$$

An easy exercise shows that C is bounded, even compact, yet not Hilbert–Schmidt.

REMARK 1.1. Note that the Cauchy integral operator on curves is well-known but it is not compact. The above Cauchy operator is compact and reproduces all C_0^∞ functions

$$\varphi(z) = -\frac{1}{\pi} \int_\Omega \frac{\partial\varphi}{\partial\bar{\xi}} \frac{1}{\zeta - z} dA(\xi).$$

2010 *Mathematics Subject Classification.* Primary 31B10, 31B20, 31B99.
The author gratefully acknowledges the NSF support of the Summer School under the grant DMS 1653027.

QUESTION 1. (i) What is the precise norm of C, for domains that are very simple, say a disk? Can we estimate the norm of C for general domains in terms of area, lowest frequency, perimeter, etc. (ii) What can be said about the asymptotics of eigenvalues of the symmetrized version of C, namely, the asymptotics of the eigenvalues of $|C|$?

As in the classical studies of the spectra of the Laplacian, there is a rich variety of questions one may ask here. Here is a sample: from the information about the spectrum, can one derive some information regarding geometry of Ω? Can one characterize disks among the domains of the same area based on the spectral information of the operator? Can we "hear the shape of a drum", i.e., are there two different domains with the same spectrum, and so on?

Two other operators L, N naturally come to mind:

$$(Lf)(z) = \frac{1}{2\pi}\int_\Omega \log\frac{1}{|z-\xi|} f(\xi)\, dA(\xi);$$
$$(Nf)(x) = \frac{1}{(n-2)\omega_n}\int_\Omega \frac{1}{|x-y|^{n-2}} f(y)\, dV(y).$$

Here, ω_n denotes the area of the unit sphere. dV is the Lebesgue measure in $\mathbb{R}^n$. The operators L and N, of course, have an obvious physical meaning. They represent the graviational (or, electrostatic) potentials in the plane or, higher dimensional space, of a mass density (charge density respectively) f distributed over Ω. All of the above questions can legitimately be asked for L, N. Also, one can certainly consider comparison (in, say, norm) of C, L, N. In 1989 Anderson and Hinkkanen [**1**] showed that

$$\|C\|_{L^2(\mathbb{D})} = \frac{2}{j_0},$$

where $j_0 \equiv 2.408\ldots$, is the smallest positive zero of the zero Bessel function

$$J_0(x) = \sum_{k=0}^{\infty}\frac{(-1)^k}{(k!)^2}\left(\frac{x}{2}\right)^{2k};$$
$(\mathbb{D}) = \{z : |z| < 1\}$ is the unit disk in $\mathbb{C}$.

The proof was a rather ingenious computation and application of an inequality of Hardy type and an ODE result due to P. Boyd (1969).

However, if one notices ([**3**]) that $j_0^2 = \lambda_1$ is the smallest eigenvalue for the Dirichlet Laplacian in $\mathbb{D}$, i.e., the smallest positive $\lambda > 0$ such that

$$\Delta\varphi + \lambda\varphi = 0 \text{ in } \Omega,\ \varphi = 0 \text{ on } \partial\Omega,$$

the problem immediately connects to mathematical physics. Furthermore, let

$$g(x,y) = \begin{cases} \dfrac{1}{(n-2)w_{n-1}}\dfrac{1}{|x-y|^{n-2}} + u_x(y), & x,y\in\Omega \quad n\geq 3 \\ \dfrac{1}{2\pi}\log\dfrac{1}{|x-y|} + u_x(y), & x,y\in\Omega \\ 0, \text{ elsewhere} \end{cases}$$

be Green's function [**14**] of Ω, $u_x(y)$ is harmonic in Ω and, $g(x,y)\mid_{y\in\partial\Omega} = 0$, and consider the operator

$$(Gf)(x) = \int_\Omega g(x,y) f(y)\, dV(y).$$

Then, $G^{-1} = -\Delta$ on the Sobolev space $W_0^{1,2}$. The eigenvalues of G are reciprocals of those of Δ. Hence, the Anderson–Hinkkanen results can be read as

$$\left\|\frac{1}{4}C^*C\right\|_{\mathbb{D}} = \frac{1}{j_0^2} = \frac{1}{\lambda_1} = \mu_1 := \|G\|_{\mathbb{D}}$$

It turns out that not all of these are coincidences.

THEOREM 1.2 ([**2**, **6**]).

$$\left\|\frac{1}{4}C^*C\right\|_{\Omega} = \left(\frac{1}{2}\|C\|\right)^2_{\Omega} = \|G\|_{\Omega} = \frac{1}{\lambda_1} = \mu_1.$$

Moreover, in view of the Raleyigh–Faber–Krahn theorem [**4**], we have

COROLLARY 1.3. *Among the domains of equal area, the norm for C (and G) is maximized over a disk, i.e.*

$$\|C\| \leq \frac{2}{j_0}\sqrt{\frac{\operatorname{Area}\Omega}{\pi}}$$

Moreover, combining the results from [**2**], [**3**] and [**5**], we have the following results on spectral asymptotics:

(i) The operator C, or, rather, $\frac{C^*C}{4}$: The singular numbers s_n, i.e., the eigenvalues of the self-adjoint operator $\left(\frac{C^*C}{4}\right)^{1/2}$ behave asymptotically as $O\left(\frac{1}{\sqrt{n}}\right)$.

All the eigenvalues are of double multiplicities so, $C \in S_{2,\infty}$, in particular, $C \subset S_p$, $p > 2$ but C is not Hilbert–Schmidt (S_p, $S_{2,\infty}$ stand for the corresponding Schatten classes — cf. [**3**]). (Recall briefly, that the Schatten classes $S_p, p \geq 1$ denote operators with l^p- summable singular numbers. $S_{p,\infty}$ denotes the space of operators whose singular numbers belong to the weighted space l^∞ with the weights $n^{1/p}$ - cf. [**3**].)

Remarkably, if we restrict C to the Bergman space, i.e.. the space of square-integrable with respect to area analytic functions in Ω, the singular numbers of CP, where P denotes the orthogonal projection onto the Bergman space, become $O\left(\frac{1}{n}\right)$, hence $CP \in S_{1,\infty}$, i.e., "2 times better than C".

(ii) The operator L. The spectra of L and $\frac{1}{4}C^*C$ are quite similar (and coincide for the disk) although in the disk L has the whole series of triple eigenvalues. The singular numbers are $O\left(\frac{1}{n}\right)$ and $L \in S_{1,\infty}$. Again the operator PLP is "twice as nice" ($\in S_{1/2,\infty}$), the eigenfunctions in the disk are simply monomials.

In 2007, M. Dostanić [**7**] established remarkable further refinement on asymptotics of C and L.

$$s_n(C) = \sqrt{\frac{\operatorname{Area}(\Omega)}{\pi}\frac{1}{n}} + O\left(\frac{1}{n}\right);$$

$$\lambda_n(L) = \frac{\operatorname{Area}(\Omega)}{4\pi}\cdot\frac{1}{n} + O\left(\frac{1}{n^{3/2}}\right).$$

Thus, as in the case of the Dirichlet Laplacian, the spectral behavior of C, L allow us to determine the area of Ω. Note that L is "two times better" than C.

As noted above, the norms of C, L allow us to determine when the domain is a disk (or, not — more often) in view of the Faher–Krahn theorem. The following questions certainly pose themselves:

(i) Are there domains with identical spectra for both operators C_Ω and L_Ω ? (Most likely "yes", but virtually nothing has been done.)

(ii) The eigenfunctions for disks, balls for the operators C, L, N are products of Bessel functions and spherical harmonics. Thus entire functions of exponential type. Does this property characterize balls? Probably not, since some of these statements hold for ellipsoids as well. Hence, it is true that the growth of the eigenfunctions characterize ellipsoids. For the operator G, or equivalently, for the Dirichlet Laplacian the eigenfunctions for ellipsoids are also entire functions of exponential type (Bessel functions combined with spherical harmonics for balls). It is tempting to suggest that that property does characterize ellipsoids, in other words for all other domains some eigenfunctions develop singularities somewhere outside the domains. At present, there is virtually no progress on the problem. A closely-related problem is associated with the so-called Khavinson–Shapiro conjecture — cf. [**11**, **12**]. The conjecture asserts that ellipsoids are the only domains for which all solutions of the Dirichlet problem $\Delta u = 0$ in Ω, $u \mid_{\Gamma:=\partial\Omega}= f$, with entire data f are themselves entire functions.

The problem where do the singularities of solutions of the Dirichlet problem with a "nice", entire or polynomial, data appear from is deep, difficult and, mostly, widely open. Some recent progress has been achieved mostly by H. Render, and , also, by S. Bell–P. Ebenfelt–D. Khavinson–H. S. Shapiro and D. Khavinson–E. Lundberg – cf. [**11**, **12**] and references therein. In two dimensions the problem leads very quickly to some intriguing and non-standard algebraic geometry: complex "lightning bolts" on Riemann surfaces, etc. – cf. [**11**].

(iii) Do the spectra of C, L, N transmit the information about the geometry of the base domain Ω? For example, can we detect corners, or cusps on $\Gamma = \partial\Omega$ from the spectral properties of those operators. Since the crude asymptotics ($O\left(\frac{1}{\sqrt{n}}\right)$ for C and $O\left(\frac{1}{n}\right)$ for L) are the same for all bounded domains, the problem appears quite delicate. It appears as a worthy task to investigate in detail the situation for C in cardioids since the conformal map from the disk onto cardioids is just a quadratic polynomial. Hence, via a simple change of variables, the problem could be moved back to the disk. Similarly, the example of a square should be possible to handle directly for the operator L. Note in passing , that similar themes regarding sensitivity of spectral asymptotics to the singularities of the boundary in the context of differential operators, e.g., the Laplacian, have been intensely studied over the years. Recently, a lot of progress has been achieved in particular in studying the asymptotics of the so-called Steklov eigenvalue problem - cf. the survey [**10**] for further details and references.

2. Single-layer potentials

Set $\Gamma = \partial\Omega$, Ω is as above. Let

$$u(x) = \int_\Gamma E_n(x-y) f(y)\, dS(y),$$

where

$$E_n(x) = \begin{cases} -\dfrac{1}{2\pi} \log|x|, & n = 2; \\ \dfrac{1}{(n-2)\omega_n} |x|^{2-n}, & n > 2, \end{cases}$$

where as above ω_n is the area of the unit sphere S^{n-1} in $\mathbb{R}^n$. dS is Lebesgue measure on Γ. Obviously, u is harmonic in $\mathbb{R}^n \setminus \Gamma$ and is continuous in $\mathbb{R}^n$. The physical meaning of the single-layer potential is simple: it represents the electrostatic (gravitational) potential of the charge distribution (mass density) f on Γ. The operator $S : f \mapsto u|_\Gamma$ is self-adjoint on $L^2(\Gamma)$ and it is not difficult to see that S is compact, even Hilbert–Schmidt. S is injective in $\mathbb{R}^n$, $n \geq 3$, and has at most 1 dimensional kernel in $\mathbb{R}^2$. It is not difficult to calculate the spectrum of S in $\mathbb{D}$, $n = 2$ or $B := \{|x| < 1\}$ in $\mathbb{R}^n$, $n \geq 3$. The eigenfunctions are spherical harmonics, the multiplicity of every eigenvalue λ_n in $\mathbb{R}^2$ is 2 while for $n \geq 3$ it is $A(n,m) - A(n,m-2)$, $A(n,m) = \frac{n(n+1)\cdots(n+m-1)}{m!}$, the dimension of the space of spherical homogeneous harmonics of degree m. The eigenvalues are $\frac{1}{2m}$ for $n = 2$ and $\frac{1}{2m+n-2}$ for $n > 2$. This observation provides us with crude asymptotics of the spectra for all bounded domains in $\mathbb{R}^n$.

Note that the constant functions are eigenfunctions for S in all dimensions on the ball.

QUESTION 2. Does this properly characterize balls?

The answer is "Yes" ([**9**]), provided that Γ satisfies certain mild smoothness assumptions. (For example, for $n = 2$, $\Omega^e = \mathbb{C} \setminus \overline{\Omega}$ is assumed to be a Smirnov domain, i.e., the logarithm of the modulus of the derivative of the conformal mapping from the disk to Ω^e is representable by the Poisson integral of its boundary values - cf. [**9**] and the function theory references there.) The result is proved in two dimensions by function theory via the Riemann mapping.

2.1. Quadrature identity. One can show that S has 1 as an eigenfunction if and only if the domain Ω' obtained from $\mathbb{R}^n \setminus \Omega$ by inversion has the following quadrature property

$$u(0) = c \int_{\Gamma' := \partial\Omega'} |x|^{-n} u(x)\, dS, \text{ for all harmonic } u \text{ in } \Omega.$$

Then it is indeed true that such domain must be a ball (W. Reichel, S. Shahgholian $\sim$ '97 – cf. references in [**9**]) but it is only known for $C^{2\alpha}$ boundaries in $\mathbb{R}^n$, $n \geq 3$.

Perhaps, one wonders, even a more ambitious conjecture is true: If for $\alpha \in \mathbb{R}$

$$(*) \qquad u(0) = c \int_\Gamma |x|^\alpha u(x)\, dS(x)$$

for all u harmonic in $\overline{\Omega}$, then Ω is a ball. $(*)$ of course, means that the harmonic measure ω at 0 is represented as $\omega(0, \Omega, \Gamma) = c|x|^\alpha\, dS(x)$.

THEOREM 2.1 ([**9**]).

(i) *For* $n = 2$ *and* $\alpha = -2$, $(*)$ *holds for ALL disks containing the origin.*

(ii) *For* $n = 2$, $\alpha = -3, -4, -5, \ldots \exists$ *solutions (domains)* Ω *which are NOT disks (!)*

(iii) *For all other* α, Ω *is a disk centered at the origin.*

(For (ii), Ω *is obtained from* $\mathbb{D}$ *via conformal maps* $\varphi(w) = \frac{w}{(A+Bw^k)^{1/k}}$, $k = 2, 3, \ldots$ *with appropriate* A, B.*)*

QUESTIONS.

(i) How can the above theorem be extended to $n \geq 3$?

(ii) Can one extend the results of Reichel, Shahgholian, etc., to less-smooth boundaries?

(iii) All the questions from the previous section, e.g., isospectrality, information about the boundary, e.g., cusps, corners, etc., make sense for the single-layer operator S as well. But virtually nothing is known.

(iv) We have shown that under some mild restrictions only balls have constants as eigenfunctions for S. What about higher-degree harmonic polynomials?

S. Zoalroshd [**16**] showed that if in $\mathbb{R}^2$ a harmonic polynomial $h = Re\, p(z)$ is an eigenfunction for S and all but one zeros of p are inside Ω, then $\Gamma = \partial\Omega$ is a circle. It is natural to conjecture that the latter hypothesis regarding the placement of zeros of p is not necessary. Is it true? Is his result true in $\mathbb{R}^n$, $n \geq 3$?

3. Double-layer potentials

Let as above $E(x,y) := E_n(x,y) = c_n \begin{cases} |x|^{2-n}, & n \geq 3 \\ \log|x| & n = 2 \end{cases}$ be the usual kernel in potential theory.

Let $K : L^2(\Gamma) \to L^2(\Gamma)$ be defined

$$(Kf)(x) = 2\int_\Gamma \frac{\partial}{\partial n_y} E(x-y) f(y)\, d\sigma(y),$$

K is called the Neumann–Poincaré operator. If we let $x \in \Omega$, we have a harmonic function whose boundary values equal $f - Kf = (I - K)f$ on Γ. Thus, proving that the operator $(I \pm K)$ is surjective allowed Fredholm to solve the Dirichlet problem (Fredholm's alternative) in terms of double layer potentials in rather general domains provided that $I \pm K$ is injective.

Unfortunately, K is never self-adjoint unless Ω is a ball, cf. [**13**]. However, using the so-called symmetrization procedure one can show that K indeed has a real spectrum and its eigenfunctions do span the L^2- space on the boundary. Poincaré has conjectured (by analogy with the sphere) that the spectrum of K is always nonnegative in $\mathbb{R}^n$, $n \geq 3$. This is now proven to be false ([**13**]). There is enormous literature on eigenvalues and eigenfunctions for K — cf. the references in [**13**]. Let me just touch on the most fundamental property of K:

QUESTION 3. Is K injective, i.e., does $\ker K = \{0\}$?

Obviously, it is true for $\Gamma = S^{n-1}$, $n \geq 3$, where the operator K, as is easily verified, is a scalar multiple of S, which is obviously injective. So, is it true for other smooth surfaces Γ in $\mathbb{R}^n$, $n \geq 3$? The answer is probably "no", but it is still unknown. In 2 dimensions the situation is even more intriguing. On $\mathbb{T} := \partial\mathbb{D}$,

$$Kf = \frac{1}{\pi} \int_0^{2\pi} f\left(e^{i\theta}\right) ds$$

has rank 1. So $\ker K = L^2(\mathbb{T}) \ominus \mathbb{C}$ is virtually almost everything. In the 1980's, D. Khavinson and H. S. Shapiro showed (unpublished) that $\ker K$ cannot be finite dimensional.

In [**8**] it was proven that not only if $\ker K \neq \{0\}$, then $\dim K = \infty$, but moreover it is an algebra! Namely, $F \in \ker K \Leftrightarrow \exists f, g$ analytic in Ω, Ω^e respectively, $g(\infty) = 0$ and such that

$$(**) \qquad f = \bar{g} \text{ on } \Gamma.$$

EXAMPLE 3.1.

(i) $\Omega = \mathbb{D}$, $f = z^n$, $n \geq 1$.

(ii) Yet, the "matching problem" $(**)$ has other solutions!

Rational lemniscates ([**8**]) are defined as Jordan curves Γ such that

$$\Gamma = \{|R(z)| = c,\ c > 0\},$$

where R is a rational function, R has no zeros in Ω^e, and no poles in Ω. The pair R, $\frac{C}{R}$ obviously solves $(**)$.

CONJECTURE 3.2. *Rational lemniscates are the only curves on which the Neumann-Poincaré is not injective .*

REMARK 3.3. It is truly important to identify the $\ker K$ since these are precisely the densities whose double layer potentials give directly explicit solutions of the Dirichlet problem.

Once again, all the questions along the lines discussed in the previous sections apply to K as well (cf. [**15**]).

Another open venue for research is to study in depth the relationships between the spectra of C, L, N and their interplay. Even in $\mathbb{D}$ (or, $B = \{|x| < 1\}$, $n \geq 3$) those spectra are close, but not identical. So understanding precisely, in geometric terms, the "corrections" between the eigenvalues is already a worthy project.

Acknowledgments

The author is grateful to the National Science Foundation for supporting the Summer School and to the Simons Foundation for supporting the author's research. The author is indebted to the anonymous referee for many thoughtful suggestions regarding the exposition.

References

[1] J. M. Anderson and A. Hinkkanen, *The Cauchy transform on bounded domains*, Proc. Amer. Math. Soc. **107** (1989), no. 1, 179–185, DOI 10.2307/2048052. MR972226

[2] J. M. Anderson, D. Khavinson, and V. Lomonosov, *Spectral properties of some integral operators arising in potential theory*, Quart. J. Math. Oxford Ser. (2) **43** (1992), no. 172, 387–407, DOI 10.1093/qmathj/43.4.387. MR1188382

[3] J. Arazy and D. Khavinson, *Spectral estimates of Cauchy's transform in $L^2(\Omega)$*, Integral Equations Operator Theory **15** (1992), no. 6, 901–919, DOI 10.1007/BF01203120. MR1188786

[4] Catherine Bandle, *Isoperimetric inequalities and applications*, Monographs and Studies in Mathematics, vol. 7, Pitman (Advanced Publishing Program), Boston, Mass.-London, 1980. MR572958

[5] Milutin Dostanić, *Asymptotic behavior of eigenvalues of certain integral operators*, Publ. Inst. Math. (Beograd) (N.S.) **59(73)** (1996), 95–113. MR1444569

[6] Milutin R. Dostanić, *The properties of the Cauchy transform on a bounded domain*, J. Operator Theory **36** (1996), no. 2, 233–247. MR1432117

[7] Milutin R. Dostanić, *Estimate of the second term in the spectral asymptotic of Cauchy transform*, J. Funct. Anal. **249** (2007), no. 1, 55–74, DOI 10.1016/j.jfa.2007.04.007. MR2338854

[8] Peter Ebenfelt, Dmitry Khavinson, and Harold S. Shapiro, *An inverse problem for the double layer potential*, Comput. Methods Funct. Theory **1** (2001), no. 2, [On table of contents: 2002], 387–401, DOI 10.1007/BF03320998. MR1941133

[9] P. Ebenfelt, D. Khavinson, and H. S. Shapiro, *A free boundary problem related to single-layer potentials*, Ann. Acad. Sci. Fenn. Math. **27** (2002), no. 1, 21–46. MR1884347

[10] Alexandre Girouard and Iosif Polterovich, *Spectral geometry of the Steklov problem (survey article)*, J. Spectr. Theory **7** (2017), no. 2, 321–359, DOI 10.4171/JST/164. MR3662010

[11] Dmitry Khavinson and Erik Lundberg, *The search for singularities of solutions to the Dirichlet problem: recent developments*, Hilbert spaces of analytic functions, CRM Proc. Lecture Notes, vol. 51, Amer. Math. Soc., Providence, RI, 2010, pp. 121–132. MR2648870

[12] Dmitry Khavinson and Erik Lundberg, *Linear holomorphic partial differential equations and classical potential theory*, Mathematical Surveys and Monographs, vol. 232, American Mathematical Society, Providence, RI, 2018. MR3821527

[13] Dmitry Khavinson, Mihai Putinar, and Harold S. Shapiro, *Poincaré's variational problem in potential theory*, Arch. Ration. Mech. Anal. **185** (2007), no. 1, 143–184, DOI 10.1007/s00205-006-0045-1. MR2308861

[14] N. S. Landkof, *Foundations of modern potential theory*, Springer-Verlag, New York-Heidelberg, 1972. Translated from the Russian by A. P. Doohovskoy; Die Grundlehren der mathematischen Wissenschaften, Band 180. MR0350027

[15] Yoshihisa Miyanishi and Takashi Suzuki, *Eigenvalues and eigenfunctions of double layer potentials*, Trans. Amer. Math. Soc. **369** (2017), no. 11, 8037–8059, DOI 10.1090/tran/6913. MR3695853

[16] Seyed M. Zoalroshd, *On Spectral Properties of Single Layer Potentials*, ProQuest LLC, Ann Arbor, MI, 2016. Thesis (Ph.D.)–University of South Florida. MR3579454

Department of Mathematics & Statistics, University of South Florida, Tampa, Florida 33620-5700

Email address: dkhavins@usf.edu

Selected Published Titles in This Series

720 **Alexandre Girouard, Editor,** Spectral Theory and Applications, 2018

716 **Alex Martsinkovsky, Kiyoshi Igusa, and Gordana Todorov, Editors,** Surveys in Representation Theory of Algebras, 2018

715 **Sergio R. López-Permouth, Jae Keol Park, S. Tariq Rizvi, and Cosmin S. Roman, Editors,** Advances in Rings and Modules, 2018

714 **Jens Gerlach Christensen, Susanna Dann, and Matthew Dawson, Editors,** Representation Theory and Harmonic Analysis on Symmetric Spaces, 2018

713 **Naihuan Jing and Kailash C. Misra, Editors,** Representations of Lie Algebras, Quantum Groups and Related Topics, 2018

712 **Nero Budur, Tommaso de Fernex, Roi Docampo, and Kevin Tucker, Editors,** Local and Global Methods in Algebraic Geometry, 2018

711 **Thomas Creutzig and Andrew R. Linshaw, Editors,** Vertex Algebras and Geometry, 2018

710 **Raphaël Danchin, Reinhard Farwig, Jiří Neustupa, and Patrick Penel, Editors,** Mathematical Analysis in Fluid Mechanics, 2018

709 **Fernando Galaz-García, Juan Carlos Pardo Millán, and Pedro Solórzano, Editors,** Contributions of Mexican Mathematicians Abroad in Pure and Applied Mathematics, 2018

708 **Christian Ausoni, Kathryn Hess, Brenda Johnson, Ieke Moerdijk, and Jérôme Scherer, Editors,** An Alpine Bouquet of Algebraic Topology, 2018

707 **Nitya Kitchloo, Mona Merling, Jack Morava, Emily Riehl, and W. Stephen Wilson, Editors,** New Directions in Homotopy Theory, 2018

706 **Yeonhyang Kim, Sivaram K. Narayan, Gabriel Picioroaga, and Eric S. Weber, Editors,** Frames and Harmonic Analysis, 2018

705 **Graham J. Leuschke, Frauke Bleher, Ralf Schiffler, and Dan Zacharia, Editors,** Representations of Algebras, 2018

704 **Alain Escassut, Cristina Perez-Garcia, and Khodr Shamseddine, Editors,** Advances in Ultrametric Analysis, 2018

703 **Andreas Malmendier and Tony Shaska, Editors,** Higher Genus Curves in Mathematical Physics and Arithmetic Geometry, 2018

702 **Mark Grant, Gregory Lupton, and Lucile Vandembroucq, Editors,** Topological Complexity and Related Topics, 2018

701 **Joan-Carles Lario and V. Kumar Murty, Editors,** Number Theory Related to Modular Curves, 2018

700 **Alexandre Girouard, Dmitry Jakobson, Michael Levitin, Nilima Nigam, Iosif Polterovich, and Frédéric Rochon, Editors,** Geometric and Computational Spectral Theory, 2017

699 **Mark L. Agranovsky, Matania Ben-Artzi, Catherine Bénéteau, Lavi Karp, Dmitry Khavinson, Simeon Reich, David Shoikhet, Gilbert Weinstein, and Lawrence Zalcman, Editors,** Complex Analysis and Dynamical Systems VII, 2017

698 **Alexander M. Blokh, Leonid A. Bunimovich, Paul H. Jung, Lex G. Oversteegen, and Yakov G. Sinai, Editors,** Dynamical Systems, Ergodic Theory, and Probability: in Memory of Kolya Chernov, 2017

697 **Fabrizio Broglia, Françoise Delon, Max Dickmann, Danielle Gondard-Cozette, and Victoria Ann Powers, Editors,** Ordered Algebraic Structures and Related Topics, 2017

696 **Ara S. Basmajian, Yair N. Minsky, and Alan W. Reid, Editors,** In the Tradition of Ahlfors–Bers, VII, 2017

For a complete list of titles in this series, visit the
AMS Bookstore at **www.ams.org/bookstore/conmseries/**.